WAVES IN ANISOTROPIC PLASMAS

WAVES IN ANISOTROPIC PLASMAS

PUBLISHED 1963 BY THE M.I.T. PRESS, CAMBRIDGE, MASSACHUSETTS

WILLIAM P. ALLIS
Professor of Physics, Massachusetts Institute of Technology

SOLOMON J. BUCHSBAUM
Member of Technical Staff, Bell Telephone Laboratories, Inc., Murray Hill, New Jersey

ABRAHAM BERS
Assistant Professor of Electrical Communications, Massachusetts Institute of Technology

Library of Congress Catalog Card Number: 62-19756

Printed in the United States of America

FOREWORD

There has long been a need in science and engineering for systematic publication of research studies larger in scope than a journal article but less ambitious than a finished book. Much valuable work of this kind is now published only in a semiprivate way, perhaps as a laboratory report, and so may not find its proper place in the literature of the field. The present contribution is the seventeenth of the M.I.T. Press Research Monographs, which we hope will make selected timely and important research studies readily accessible to libraries and to the independent worker.

J. A. Stratton

ACKNOWLEDGMENT

This is Special Technical Report Number 8 of the Research Laboratory of Electronics of the Massachusetts Institute of Technology.

The Research Laboratory of Electronics is an interdepartmental laboratory in which faculty members and graduate students from numerous academic departments conduct research.

The research reported in this document was made possible in part by support extended the Massachusetts Institute of Technology, Research Laboratory of Electronics, jointly by the U.S. Army, the U.S. Navy (Office of Naval Research), and the U.S. Air Force (Office of Scientific Research) under Contract DA36-039-sc-78108, Department of the Army Task 3-99-25-001-08; and in part by Contract DA-SIG-36-039-61-G14; additional support was furnished by the National Science Foundation under Grant G-9330.

PREFACE

This work originated in attempts to understand the literature on waves in plasmas, in which terms such as "longitudinal" and "transverse," "ordinary" and "extraordinary," and "resonance" have ambiguous meanings, and in which the assumptions are assumed to be obvious. This led us to the realization that sketches of the phase-velocity surface for arbitrary but ordered plasma parameters — the (α^2, β^2) diagram — are sufficient to locate the domain of various approximations. Also, they provide a graphic description of the local properties of the waves. The theory was kept simple: plane monochromatic waves with no collisions, no damping, no boundaries, no shocks, not much temperature. Even so, there was sufficient variety in the results to make it difficult to see the whole picture, and we were encouraged to write it out. This constitutes the first part of this monograph. It is incomplete in several ways, perhaps the most obvious omission is the absence of a whole class of solutions that are merely mentioned in the section on the $(n_{\parallel}^2, n_{\perp}^2)$ diagram.

The second part of this monograph originated independently from a need to interpret observations in bounded plasmas. Energy and power flow relations were found to contain many of the important properties of both free and guided waves. Although the field analysis of plasma waveguides is rather formal, many aspects of the dispersion relation such as the resonances and cutoffs — the poles and zeros of the propagation constant — are easily related to the properties of the free waves treated in the first part. A remaining gap between the two parts of this monograph is the lack of a more complete study of surface waves on a semi-infinite plasma.

We are indebted to many authors for the ideas contained in this monograph and have chosen not to try to trace each idea back to its source. We prefer to consider this as a new arrangement of old ideas. We are indebted to many on the staffs of the Research Laboratory of Electronics and of the Bell Telephone Laboratories for discussions, for computations, and for the preparation of the manuscript and of the figures.

W. P. Allis
S. J. Buchsbaum
A. Bers

Cambridge, Massachusetts
December, 1962

CONTENTS

PART II ENERGY-POWER THEOREMS AND GUIDED WAVES
Abraham Bers

INTRODUCTION

Ever since 1902, when Oliver Heaviside postulated the existence of an ionized layer in the atmosphere, a great deal of attention has been paid to the study of electromagnetic waves in the ionosphere. Appleton and Hartree in 1931 developed a theory based on Lorentz's Theory of Electrons (1909) applied to the motions of free electrons in a static magnetic field and an alternating electric field. The results of this magnetoionic theory are well presented in The Magnetoionic Theory and Its Applications to the Ionosphere by Ratcliffe in 1959. Experiments in this field are performed by observing waves reflected from the various layers of the ionosphere. The experimental variable is the frequency $\nu = \omega/2\pi$ † of the transmitter. Thus the results of the theory are appropriately represented by plots of phase or group velocity (u or u_g) of the wave or of propagation constant ($k = \omega/u = 2\pi/\lambda$) against frequency. Two "waves" are recognized on these plots, one of which is qualitatively independent of the magnetic field and was, unfortunately, named "ordinary" (denoted by subscript o on our plots). The second wave was therefore "extraordinary" (denoted by subscript $\times$). The ordinary wave has a low-frequency cutoff $u \to \infty$ at the "critical frequency" $\omega_p = \sqrt{Ne^2/m\epsilon_0} = 56.5\sqrt{N}$ radians/second.‡ This formula was first written by Lord Rayleigh in 1906 to express the resonance frequencies of the pumpkin model of the atom. It is a parallel resonance between conduction and displacement currents, but we reserve the term "resonance" for the series resonance $u \to 0$.

The extraordinary wave has two low-frequency cutoffs displaced by the magnetic field to both sides of the critical frequency and a resonance $u \to 0$ related to the cyclotron frequency $\omega_b = eB_0/m$. The two "waves" are therefore defined in terms of the frequency dependence of their index of refraction $n = c/u$.

In optics the term "ordinary" is used for a wave whose index of refraction n is independent of the direction of propagation, and which therefore obeys Snell's law of refraction, whereas the "extraordinary" wave is anisotropic and does not obey Snell's law. As the optical definitions are different from the magnetoionic ones, some confusion is inevitable. We shall use the term "ordinary" in

† We shall generally refer to ω as the frequency.

‡ Rationalized mks (meter-kilogram-second) units are used throughout, so that N is the number of electrons per cubic meter.

the strictly magnetoionic sense (n independent of B_0) and shall show that only one of the waves that are propagated across the magnetic field is strictly ordinary.

In 1929, Langmuir and Tonks observed that electron beams are scattered by a plasma in a distance much shorter than can be explained by collisions, that is, by Rutherford scattering. They postulated "plasma-electron oscillations" at the "plasma frequency" ω_p as the mechanism to explain the observations. Making use of the term $\underset{\sim}{v} \cdot \underset{\sim}{\nabla} f$ in the Boltzmann equation, which is equivalent to introducing the pressure $p = NeT$,† Vlasov in 1938 showed that plasma oscillations are the cutoff condition of a plasma wave, but both theory (Landau, 1946) and experiment (Looney and Brown, 1954) relative to plasma waves proved difficult. In 1929, Tonks and Langmuir also introduced another plasma wave propagating like a sound wave and involving the motion of ions as well as electrons.

We now have four waves, two of which depend on Maxwell's equations for their propagation and go over, in the high-frequency limit, to the vacuum electromagnetic waves; and two of which depend on the pressure for their propagation and have phase velocities u of the order of the thermal velocities $v_T = \sqrt{eT/m}$ of the particles. Let us remind the reader that an anisotropic two-element, ionic crystal transmits six elastic waves. If we eliminate the shear waves, which should not be transmitted by a plasma, there remain two compression waves: one in the "acoustic" branch in which neighboring ions of opposite signs move in phase, and one in the "optical" branch in which neighboring but oppositely charged ions move 180° out of phase. The Vlasov, or plasma-electron, wave corresponds to the "optical" branch of the compression waves, the Thomson or plasma-ion wave to the "acoustic" branch. These are "electrodynamic" waves, because the compression forces are transmitted by electric fields rather than by collisions as in the ordinary acoustic wave, but no electromagnetic interactions are required. Adding the two electromagnetic waves, we have the correct total of four waves for the medium.

In 1942, the interest shifted to cosmic plasmas, and Alfvén made the appropriate assumptions: an incompressible, infinitely conducting fluid on which the Lorentz force $\underset{\sim}{J} \times \underset{\sim}{B}_0$ acts. He found a wave, whose velocity u is related to the Alfvén speed $u_a = B_0/\sqrt{\mu_0 d}$, where d is the mass density of the fluid. This is not an electromagnetic wave, as it does not depend on the displacement current. It is a hydromagnetic wave. However, real and displacement currents are merely additive and not readily distinguished in the equations, so that the hydromagnetic wave is of the electromagnetic family. The velocity of light c is hidden in the permeability μ_0,

---

† We express temperatures in electron volts; one electron volt is equivalent to 11,600°K.

In 1950, Herlofson introduced compressibility and found three magnetohydrodynamic waves, one of which is nearly spherical and hence "ordinary" by optical terminology, but "extraordinary" by radio terminology, as it depends critically on B_0. The radio "ordinary" wave is cut off when conduction current exceeds the displacement current.

In the magnetohydrodynamic limit only three waves appear because the inertia of the electrons is negligible.

The interest in guided waves in plasmas dates back to the time of the Second World War. Soon after the war was over, the then highly developed microwave techniques were applied to experimental studies of gaseous discharges (Brown, et al.). At that time the theory of wave propagation in plasma and, in particular, in bounded plasmas, had not yet received much attention. Most experiments were designed so that the plasma would act only as a perturbation on standard waveguide and resonator structures, and perturbation theories were developed for interpreting the measurements.

At about the same time the increased interest in passive nonreciprocal microwave devices led some workers to a study of the plasma waveguide in the presence of a magnetic field (Goldstein). This was, however, soon abandoned in favor of the ferrite medium, and much emphasis was placed on the study of guided wave propagation in anisotropic ferrite media. Some of the workers in this field (Suhl and Walker, and Van-Trier) allowed their mathematical treatment to be general enough to include also the simplest anisotropic plasma medium. But, with few exceptions (Schumann), detailed solutions were carried out only for the anisotropic ferrite waveguide.

More recently, workers in the field of microwave electron-beam tubes became interested in the interaction between an electron beam and a stationary plasma. Interest in the interaction with slow waves in a bounded plasma lead to quasistatic theories of wave propagation in anisotropic plasma waveguides (Smullin, Gould). The slow-wave regions near the cyclotron and plasma resonances have been used for interaction with an electron beam.

In the low-frequency regime, waveguide solutions have been obtained for the infinitely conducting fluid model of a plasma (Newcomb, Gajewski).

Government sponsorship of controlled fusion research in 1952, and its declassification in 1958, lead to a vast increase in the number of papers on plasmas. With the number of papers, the number of "waves" increased. We read of electron and ion cyclotron waves, and of magnetoacoustic waves. Obviously we need a definition of what constitutes "a wave," as there are only four modes in a two-component fluid medium. Furthermore, each paper relates to an experimental field and makes the appropriate assumptions: "Terms of order m_-/m_+ are neglected." "The displacement current is negligible compared to real currents in high-density plasmas." "The electrons are tied to the magnetic lines of force." These statements are all true in appropriate domains, but the domains of validity hardly overlap. The small ratio m_-/m_+ is neglected by both Appleton and Alfvén, with the result that only m_- appears in Appleton's theory, only m_+ in Alfvén's!

In this monograph we try to bring order intc this diversity. Terms are defined carefully, and no limits are placed on the principal variables: frequency ω, plasma density N, and magnetic field $\underset{\sim}{B}_0$. Nor is the ratio m_-/m_+ neglected. Thus the range of validity covers the magnetoionic, the acoustic, and the magnetohydrodynamic. On the other hand, v^2/c^2 is neglected so that the temperature T is introduced only to first order and the treatment is nonrelativistic. The major limitations are the neglect of collisions ($\nu_c = 0$) and of Landau damping. This is done intentionally because these phenomena mask many details of the wave structure which are interesting and helpful in understanding the entire scheme of waves. To keep these features complicates the equations inordinately. It is believed that, unlike m_-/m_+ or displacement current, collisions may be added later, and their effect is to some extent intuitively obvious. Furthermore, Landau damping is sensitive to the form of the velocity distribution function, and the interaction between the wave amplitude distribution and the electron velocity distribution has not yet been fully worked out.

In the treatment of guided waves, we have similarly tried to avoid the usual assumptions of infinite ion mass or infinite conductivity. Nevertheless, the problem still remains complicated in detail even for the simplest plasma. To aid in understanding the nature of the solutions, we have included two chapters devoted to conservation principles for the first-order fields. From these, power flow and energy relations are established and related to wave solutions. The last chapter illustrates the details of setting up some boundary-value problems and gives the quasi-static approximation that has proved so useful.

Recently (1961) a book by Denisse and Delcroix has appeared which has much the same objective as ours, but does not treat guided waves. They retain the radio point of view, judging waves by their frequency characteristics. We have adopted the optical criteria of judging waves by their local characteristics: the shape of the phase velocity surface and the polarization.

A selected literature list is given at the end of this book. We have not tried, however, to ascribe in each instance each fact to its original author. No claim is made to the discovery of new facts in this work. Only the mode of presentation is original, and it is hoped useful.

PART I

FREE WAVES

William P. Allis

and

Solomon J. Buchsbaum

Chapter 1

GENERAL PROPERTIES OF WAVES IN ANISOTROPIC MEDIA

1.1 The Material Equations

The solution of any theory in plasma physics is generally obtained by assuming a form for the electric field $\underset{\sim}{E}$, using this to calculate the charge density ρ and current $\underset{\sim}{J}$, and then requiring that $\underset{\sim}{E}$, ρ, and $\underset{\sim}{J}$ be consistent with Maxwell's equations. We wish to study periodic waves and shall therefore assume the time and space variation of $\underset{\sim}{E}$ to be as $\exp j(\omega t - \underset{\sim}{k} \cdot \underset{\sim}{r})$ and shall further specify that ω is real and let the propagation vector $\underset{\sim}{k}$ come out as it may. There are several choices for the next step: In Chapters 2 to 4 we use the equation of motion for a single particle of mass m_i and charge e_i with the Lorentz $(\underset{\sim}{E} + \underset{\sim}{v}_i \times \underset{\sim}{B})$ and Langevin $(-m_i \nu_{ci} \underset{\sim}{v}_i)$ forces

$$m_i \frac{\partial \underset{\sim}{v}_i}{\partial t} = e_i(\underset{\sim}{E} + \underset{\sim}{v}_i \times \underset{\sim}{B}) - m_i \nu_{ci} \underset{\sim}{v}_i \tag{1.1}$$

In Chapter 5 we use the transport equations for the particle density N_i and the particle current $\underset{\sim}{\Gamma}_i$:

$$\left.\begin{aligned} &\frac{\partial N_i}{\partial t} + \underset{\sim}{\nabla} \cdot \underset{\sim}{\Gamma}_i = 0 \\ &\frac{\partial \underset{\sim}{\Gamma}_i}{\partial t} + \underset{\sim}{\nabla} \frac{N_i e_i T_i}{m_i} - \frac{e_i}{m_i}(N_i \underset{\sim}{E} + \underset{\sim}{\Gamma}_i \times \underset{\sim}{B}) = -\nu_{ci} \underset{\sim}{\Gamma}_i \end{aligned}\right\} \tag{1.2}$$

In Chapter 6 we use the Boltzmann equation for the particle distribution function $f_i(\underset{\sim}{r}, \underset{\sim}{v}, t)$

$$\frac{\partial f_i}{\partial t} + \underset{\sim}{\nabla}_r \cdot \underset{\sim}{v} f_i + \frac{e_i}{m_i} \underset{\sim}{\nabla}_v \cdot (\underset{\sim}{E} + \underset{\sim}{v} \times \underset{\sim}{B}) f_i = \nu_{ci}(f_{0i} - f_i) \tag{1.3}$$

In each case the equations are first linearized because we do not choose to study the nonlinear effects, and are then applied separately to electrons and ions, several kinds of ions if desired (Section 4.5), and the results added to get the charges

$$\rho = \sum_i e_i N_i \quad \text{or} \quad \sum_i e_i \int f_i(\underset{\sim}{r}, \underset{\sim}{v}, t)\, d^3v \tag{1.4}$$

and currents

$$\underset{\sim}{J} = \sum_i e_i \underset{\sim}{v}_i N_i \quad \text{or} \quad \sum_i e_i \underset{\sim}{\Gamma}_i \quad \text{or} \quad \sum_i e_i \int \underset{\sim}{v} f_i(\underset{\sim}{r}, \underset{\sim}{v}, t) \, d^3v \qquad (1.4a)$$

These must satisfy the continuity equation

$$\frac{\partial \rho}{\partial t} + \underset{\sim}{\nabla} \cdot \underset{\sim}{J} = 0 \qquad (1.5)$$

In this chapter we shall assume Ohm's law

$$\underset{\sim}{J} = \underset{\approx}{\sigma} \cdot \underset{\sim}{E} \qquad (1.6)$$

in which the conductivity $\underset{\approx}{\sigma}$ has tensor character because of the presence of a static magnetic field $\underset{\sim}{B}_0$ directed along the z axis. This conductivity can be made to include the effects of the spatial gradient terms of the transport or Boltzmann equations because, in the linear approximation, these are also proportional to $\underset{\sim}{E}$. Therefore, the relations obtained here have a considerable degree of generality. We shall not at this point make any assumptions about the dependence of $\underset{\approx}{\sigma}$ on the density N, temperature T, circular frequency ω, or other parameters, but we shall use the symmetry that follows from a special choice of coordinate axes and from the axial nature of the static magnetic field vector $\underset{\sim}{B}_0$

According to the Onsager relations, the subscripts on the conductivity tensor are interchanged when the sign of an axial vector ($\underset{\sim}{B}_0$) is changed but are not affected by a change in sign of a polar vector such as the propagation vector $\underset{\sim}{k}$:

$$\sigma_{ij}(\underset{\sim}{B}_0, \underset{\sim}{k}) = \sigma_{ji}(-\underset{\sim}{B}_0, \underset{\sim}{k}) = \sigma_{ji}(-\underset{\sim}{B}_0, -\underset{\sim}{k}) \qquad (1.7)$$

Let the propagation vector $\underset{\sim}{k}$ be so oriented with respect to a Cartesian coordinate system that the z, y, x axes are along $\underset{\sim}{B}_0$, $\underset{\sim}{B}_0 \times \underset{\sim}{k}$, and $(\underset{\sim}{B}_0 \times \underset{\sim}{k}) \times \underset{\sim}{B}_0$, respectively. Changing the signs of $\underset{\sim}{B}_0$ and $\underset{\sim}{k}$ then changes the directions of the x and z axes but not of y, and hence changes the signs of σ_{xy} and σ_{yz} but not of σ_{xz} or of the diagonal elements. It follows that

$$\sigma_{xy} = -\sigma_{yx}, \quad \sigma_{yz} = -\sigma_{zy}, \quad \sigma_{zx} = \sigma_{xz} \qquad (1.8)$$

and this reduces the number of independent tensor components to six.

Starting with Section 2.4, we shall be particularly concerned with lossless plasmas. It then follows that $\underset{\approx}{\sigma}$ is anti-Hermitian, and hence that σ_{xy} and σ_{yz} are pure real, while σ_{zx} and the diagonal components are pure imaginary.

1.2 The Tensor Dielectric Coefficient

As $\underset{\approx}{\sigma}$ often turns out to be nearly pure imaginary, it is more convenient to use an equivalent dielectric coefficient defined by

$$\underset{\approx}{K} = \underset{\approx}{I} + \frac{\underset{\approx}{\sigma}}{j\omega\epsilon_0} \tag{1.9}$$

where $\underset{\approx}{I}$ is the unity matrix and ϵ_0 the permittivity of free space. The dielectric tensor has the same symmetry as the conductivity:

$$K_{xy} = -K_{yx}, \quad K_{yz} = -K_{zy}, \quad K_{zx} = K_{xz} \tag{1.10}$$

If there are no energy losses, the tensor is Hermitian, so that K_{xy} and K_{yz} are pure imaginary and K_{zx} as well as the diagonal terms is real, but no use will be made of this property at present.

It is convenient to use the left and right rotating coordinates

$$\left. \begin{array}{ll} \sqrt{2}\,\ell = x - jy, & \sqrt{2}\,E_\ell = E_x - jE_y \\[2mm] \sqrt{2}\,r = x + jy, & \sqrt{2}\,E_r = E_x + jE_y \end{array} \right\} \tag{1.11}$$

and the transformation is effected by the unitary matrix

$$U = \frac{1}{\sqrt{2}} \begin{bmatrix} 1 & -j & 0 \\ 1 & j & 0 \\ 0 & 0 & \sqrt{2} \end{bmatrix}, \quad U^{-1} = \frac{1}{\sqrt{2}} \begin{bmatrix} 1 & 1 & 0 \\ j & -j & 0 \\ 0 & 0 & \sqrt{2} \end{bmatrix} \tag{1.12}$$

The tensor $\underset{\approx}{K}$ transforms into the symmetric tensor

$$K' = UKU^{-1} = \begin{bmatrix} K_\ell & K_{\ell r} & K_{\ell p} \\ K_{\ell r} & K_r & K_{rp} \\ K_{\ell p} & K_{rp} & K_p \end{bmatrix} \tag{1.13}$$

whose components are

$$2K_\ell = K_{xx} + K_{yy} + 2jK_{xy} \qquad 2K_{xx} = K_\ell + K_r + 2K_{\ell r}$$

$$2K_r = K_{xx} + K_{yy} - 2jK_{xy} \qquad 2K_{yy} = K_\ell + K_r - 2K_{\ell r}$$

$$K_p = K_{zz} \qquad\qquad\qquad K_{zz} = K_p$$

$$2K_{\ell r} = K_{xx} - K_{yy} \qquad\qquad 2K_{xy} = j(K_r - K_\ell) = -2K_{yx}$$

$$\sqrt{2}\,K_{\ell p} = K_{xz} - jK_{yz} \qquad \sqrt{2}\,K_{xz} = K_{rp} + K_{\ell p} = \sqrt{2}\,K_{zx}$$

$$\sqrt{2}\,K_{rp} = K_{xz} + jK_{yz} \qquad \sqrt{2}\,K_{yz} = K_{\ell p} - K_{rp} = \sqrt{2}\,K_{zy}$$

$$\tag{1.14}$$

The greatest advantage of this representation occurs when $K_{xx} = K_{yy}$ and $K_{xz} = K_{yz} = 0$, so that Equation 1.13 is a diagonal tensor indicating that the three waves represented by (E_ℓ, E_r, E_z) propagate independently of each other.

1.3 The Field Equations

Maxwell's equations are

$$\left.\begin{aligned} \nabla \cdot \mathbf{E} &= \frac{\rho}{\epsilon_0}, & \nabla \cdot \mathbf{B} &= 0 \\ \nabla \times \mathbf{E} &= -\mu_0 \frac{\partial \mathbf{H}}{\partial t}, & \nabla \times \mathbf{H} &= \epsilon_0 \frac{\partial \mathbf{E}}{\partial t} + \mathbf{J} \end{aligned}\right\} \tag{1.15}$$

in which the plasma appears through the space charge ρ and the conduction current $\mathbf{J}$.

In terms of the dielectric coefficient $\mathbf{K}$ and for fields that vary as $\exp j(\omega t - \mathbf{k} \cdot \mathbf{r})$, Maxwell's equations become

$$\mathbf{k} \cdot \mathbf{D} = \epsilon_0 \mathbf{k} \cdot \mathbf{K} \cdot \mathbf{E} = 0 \tag{1.16}$$

$$\mathbf{k} \cdot \mathbf{B} = 0 \tag{1.17}$$

$$\mathbf{k} \times \mathbf{E} = \omega \mathbf{B} \tag{1.18}$$

$$\mathbf{k} \times \mathbf{H} = -\omega \mathbf{D} = -\epsilon_0 \omega \mathbf{K} \cdot \mathbf{E} \tag{1.19}$$

In Equation 1.16, use was made of the continuity equation (1.5), and the total displacement $\mathbf{D}$ was defined as

$$\mathbf{D} = \epsilon_0 \mathbf{K} \cdot \mathbf{E} = \epsilon_0 \mathbf{E} + \frac{\mathbf{J}}{j\omega} \tag{1.20}$$

From Equations 1.16, 1.17, and 1.19, it follows that for a plane wave the vectors $\mathbf{k}$, $\mathbf{H}$, and $\mathbf{D}$ form an orthogonal set. The fields $\mathbf{H}$ and $\mathbf{D}$ are then always transverse to $\mathbf{k}$, but $\mathbf{E}$ and $\mathbf{J}$ are not necessarily so (Figure 1.1). In fact, the more interesting waves will be those in which $\mathbf{E}$ is partly (or wholly) longitudinal.

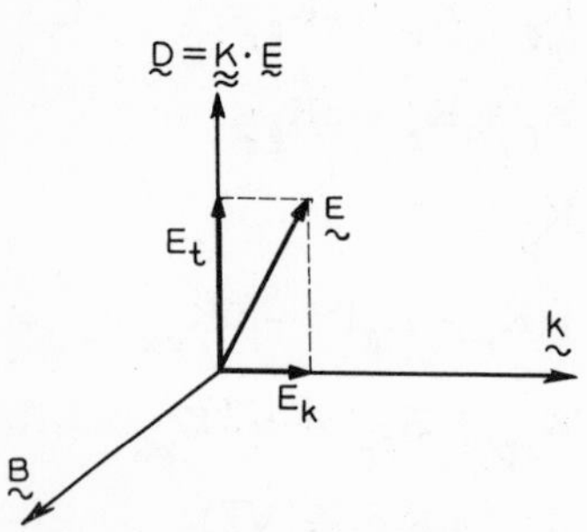

Figure 1.1. Directions of field vectors ($\mathbf{E}$, $\mathbf{B}$, $\mathbf{D}$) and of propagation vector $\mathbf{k}$ in a plane wave.

The relation between the propagation constant $\mathbf{k}$ and the dielectric coefficient $\mathbf{K}$ is obtained by iterating Equations 1.18 and 1.19. The result is

$$\mathbf{k} \times (\mathbf{k} \times \mathbf{E}) + k_0^2 \mathbf{K} \cdot \mathbf{E} = 0 \tag{1.21}$$

where $k_0 = \omega/c$ is the propagation constant in free space.

For some plasma models, it is easier to obtain the field $\mathbf{E}$ in terms of the current $\mathbf{J}$. In this case we write

$$j\omega\epsilon_0 \mathbf{E} = \mathbf{R} \cdot \mathbf{J} \tag{1.22}$$

where $\mathbf{R} = j\omega\epsilon_0 \boldsymbol{\sigma}^{-1}$ is the normalized plasma resistivity tensor. Then the wave equation is

$$\underset{\sim}{k} \times \underset{\sim}{k} \times (\underset{\approx}{R} \cdot \underset{\sim}{J}) + k_0^2 (\underset{\approx}{R} + \underset{\approx}{I}) \cdot \underset{\sim}{J} = 0 \tag{1.23}$$

where $\underset{\approx}{I}$ is the identity matrix.

1.4 The Dispersion Relation

Equation 1.21 represents a set of three linear homogeneous equations for the three field components (E_x, E_y, E_z). A nonzero solution exists only when the determinant of the coefficients vanishes.

$$\begin{vmatrix} k_0^2 K_{xx} - k_y^2 - k_z^2 & k_x k_y + k_0^2 K_{xy} & k_x k_z + k_0^2 K_{xz} \\ k_y k_x + k_0^2 K_{yx} & k_0^2 K_{yy} - k_z^2 - k_x^2 & k_y k_z + k_0^2 K_{yz} \\ k_z k_x + k_0^2 K_{zx} & k_z k_y + k_0^2 K_{zy} & k_0^2 K_{zz} - k_x^2 - k_y^2 \end{vmatrix} = 0 \tag{1.24}$$

Equation 1.24 can be solved for the propagation constant

$$k = \sqrt{k_x^2 + k_y^2 + k_z^2}$$

in terms of the direction cosines $(k_x : k_y : k_z)$ of $\underset{\sim}{k}$ and the components $\underset{\approx}{K}$. The properties of the vector $\underset{\sim}{k}$ determine the propagation and attenuation of the waves.

Let the y axis be taken along $\underset{\sim}{B}_0 \times \underset{\sim}{k}$ and let θ be the angle that $\underset{\sim}{k}$ makes with the direction of the static magnetic field, so that $k_x = k \sin\theta$, $k_y = 0$, $k_z = k\cos\theta$ (Figure 1.2). Then with the

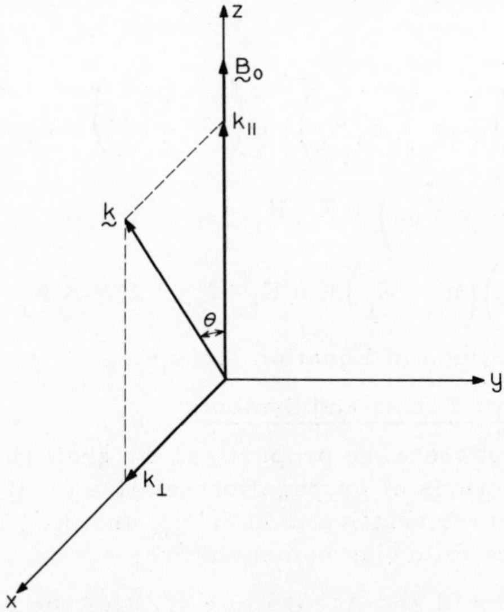

Figure 1.2. Orientation of coordinate axes with respect to static magnetic field $\underset{\sim}{B}_0$ and propagation vector $\underset{\sim}{k}$.

dielectric coefficient of the form given by Equation 1.10, the dispersion relation (1.24) can be put in the form

$$An^4 - Bn^2 + C = 0 \tag{1.25}$$

where $\underset{\sim}{n} = \underset{\sim}{k}/k_0$ is the vector index of refraction of the wave and where

$$A = K_{xx} \sin^2 \theta + 2K_{xz} \cos \theta \sin \theta + K_{zz} \cos^2 \theta \tag{1.26a}$$

$$B = K_{xx}K_{zz} - K_{xz}^2 + \left(K_{xx}K_{yy} + K_{xy}^2\right) \sin^2 \theta + 2\left(K_{yy}K_{xz} - K_{xy}K_{yz}\right) \sin \theta \cos \theta + \left(K_{yy}K_{zz} + K_{yz}^2\right) \cos^2 \theta \tag{1.26b}$$

$$C = |K| \tag{1.26c}$$

Equation 1.25 is quadratic in n^2, so that in general two distinct waves, propagating in either direction, can exist. However, when the dielectric coefficient $\underset{\approx}{K}$ itself is a function of the index of refraction, more than two waves may be possible. We shall find that in a plasma where thermal motions of the particles can be neglected, only two waves can propagate, but that in an N-component plasma with the thermal motion taken into account, N+2 distinct waves can exist.

It is often more convenient to solve the determinantal equation for tan θ instead of for the index n. The equation we obtain is then

$$a \tan^2 \theta + b \tan \theta + c = 0 \tag{1.27}$$

where

$$\left.\begin{aligned}
a &= \left(n^2 - K_{\parallel}\right)\left(K_{xx}n^2 - K_rK_\ell\right) + n^2\left(\frac{k_{xz}^2 + K_{r\ell}^2}{2}\right) + C - K_{\parallel}K_rK_\ell \\
b &= 2n^2\left[K_{xz}\left(n^2 - K_{yy}\right) + K_{xy}K_{yz}\right] \\
c &= K_{\parallel}\left(n^2 - K_r\right)\left(n^2 - K_\ell\right) + n^2K_{\ell p}K_{rp} + C - K_{\parallel}K_rK_\ell
\end{aligned}\right\} \tag{1.28}$$

where C is defined in Equation 1.26c.

1.5 Definition of Terms and Symbols

In the study of the wave properties, we shall find it necessary to talk about directions of propagation relative to $\underset{\sim}{B}_0$ and components of the field vectors relative to both $\underset{\sim}{B}_0$ and $\underset{\sim}{k}$. To avoid confusion, we shall use the following nomenclature:

Along ($\theta = 0$) and Across ($\theta = \pi/2$) for the principal directions of propagation relative to the static magnetic field $\underset{\sim}{B}_0$.

Parallel (‖ or p) and Perpendicular (⊥) for the Cartesian components of any field vector relative to $\underset{\sim}{B}_0$. Left (ℓ) and Right (r) for rotating components of any field vector relative to $\underset{\sim}{B}_0$.

Longitudinal (k) and Transverse (t) for the components of any field vector relative to the direction of propagation. The transverse component is further decomposed into its polar (θ) and azimuthal (φ) components.

The static magnetic field will always be indicated by $\underset{\sim}{B}_0$, and the wave magnetic field by $\underset{\sim}{B}$ (or $\underset{\sim}{H}$).

Principal waves are waves that propagate either along or across $\underset{\sim}{B}_0$.

1.6 Cutoffs and Resonances

The term cutoff will be used for the condition in which the phase velocity of the wave $u = \omega/k$ is infinite ($u = \infty$, $n^2 = 0 = k^2$), and the term resonance for the condition in which the phase velocity is zero ($u = 0$, $n^2 = \infty = k^2$). When the plasma parameters and the frequency are varied, the index n^2 moves about the complex plane and may pass through, or close to, these two points. When n^2 is real, so that it passes through these points, the cutoffs and resonances are sharp, otherwise they are diffuse. We shall in what follows adopt a model in which they are sharp, that is, the absence of collisions.

From Maxwell's equation (1.18), it follows that

$$\frac{\left| E_t \right|}{\left| B \right|} = \frac{\omega}{k} = u \tag{1.29}$$

As $\underset{\sim}{B}$ is always transverse, the phase velocity measures the impedance of the medium. In order to transfer energy efficiently to or from a wave, it is necessary to match impedances. Electron beams can be used effectively near a resonance. It is difficult to match to a wave near cutoff.

At a cutoff, then $\underset{\sim}{B}$ and $\underset{\sim}{J} + \epsilon_0 \dot{\underset{\sim}{E}} = -jk \times \underset{\sim}{H} = 0$. The real and displacement currents, which always cancel in the longitudinal direction (because $\underset{\sim}{D}$ is transverse), now cancel in the transverse direction as well. At a resonance, $E_t = 0$, so that the electric vector is purely longitudinal, $\underset{\sim}{E} = \underset{\sim}{E}_k$, or is zero. The transverse current $\underset{\sim}{J}_t = -jk \times \underset{\sim}{H}$ is then infinite when $\underset{\sim}{H} \neq 0$ or is finite when $\underset{\sim}{H} = 0$. We shall find both situations.

In inhomogeneous plasmas a wave in general will be reflected from a cutoff surface but will be absorbed at a resonant surface. Consider the limit of geometrical optics. If the index of refraction n is a function of position, say, of x, then for the laws of geometrical optics to hold, the relative change in the refractive index per wavelength in the medium must be small; that is,

$$\frac{1}{n}\left|\frac{dn}{dx}\right| \lambda << 1 \tag{1.30}$$

Since $\lambda = \lambda_0/n$, where λ_0 is the free-space wavelength, Condition 1.30 becomes

$$\frac{1}{n^2}\left|\frac{dn}{dx}\right| \lambda_0 << 1 \tag{1.31}$$

When Condition 1.31 is satisfied, the ray is determined by the principle of least time $\delta \int \underset{\sim}{k} \cdot d\underset{\sim}{s} = 0$, and the field vectors are given by the WKB approximation

$$E \sim \frac{E}{\sqrt{n(x)}} \exp j\left(\omega t \pm k_0 \int n(x)dx\right) \tag{1.32}$$

The two waves given by the plus and minus signs in Equation 1.32 refer to waves that travel in opposite directions and are independent of each other. Should there be a sharp boundary where Condition 1.31 fails, the two waves are no longer independent: One transforms into the other, and reflection occurs. Consider a ray approaching a cutoff, $n \to 0$; then n is decreasing, and the ray is refracted away from the cutoff surface and may be reflected in this way (Figure 1.3). If it is not refracted away, it will necessarily violate Condition 1.31 and thus be at least partially reflected. On the other hand, for a wave approaching a resonance, $n \to \infty$. With

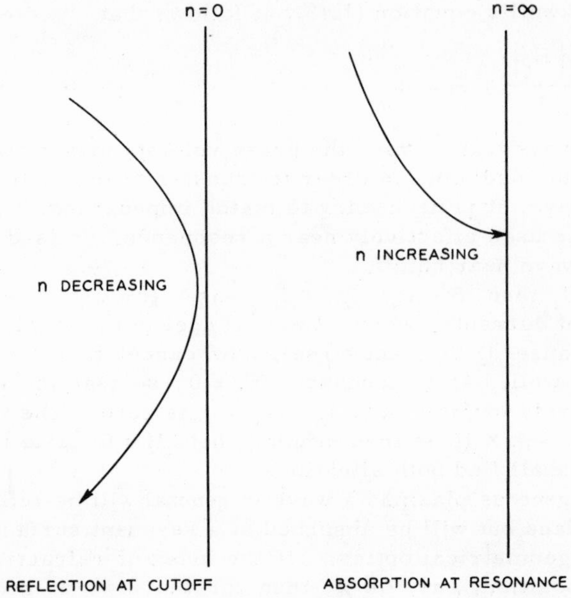

Figure 1.3. Rays in inhomogeneous plasma.

n increasing, it is refracted toward the resonant surface and reaches it normally (Figure 1.3). As it satisfies Condition 1.31 increasingly well, there is no reflection, the energy of the incoming wave being completely stored in the very large currents of the medium. In the presence of damping, however small, the stored energy is dissipated as heat. In the absence of any damping mechanism, there is no steady state and the currents increase indefinitely.

Cutoffs and resonances separate values of the plasma parameters in which n^2 is positive or negative and hence regions of propagation and of nonpropagation. The attenuation $\sqrt{-k^2}$ is small just beyond a cutoff but large just beyond a resonance. The characteristics of cutoffs and resonances are listed in Table 1.1.

Table 1.1. Characteristics of Cutoffs and Resonances

Cutoff	Resonance
$u = \infty$	$u = 0$
$k = n = 0$	$k = n = \infty$
$\underset{\sim}{H} = 0$	$E_t = 0$
$\underset{\sim}{J} + \epsilon_0 \dot{\underset{\sim}{E}} = 0$	$J/H = \infty$
Reflection	Absorption
All directions	Resonance cone

It can be seen from Equation 1.25 that if $C = 0$ and either A or $B \neq 0$, at least one root of the equation is zero. This represents, then, the cutoff condition. As C is independent of θ, the cutoff condition does not depend on the direction of propagation.

Similarly, $A = 0$ represents the resonance condition. As A does depend on θ, a resonance cone is defined by

$$K_{xx} \tan^2 \theta_{res} + 2K_{xz} \tan \theta_{res} + K_{zz} = 0 \qquad (1.33)$$

At angles near θ_{res}, the phase velocity of the wave is much below that of light in free space, and therefore the conditions of Cerenkov radiation are readily satisfied. An electron beam in any direction within the resonance cone, in particular, one along $\underset{\sim}{B}_0$, will produce Cerenkov radiation along the resonance cone[1] (Figure 1.4).

The direction of propagation along $\underset{\sim}{B}_0$ is, however, singular, and as we shall see shortly, the cutoff and resonance conditions given above do not necessarily hold in this direction.

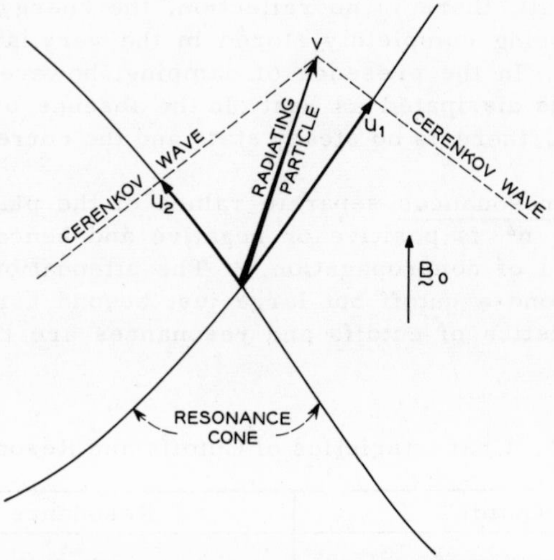

Figure 1.4. Cerenkov radiation at resonance cone.

1.7 Polarization of the Waves

When the determinant (1.24) vanishes, the relative magnitude of the components of the electric field vector $E_x : E_y : E_z$ are given by the ratio of the cofactors of the coefficients in Equation 1.21. Taking cofactors of the top row, we find in Cartesian coordinates

$$\left.\begin{aligned} E_x &\approx \left(n^2 - K_{yy}\right)\left(n^2 \sin^2\theta - K_{zz}\right) + K_{yz}^2 \\ E_y &\approx K_{yx}\left(n^2 \sin^2\theta - K_{zz}\right) + K_{yz}\left(n^2 \sin\theta\cos\theta + K_{xz}\right) \\ E_z &\approx \left(n^2 - K_{yy}\right)\left(n^2 \sin\theta\cos\theta + K_{xz}\right) + K_{xy}K_{yz} \end{aligned}\right\} \quad (1.34)$$

in rotating coordinates

$$\left.\begin{aligned} \sqrt{2}\,E_{r,\ell} = &\left(n^2 \sin^2\theta - K_{zz}\right)\left(n^2 - K_{yy} \pm jK_{yx}\right) \\ &\pm jK_{yz}\left(n^2 \sin\theta\cos\theta + K_{xz} \mp jK_{yz}\right) \end{aligned}\right\} \quad (1.35)$$

and the components of $\underset{\sim}{E}$ in the spherical coordinate system (k, θ, ϕ) are proportional to

$$\left.\begin{aligned}
E_\theta &\approx \left(K_{yy} - n^2\right)\left(K_{zx} \sin\theta + K_{zz} \cos\theta\right) \\
&\qquad + K_{yz}\left(K_{yx} \sin\theta + K_{yz} \cos\theta\right) \\
E_\phi &\approx n^2 \sin\theta\left(K_{yx} \sin\theta + K_{yz} \cos\theta\right) \\
&\qquad + K_{yz}K_{zx} + K_{xy}K_{zz} \\
E_k &\approx \left(K_{yy} - n^2\right)\left[\left(K_{zz} - n^2\right) \sin\theta - K_{zx} \cos\theta\right] \\
&\qquad + K_{yz}\left(K_{yz} \sin\theta - K_{yx} \cos\theta\right)
\end{aligned}\right\} \tag{1.36}$$

Then E_k is the longitudinal component of the wave. We note that E_k contains n^4, whereas E_θ and E_ϕ contain only n^2, confirming that near a resonance, and except when $\theta = 0$, the electric field becomes longitudinal.

Chapter 2

PARTICLE DISPLACEMENTS IN THE ELECTRIC FIELD

2.1 The "Temperate" Plasma

There are two main difficulties in the solution of Equations 1.1 through 1.3. First, the product $\underset{\sim}{E} \cdot \partial f/\partial \underset{\sim}{v}$, occurring in the Boltzmann equation renders this equation nonlinear in $\underset{\sim}{E}$. The condition for linearization may best be seen by considering the induced velocity $\underset{\sim}{v}_E$ of the average particle in the absence of a magnetic field

$$\underset{\sim}{v}_E = \frac{e\underset{\sim}{E}}{m} \frac{1}{(\nu_c + j\omega)} \tag{2.1}$$

Since the collision frequency ν_c depends on the total velocity, $\underset{\sim}{v}_E$ plus $\underset{\sim}{v}_{thermal}$, this equation is nonlinear unless the thermal speed

$$v_T = \sqrt{\frac{eT}{m}} \tag{2.2}$$

is much larger than v_E $(v_E << v_T)$. This is also the condition for convergence of small-signal approximations of the Boltzmann equation even in the absence of collisions.

Second, Equation 2.1 is itself not correct because the induced velocity v_E should be obtained by an integral of $\underset{\sim}{E}$ over the particle's trajectory, and the spatial variation of $\underset{\sim}{E}$ has not been taken into account in deriving Equation 2.1. This is acceptable only if the wavelength is sufficiently long or, more specifically, if $v_T << u$, where $u = \omega/k$ is the phase velocity of the wave. Thus the analysis, and also the phenomena, are peculiarly simple if the temperature is bracketed:

$$v_E << v_T << u \tag{2.3}$$

and such a plasma will be termed "temperate."

Most plasmas are, in fact, temperate except at resonances. Phase velocities are generally of the order of 3×10^8 m/sec. The electron thermal velocity of a 10-ev plasma is $v_T = 2 \times 10^6$ m/sec, which then requires that $E/|\nu_c + j\omega| << 10^{-5}$ volt-sec/m, which, if ν_c or ω, whichever is larger, is 10^7/sec, requires that $E << 1$ volt/cm; ν_c is about 10^7/sec in weakly ionized argon or mercury at a pressure of 1 micron.

2.2 Equations of Motion

For a temperate plasma, the induced motion is given by Newton's law with the Lorentz and Langevin forces

$$m_i \frac{d\underset{\sim}{v}_E}{dt} = e_i(\underset{\sim}{E} + \underset{\sim}{v}_E \times \underset{\sim}{B}) - m_i \nu_c \underset{\sim}{v}_E \tag{2.4}$$

By Inequality 2.3, ν_c is an average over the thermal motions, the precise average to be given in a later section, and therefore ν_c is temperature- and not field-dependent. We also linearize by including in $\underset{\sim}{B}$ only the static applied magnetic field $\underset{\sim}{B}_0$ and neglecting the alternating magnetic field of the wave. This amounts to neglecting v_E/u, which by Inequality 2.3 is a fortiori satisfied.

We define the cyclotron frequency vector for the ith constituent by

$$\underset{\sim}{\omega}_{b_i} = -\frac{e_i}{m_i} \underset{\sim}{B}_0 \tag{2.5}$$

where $\underset{\sim}{B}_0$ is the static magnetic field. Inserting Equation 2.5 in Equation 2.4 and dropping the subscripts give

$$(\nu_c + j\omega - \underset{\sim}{\omega}_b \times)\underset{\sim}{v}_E = \frac{e\underset{\sim}{E}}{m} \tag{2.6}$$

where we have assumed that both the field and the drift velocity vary harmonically with time as exp $(j\omega t)$, but have not used the factor exp $(-j\underset{\sim}{k} \cdot \underset{\sim}{r})$ because of Inequality 2.3.

The solution of the vector Equation 2.6 in vector form is

$$\underset{\sim}{v}_{E\perp} = \frac{(\nu_c + j\omega + \underset{\sim}{\omega}_b \times)\dfrac{e\underset{\sim}{E}_\perp}{m}}{[\nu_c + j(\omega - \omega_b)][\nu_c + j(\omega + \omega_b)]}\ ; \quad v_{E\parallel} = \frac{\dfrac{eE_\parallel}{m}}{\nu_c + j\omega} \tag{2.7}$$

and in tensor form

$$\underset{\sim}{v}_E = \begin{vmatrix} \ell + r & j(\ell - r) & 0 \\ j(r - \ell) & \ell + r & 0 \\ 0 & 0 & 2p \end{vmatrix} \cdot \frac{e\underset{\sim}{E}}{2m} \tag{2.8}$$

where

$$\ell, r = \frac{1}{j(\omega \pm \omega_b) + \nu_c}\ ; \quad p = \frac{1}{j\omega + \nu_c} \tag{2.9}$$

are times that correspond to left circular, right circular, and parallel motions relative to $\underset{\sim}{B}_0$. In Equations 2.7, $\underset{\sim}{E}_\perp$ and $\underset{\sim}{E}_\parallel$ are the components of $\underset{\sim}{E}$ perpendicular to and parallel with the static magnetic field, which is in the z direction.

When ν_c is a function of velocity, the quantities ℓ, r, and p are proper averages over particle distribution functions. For electrons these are [2]

$$\left.\begin{aligned} \ell, r &= -\frac{4\pi}{3}\int_0^\infty \frac{v^3 \dfrac{\partial f_0}{\partial v}\, dv}{j(\omega \pm \omega_b) + \nu_c(v)} \\ p &= -\frac{4\pi}{3}\int_0^\infty \frac{v^3 \dfrac{\partial f_0}{\partial v}\, dv}{j\omega + \nu_c(v)} \end{aligned}\right\} \tag{2.10}$$

where $f_0(v)$ is the spherically symmetric part of electron distribution function. The quantity $\nu_c(v)$ that appears in these integrals represents an expansion in powers of (m_-/M_s) of the collision integral that appears in the Boltzmann equation, where m_- is the electron mass and M_s is the effective mass of the scattering center (atom, molecule, or ion). Thus the integrals can properly describe the effect of electron-atom or electron-ion collisions provided $f_0(v)$ is known. No corresponding expressions have yet been derived for ions where (m_+/M_s) is of the order of unity.[3] In this monograph we shall not consider in detail the effect of collisions on wave propagation. We shall thus assume that Equations 2.8 and 2.9 are sufficiently accurate to represent the effect of collisions of both electrons and ions.

The significance of the terms ℓ, r, and p in Equation 2.8 is best shown by transforming to a rotating system of coordinates through the unitary transformation (1.12). Then

$$\underset{\sim}{v}_E = \frac{e}{m}\begin{bmatrix} \ell & 0 & 0 \\ 0 & r & 0 \\ 0 & 0 & p \end{bmatrix} \cdot \underset{\sim}{E} \tag{2.11}$$

From $\underset{\sim}{v}_E$ we obtain the conductivity by adding the currents:

$$\sum N_i e_i \underset{\sim}{v}_E = \underset{\approx}{\sigma} \cdot \underset{\sim}{E} \tag{2.12}$$

where N_i is the density of ith charged-particle species. Hence, in Cartesian coordinates the dielectric coefficient $\underset{\approx}{K}$ has the form

$$\underset{\approx}{K} = \begin{bmatrix} K_\perp & -K_\times & 0 \\ K_\times & K_\perp & 0 \\ 0 & 0 & K_\parallel \end{bmatrix} \tag{2.13}$$

where the subscript $\times$ refers to the "cross-product" component of the tensor. In rotating coordinates,

$$\underset{\approx}{K} = \begin{bmatrix} K_\ell & 0 & 0 \\ 0 & K_r & 0 \\ 0 & 0 & K_p \end{bmatrix} \tag{2.14}$$

where $K_p \equiv K_{\|}$, $K_\perp = (K_\ell + K_r)/2$, $K_\times = j(K_\ell - K_r)/2$.

2.3 The Tensor Components

Let us define for electrons (-) the parameters†

$$\left.\begin{aligned} \omega_{p_-}^{\ 2} &= \frac{Ne^2}{\epsilon_0 m_-}, & \alpha_- &= \frac{\omega_{p_-}}{\omega} \\ \omega_{b_-} &= \frac{eB_0}{m_-}, & \beta_- &= \frac{\omega_{b_-}}{\omega} \\ & & \gamma_- &= \frac{\nu_{c_-}}{\omega} \end{aligned}\right\} \tag{2.15}$$

and consider first waves of frequency high enough so that ions do not contribute appreciably to the current. We then have

$$\left.\begin{aligned} K_{\|} &= 1 - \frac{\alpha_-^2}{1 - j\gamma_-} \\ K_\ell &= 1 - \frac{\alpha_-^2}{1 + \beta_- - j\gamma_-} \\ K_r &= 1 - \frac{\alpha_-^2}{1 - \beta_- - j\gamma_-} \end{aligned}\right\} \tag{2.16}$$

There exists a simple geometrical construction[4] that represents these coefficients in the complex plane (Figure 2.1).

The figure is centered at the point $K = 1$, because all points of the figure move outward along radii from here as α_-^2 increases.

The point $K_{\|}$ is on a radius that passes through the point $-j\gamma_-$, and as the vectors

$$\left.\begin{aligned} K_{\|} - 1 &= -\frac{\alpha_-^2}{(1 - j\gamma_-)} \\ K_{\|} - 1 + \alpha_-^2 &= -\frac{j\gamma_-\alpha_-^2}{(1 - j\gamma_-)} \end{aligned}\right\} \tag{2.17}$$

are orthogonal, the circle for which 1 and $1 - \alpha_-^2$ are diametrical points passes through $K_{\|}$. Thus the coordinate lines corresponding

† Our symbols α_-^2, β_-, and γ_- have the same meanings as the U.R.S.I. symbols X, Y, and Z.

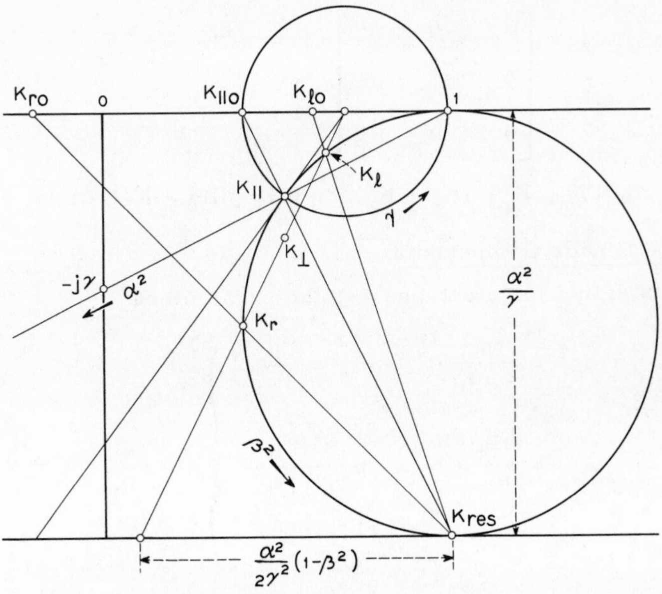

Figure 2.1. Geometrical representation of the components of dielectric tensor in the complex dielectric-coefficient plane.

to the parameters α_-^2 and γ_- are the lines $(1, -j\gamma_-)$ and the circles $(1, 1 - \alpha_-^2)$.

Consider now the quantities

$$K_{\ell,r} - 1 + \frac{j\alpha_-^2}{2\gamma_-} = \frac{j\alpha_-^2}{2\gamma_-}\,\frac{1 \pm \beta_- + j\gamma_-}{1 \pm \beta_- - j\gamma_-} \tag{2.18}$$

These quantities, as also $K_{\parallel} - 1 + (j\alpha_-^2/2\gamma_-)$, have the same absolute value $\alpha_-^2/2\gamma_-$. Therefore, the points K_ℓ, K_r, $K_{\parallel}$, and 1 are all on a circle centered at $1 - (j\alpha_-^2/2\gamma_-)$. The points K_ℓ and K_r move along this circle when β_-^2 varies at constant α_-^2/γ_-. How they move is determined by the vectors

$$K_{\ell,r} - 1 + \frac{\alpha_-^2}{2} = \frac{\alpha_-^2}{2}\,\frac{\beta_-^2 - (1 + j\gamma_-)^2}{(1 \pm \beta_-^2) + \gamma_-^2} \tag{2.19}$$

These vectors have the same direction [slope $-2j\gamma_-/(\beta_-^2 + \gamma_-^2 - 1)$] and hence the points $1 - (\alpha_-^2/2)$, K_ℓ, and K_r are on the same line. This line passes through the point $1 + [(\beta_-^2 - 1)\alpha_-^2/2\gamma_-]$, $-j\alpha_-^2/\gamma_-$ that moves linearly with β_-^2 to the right. Thus K_ℓ and K_r are the intersections of the β_-^2 circle and a radius vector of the γ_- circle. This radius vector rotates counterclockwise as β_-^2 increases. It is tangent to the β_-^2 circle for $\beta_-^2 = 0$ and $\beta_-^2 = \infty$ and passes through the point $K_{res} = 1 - (j\alpha_-^2/\gamma_-)$ at cyclotron resonance $(\beta_-^2 = 1)$.

The point $K_\perp$ is on the same radius vector halfway between K_ℓ and K_r. The point $K_\times$ is not easily represented on this diagram.

The diagram of Figure 2.1 cannot be drawn if $\gamma_- = 0$, because all the points K_ℓ, $K_\parallel$, and K_r fall on the real axis. However, it is readily shown that these points, represented by K_{ℓ_0}, $K_{\parallel_0}$, and K_{r_0} on the diagram, are obtained by projecting the points K_ℓ, $K_\parallel$, and K_r on the real axis from the point K_{res}. This is so because K_ℓ, $K_\parallel$, and K_r all move on γ_- circles as γ_- decreases.

We shall henceforth neglect collisions entirely, setting $\gamma_- = 0$. They can be retained in a perfectly straightforward manner, but to do so results in complex expressions that are otherwise real or pure imaginary, and this complicates the discussion unnecessarily

For a neutral plasma with electrons (-) and a single kind of ion (+),

$$\left.\begin{aligned} K_\ell &= 1 - \frac{\alpha_-^2}{1+\beta_-} - \frac{\alpha_+^2}{1-\beta_+} = 1 - \frac{\alpha^2}{(1+\beta_-)(1-\beta_+)} \\ K_r &= 1 - \frac{\alpha^2}{(1-\beta_-)(1+\beta_+)} \\ K_\parallel &= 1 - \alpha_-^2 - \alpha_+^2 = 1 - \alpha^2 \end{aligned}\right\} \tag{2.20}$$

$$\left.\begin{aligned} K_\perp &= \frac{(K_\ell + K_r)}{2} = 1 - \frac{\alpha^2(1-\beta_+\beta_-)}{(1-\beta_-^2)(1-\beta_+^2)} \\ K_\times &= \frac{j(K_\ell - K_r)}{2} = j\,\frac{\alpha^2(\beta_- - \beta_+)}{(1-\beta_-^2)(1-\beta_+^2)} \end{aligned}\right\} \tag{2.21}$$

where

$$\alpha^2 = \frac{\omega_p^2}{\omega^2} = \frac{Ne^2}{\epsilon_0\omega^2}\,\frac{m_+ + m_-}{m_+ m_-} \tag{2.22}$$

Note that

$$K_\perp^2 + K_\times^2 = K_r K_\ell \tag{2.23}$$

When α is written without a subscript, the reduced mass is to be used in the plasma frequency formula.

2.4 The Dispersion Relation

The dispersion relation

$$F(\underset{\sim}{n}) = An^4 - Bn^2 + C = 0 \tag{2.24}$$

now has the coefficients

$$\left.\begin{aligned} A &= K_\perp \sin^2\theta + K_\parallel \cos^2\theta \\ B &= K_r K_\ell \sin^2\theta + K_\parallel K_\perp (1+\cos^2\theta) \\ C &= K_r K_\ell K_\parallel \end{aligned}\right\} \quad (2.25)$$

The discriminant of this equation is

$$D^2 = B^2 - 4AC = -4K_\parallel^2 K_\times^2 \cos^2\theta + (K_r K_\ell - K_\parallel K_\perp)^2 \sin^4\theta \quad (2.26)$$

As $K_\times^2$ is negative, the discriminant D^2 is positive. Consequently, for this plasma model, n^2 is always real, and n is either real or imaginary. This sharp distinction between conditions of propagation and nonpropagation exists by virtue of the simple assumption made in deriving the tensor $\underset{\approx}{K}$, namely, the absence of collisions and the neglect of thermal motion.

It is somewhat easier to understand the properties of the waves if Equation 2.24 is solved for the direction of propagation θ in terms of index of refraction n, as was done first by Åström:[5]

$$\tan^2\theta = -\frac{K_\parallel(n^2 - K_r)(n^2 - K_\ell)}{(n^2 - K_\parallel)(K_\perp n^2 - K_r K_\ell)} \quad (2.27)$$

The indices along the principal directions of propagation, along $\underset{\sim}{B}_0$ ($\theta = 0$) and across $\underset{\sim}{B}_0$ ($\theta = 90°$), are then particularly simple to discuss.

2.5 Geometrical Representations

The usual geometrical representations of the dielectric coefficient $\underset{\approx}{K}$ are the Fresnel and index ellipsoids. Both of these represent the equation

$$\underset{\sim}{E} \cdot \underset{\sim}{D} = 1 \quad (2.28)$$

the first being a polar plot of $\underset{\sim}{E}$ and the second a polar plot of $\underset{\sim}{D}$. These surfaces are ellipsoids of revolution with axes proportional to $K_\perp^{-\frac{1}{2}}$, $K_\perp^{-\frac{1}{2}}$, $K_\parallel^{-\frac{1}{2}}$ or $K_\perp^{\frac{1}{2}}$, $K_\perp^{\frac{1}{2}}$, $K_\parallel^{\frac{1}{2}}$, respectively. Because $K_\perp$ occurs twice in these ratios, it determines a circular diameter of the ellipsoids, and one diameter of all central sections of the ellipsoids. It follows, in crystal optics, that one wave has an index $n = K_\perp^{\frac{1}{2}}$ independently of the direction of propagation. This wave obeys Snell's law and is called the "ordinary" wave. The other wave has an index depending on both $K_\perp$ and $K_\parallel$ and on the direction of propagation and is called the "extraordinary" wave.

Students of the ionosphere have been concerned with the frequency dependence of the index of refraction, and have noticed that one of the waves propagating across the magnetic field had an index $n = K_\parallel^{\frac{1}{2}}$ which did not depend on the strength of the magnetic field, and in particular did not exhibit cyclotron resonance. They named this wave "ordinary" and the other one "extraordinary. The large volume of literature on ionospheric propagation makes it impossible now to correct this conflict in terminology, but the situation is not too serious because the crystal definition applies to nongyratory media, whereas the ionosphere definition applies to nonpolarizable media. We are adopting the ionosphere notation, but we shall restrict it to the wave that is strictly independent of $\underset{\sim}{B}_0$ according to the original ionospheric definition.

Students of magnetohydrodynamics have adopted the terminology of crystal optics. This is in direct conflict with the ionosphere terminology, as the magnetoionic waves go continuously into the magnetohydrodynamic ones.

The Fresnel and index ellipsoids are not suitable figures to represent gyratory media because the antisymmetric parts of the dielectric tensor cancel out in Equation 2.28.

There are three surfaces that are of use in representing wave propagation: (1) the "slowness surface," (2) the "phase velocity surface," and (3) the "ray surface."

William Rowan Hamilton defined the slowness surface (sometimes called the reciprocal wave surface or the index surface) as a polar plot of the vector $\underset{\sim}{k}/\omega = \underset{\sim}{n}/c$, and its equation for our case is Equation 2.24. Its importance derives from the fact that the net energy flow, that is, the time-averaged Poynting vector, is at a right angle to it (Part II, Equation 7.75; but see also Equation 8.100).

The inverse of the slowness surface is the phase velocity surface, or wave normal surface, which is a polar plot of the phase velocity

$$\underset{\sim}{u} = \frac{\omega \underset{\sim}{k}}{k^2} \tag{2.29}$$

In this monograph, we adopt the "phase velocity surface" to represent wave propagation, but we shall occasionally abbreviate it "velocity surface." Its equation is readily obtained from Equation 2.24 and can be written in the normal form

$$\frac{\cos^2\theta}{u^2 - u_0^2} + \frac{1}{2}\,\frac{\sin^2\theta}{u^2 - u_r^2} + \frac{1}{2}\,\frac{\sin^2\theta}{u^2 - u_\ell^2} = 0 \tag{2.30}$$

where

$$u_0^2 = \frac{c^2}{K_\parallel}, \quad u_r^2 = \frac{c^2}{K_r}, \quad u_\ell^2 = \frac{c^2}{K_\ell} \tag{2.31}$$

Equation 2.30 can be compared with the similar phase velocity equation in crystal optics

$$\frac{\sin^2\theta\,\cos^2\phi}{u^2 - u_1^2} + \frac{\sin^2\theta\,\sin^2\phi}{u^2 - u_2^2} + \frac{\cos^2\theta}{u^2 - u_3^2} \tag{2.32}$$

in which u_1, u_2, and u_3 are the principal velocities.

Although the phase velocity surfaces in a gyrotropic plasma have three independent velocities, they have cylindrical symmetry, which the crystal waves do not. The variety of surfaces to be displayed in the next chapter does not arise from the form of the wave surface equation but from the dependence of the coefficients K_r and K_ℓ on α^2 and β^2 which allows these coefficients to take all values from $-\infty$ to ∞. In crystal optics the principal velocities differ by minute amounts.

The envelope at time $t = 1$ of wave planes that passed the origin at time $t = 0$ is the ray surface. The phase velocity surface is the pedal surface[6] of the ray surface. The slowness surface is the polar reciprocal of the ray surface. The ray surface is the one that must be used in Huyghen's construction. Its equation is given in parametric form by

$$\underset{\sim}{v} = \frac{c\underset{\sim}{\nabla}_n F}{\underset{\sim}{n} \cdot \underset{\sim}{\nabla}_n F} \tag{2.33}$$

where $F(n)$ is the dispersion equation (2.24) and $\underset{\sim}{\nabla}_n$ denotes the gradient operator with respect to the components of the index n. The group velocity $\underset{\sim}{u}_g$ has the same direction, but not the same magnitude, as the ray velocity. It is given by

$$\underset{\sim}{u}_g = -\frac{c\underset{\sim}{\nabla}_n F}{\omega \dfrac{\partial F}{\partial \omega}} \tag{2.34}$$

Chapter 3

PHASE VELOCITY SURFACES

3.1 Principal Waves along $\underset{\sim}{B}_0$

From Equation 2.27, it is seen that for propagation along the magnetic field ($\theta = 0$) the two possible waves have the indices

$$\frac{c^2}{u_r^2} = n_r^2 = K_r \tag{3.1}$$

and

$$\frac{c^2}{u_\ell^2} = n_\ell^2 = K_\ell \tag{3.2}$$

Recourse to the discussion leading to Equation 1.13 indicates that these waves are right and left circularly polarized with respect to the static magnetic field. These waves do or do not propagate depending on the signs of K_r and K_ℓ, and these change at the cutoffs and resonances that, from Equations 2.20, are determined by the expressions in Table 3.1.

Table 3.1. Definition of Cyclotron Cutoffs and Resonances

	Cyclotron Cutoff	Cyclotron Resonance
Left-hand Wave	$\alpha^2 = (1 + \beta_-)(1 - \beta_+)$ (3.3a)	$\beta_+ = 1$ (3.4a)
Right-hand Wave	$\alpha^2 = (1 - \beta_-)(1 + \beta_+)$ (3.3b)	$\beta_- = 1$ (3.4b)

These will be called the cyclotron resonances and the cyclotron cutoffs. At low plasma densities, $\alpha^2 \ll 1$, the two cutoffs are quite close to the corresponding resonances, so that the frequency range over which these waves do not propagate is a narrow interval above the resonance frequency, but at high plasma densities the cutoff is far from the resonance.

The dispersion relation is shown by a conventional propagation constant plot (k^2c^2 versus ω^2) and ($kc = n\omega$ versus ω) in Figures 3.1 and 3.2 by curves marked r and ℓ. More information can be shown on a single diagram using the normalized variables α^2, $\beta_-\beta_+$, and n^2. Such diagrams will be referred to as (α^2, β^2) diagrams, or (α^2, β^2) planes. The cutoff and resonance conditions are shown in a plot of $\beta_-\beta_+$ against α^2 (Figure 3.3). Increasing the plasma

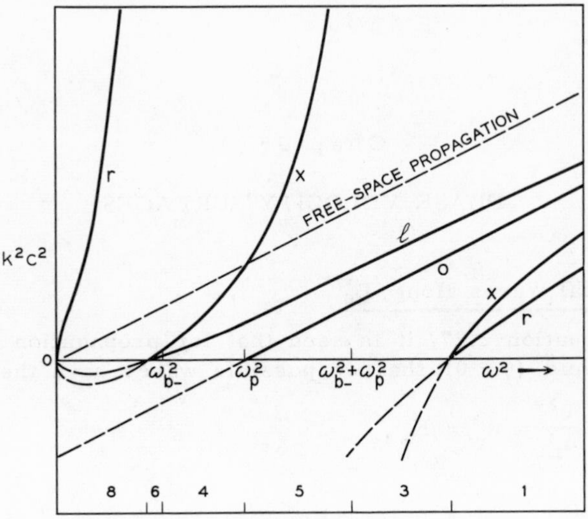

Figure 3.1. Propagation constant plot for the principal waves when ions are assumed immobile. The numbers at the bottom scale refer to regions of the (α^2, β^2) plane (Section 3.3).

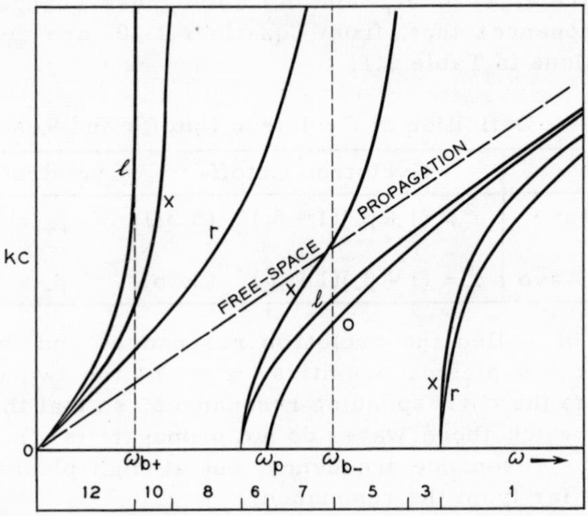

Figure 3.2. Propagation constant plot for the principal waves with ion motion included. The numbers at the bottom scale refer to regions of the (α^2, β^2) plane (Section 3.3).

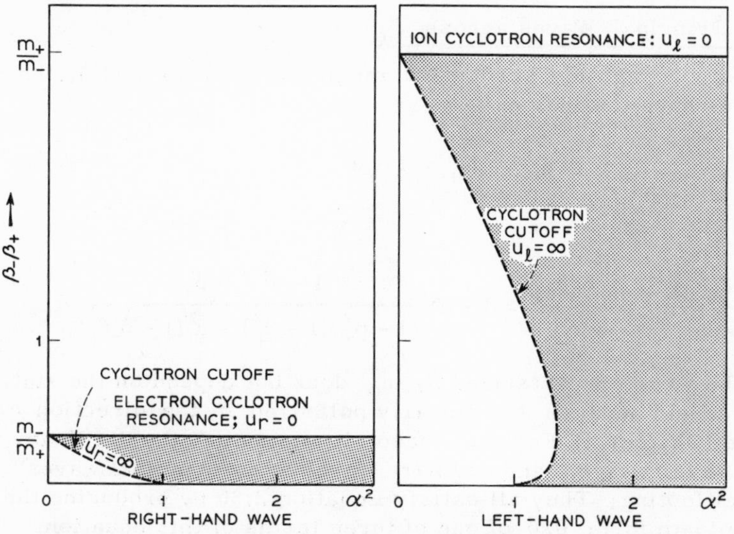

Figure 3.3. The (α^2, β^2) plot for propagation along $\underset{\sim}{B}_0$ for ion-to-electron mass ratio of 3 to 1. In the dotted regions the waves do not propagate $(u^2 < 0)$.

density N corresponds to displacement of a point on such a plot to the right, increasing the magnetic field to a displacement upward, and increasing the frequency to a radial displacement toward the origin. The two cyclotron resonances are horizontal lines whose ordinates are in the ratio $(m_+/m_-)^2$. For practical reasons, Figure 3.3 and succeeding figures are drawn for an assumed ratio $m_+/m_- = 3$. This would indeed be possible for holes and electrons in a semiconductor or a metal plasma, but for an ionized gas it must be realized that the figures are greatly off scale. The two cutoff lines belong to the same parabola whose summit is near $\alpha^2 = 1$, $\beta^2 = 0$ and whose axis is inclined at an angle $\sqrt{m_-/m_+}$ to the vertical, so that it passes through the two points $(\alpha^2 = 0, \beta_-^2 = 1)$ and $(\alpha^2 = 0, \beta_+^2 = 1)$.

It can be verified that $K_\ell < K_r$ or $u_r < u_\ell$ between the electron and ion cyclotron resonance lines, and conversely $u_\ell < u_r$ outside of these lines. The region over which these waves do not propagate has been shaded in Figure 3.3.

There is a third solution of Equation 2.27 for $\theta = 0$ which is

$$K_{\parallel} = 0 \quad \text{or} \quad \omega = \omega_p \tag{3.5}$$

This is a longitudinal oscillation at the plasma frequency. It cannot be properly called a "wave" because n and, therefore, λ are arbitrary. It is a "plasma oscillation" and does not in this model extend to directions other than those along $\underset{\sim}{B}_0$.

3.2 Principal Waves across $\underset{\sim}{B}_0$

For propagation across the magnetic field ($\theta = 90°$), the two waves have the indices

$$\frac{c^2}{u_o^2} = n_o^2 = K_{\parallel} = 1 - \alpha^2 \tag{3.6}$$

and

$$\frac{c^2}{u_\times^2} = n_\times^2 = \frac{K_r K_\ell}{K_\perp} = 1 - \alpha^2 \frac{1 - \alpha^2 - \beta_+\beta_-}{(1 - \beta_-^2)(1 - \beta_+^2) - \alpha^2(1 - \beta_-\beta_+)} \tag{3.7}$$

The wave characterized by n_o does not depend on the static magnetic field because it is linearly polarized in the direction of this field (Chapter 4). For this reason it was named the "ordinary wave."

The left, right, and ordinary waves are "principal waves" on an equal footing: They all satisfy Equation 2.30 by producing the indeterminate form 0/0 of one of three terms of this equation. The fourth principal wave is different, and has been named "extraordinary," in that it is obtained by equating the sum of the coefficients of $\sin^2\theta$ in Equation 2.30 to zero so that

$$2u_\times^2 = u_r^2 + u_\ell^2 \tag{3.8}$$

Thus the velocity of the extraordinary wave is always intermediate between the velocities of the right and left circularly polarized waves. It is "extraordinary" also, as will be seen in Chapter 4, in that it has the polarization appropriate to the right or left waves although it travels across $\underset{\sim}{B}_0$. The polarizations of all principal waves are shown in Figure 3.4.

The propagation constant plots of the ordinary and extraordinary waves are shown in Figures 3.1 and 3.2 and are denoted by o and $\times$.

The cutoff and resonance conditions for the ordinary and extraordinary waves are given in Table 3.2.

Table 3.2. Definition of Plasma Cutoffs and Resonances

	Plasma Cutoff	Plasma Resonance
Ordinary Wave	$\alpha^2 = 1$ (3.9a)	None
Extraordinary Wave	$(\alpha^2 - 1 + \beta_+\beta_-)^2 = (\beta_- - \beta_+)^2$ (3.9b)	$\alpha^2(1 - \beta_-\beta_+) = (1 - \beta_-^2)(1 - \beta_+^2)$ (3.10)

The ordinary wave cuts off at the plasma frequency (Figure 3.5). We emphasize that this is a cutoff and not a resonance frequency.

The extraordinary wave cuts off at both cyclotron cutoffs (Equations 3.3 and 3.4). Its resonance will be called "plasma resonance," and it plots on an (α^2, β^2) diagram as a hyperbola. One branch

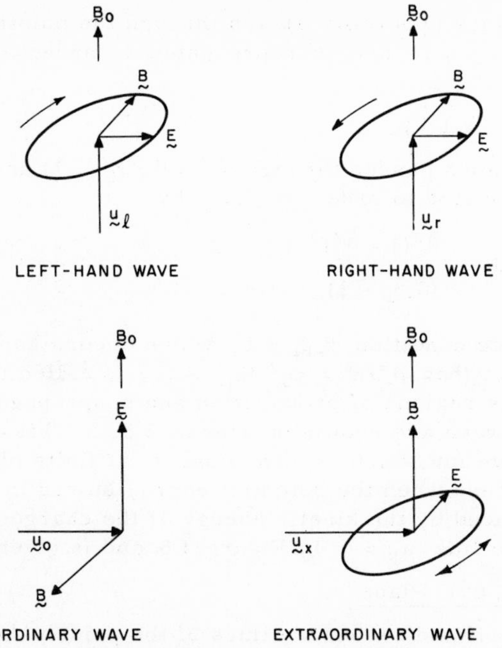

Figure 3.4. Polarization of the principal waves. The extraordinary wave can rotate either left- or right-handed with respect to $\underset{\sim}{B}_0$.

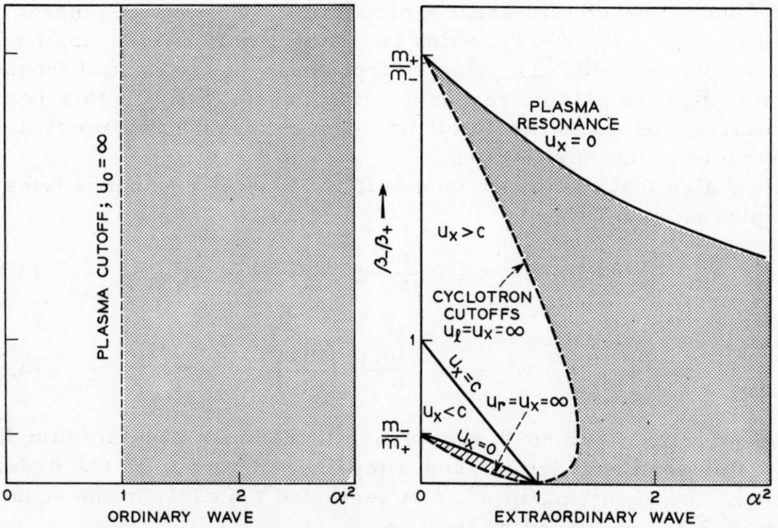

Figure 3.5. The (α^2, β^2) plot for propagation across $\underset{\sim}{B}_0$ for ion-to-electron mass ratio of 3 to 1. In the dotted regions the waves do not propagate $(u^2 < 0)$.

of the resonance hyperbola passes through the points $(\alpha^2 = 1, \beta_-^2 = 0)$ and $(\alpha^2 = 0, \beta_-^2 = 1)$ and is represented to order m_-/m_+ by the straight line

$$\alpha^2 = 1 - \beta_-^2 \tag{3.11}$$

The other branch passes through $(\alpha^2 = 0, \beta_+^2 = 1)$ and $(\alpha^2 = \infty, \beta_-\beta_+ = 1)$ and is represented to order m_-/m_+ by

$$\alpha^2 = \frac{\beta_-^2(1 - \beta_+^2)}{(\beta_-\beta_+ - 1)} \tag{3.12}$$

The resonance condition $\beta_-\beta_+ = 1$, which occurs for densities sufficiently large that $\beta_-^2/\alpha^2 = \omega_{b_-}^2/\omega_p^2 \ll 1$, is called the "hybrid resonance." The regions of propagation and nonpropagation of the extraordinary wave are shown in Figure 3.5. This is the only principal wave for which u can equal c at finite plasma densities, and this happens when the potential energy stored in the space-charge field is balanced by the kinetic energy of the charged particles. This occurs on the line $u_x = c$ in Figure 3.5, and is given by $\alpha^2 = 1 - \beta_-\beta_+$.

3.3 The (α^2, β^2) Plane

All the cutoff and resonance lines of the principal waves are shown in Figure 3.6. They divide the (α^2, β^2) plane into thirteen regions that are numbered alternately left and right of plasma cutoff, the numbers increasing with $\beta_-\beta_+$ as indicated. Part of the reason for this system of numbering is that areas whose numbers differ by 6 are found to have quite similar properties. Note the degeneracy at the point $(\beta^2 = 0, \alpha^2 = 1)$ which is removed only by the application of a magnetic field. The plasma frequency ω_p is a cutoff frequency even though the plasma resonance line passes through this point. Similarly, the cyclotron cutoff line passes through the two cyclotron resonance points for $\alpha^2 = 0$.

Note also that the cyclotron cutoff and plasma resonance lines cut the plasma cutoff line at

$$\alpha^2 = 1, \quad \beta_+ = 1 - \frac{m_-}{m_+} \tag{3.13}$$

and

$$\alpha^2 = 1, \quad \beta_+^2 = 1 - \frac{m_-}{m_+} + \left(\frac{m_-}{m_+}\right)^2 \tag{3.14}$$

respectively. Regions 9 and 11 are therefore extremely thin for a real plasma, and are missed altogether if terms of the order m_-/m_+, or even $(m_-/m_+)^2$, are neglected too early in the equations. Region 7 is subdivided by the line

$$K_{\parallel}K_{\perp} = K_r K_\ell \quad \text{or} \quad \alpha^2 = \frac{m_+}{m_-} - 1 + \frac{m_-}{m_+} - \beta_+\beta_- \tag{3.15}$$

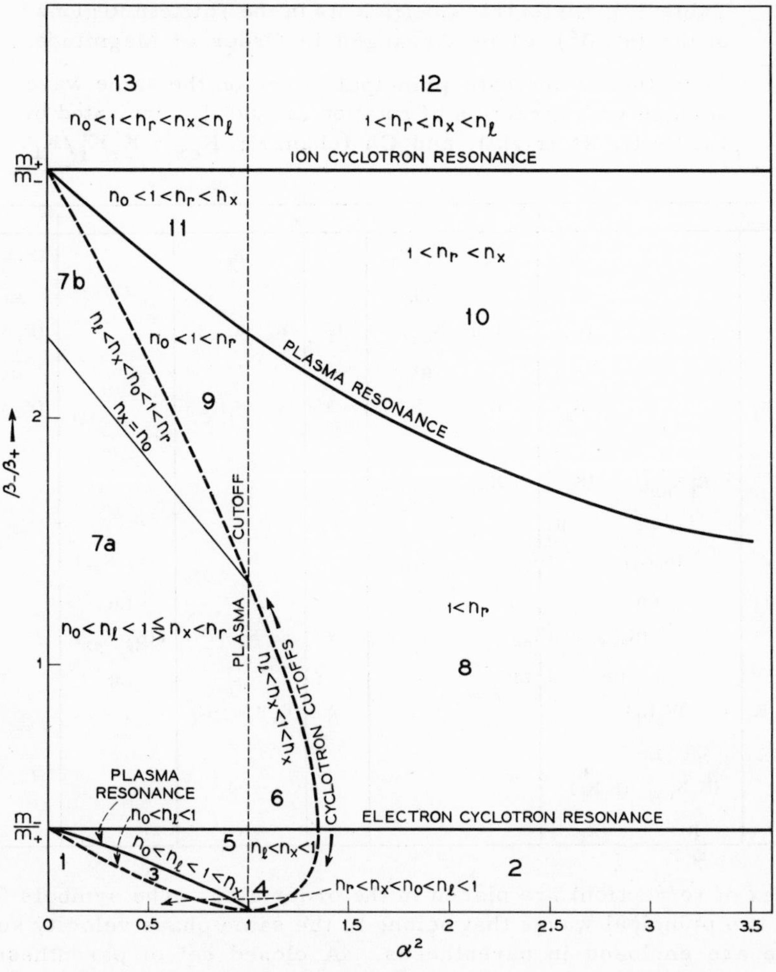

Figure 3.6. The division of the (α^2, β^2) plane into thirteen regions by the cutoff curves (dashed lines) and resonance curves (solid lines) of the principal waves. In each of the thirteen regions the waves possess different topological properties. The indices of the principal waves in each region are written in order of increasing magnitude. The absence of an index indicates that the corresponding wave does not propagate in the region.

along which $n_\times = n_o$. At this line the magnitudes of $n_\times$ and n_o cross.

In Figure 3.6 and Table 3.3, the four principal indices are ordered according to their magnitude for each of the thirteen regions of the (α^2, β^2) plane, and the values zero (cutoff) and unity (free-space

Table 3.3. Dielectric Coefficients in the Thirteen Regions of the (α^2, β^2) Plane Arranged in Order of Magnitude.

Parentheses indicate principal waves on the same wave surface with direction of rotation around $\underset{\sim}{B}_0$ indicated by Lft (left), Rt (right), and Ch (change); $K_{ex} = K_r K_\ell / K_\perp$.

	−∞ to 0	0 to 1	1 to +∞		−∞ to 0	0 to 1	1 to …
13		$(K_\parallel$	$K_r)$ $(K_{ex} K_\ell)$	12	$K_\parallel$		$(K_r K_{ex})$ $(K$
		Rt	Ch				Rt
11	K_ℓ	$(K_\parallel$	$K_r)$ $(K_{ex}$	10	K_ℓ $K_\parallel$		$(K_r K_{ex})$
		Rt	Rt				Rt
9	$K_{ex} K_\ell$	$(K_\parallel$	$K_r)$	8	K_{ex} $K_\parallel K_\ell$		$(K_r$
		Rt					Rt
7b		$(K_\ell K_{ex})$ $(K_\parallel$	K_r				
		Lft Rt					
7a		$(K_\parallel K_\ell)$	$(K_{ex}$ $K_r)$	6	$K_\parallel$	$(K_\ell\ K_{ex})$	$(K_r$
		Lft	Ch			Lft	Rt
5	K_r	$(K_\parallel K_\ell)$	$(K_{ex}$	4	K_r $K_\parallel$	$(K_\ell\ K_{ex})$	
		Lft	Lft			Lft	
3	K_{ex} K_r	$(K_\parallel K_\ell)$		2	$K_r K_\parallel K_{ex} K_\ell$		
		Lft					
1		$(K_r K_{ex})$ $(K_\parallel K_\ell)$					
		Rt Lft					

index of refraction) are placed in the order also. The symbols for the two principal waves that belong to the same phase velocity surface are enclosed in parentheses. A closed set of parentheses indicates that a complete velocity surface exists in the region. When the set is open, it indicates the presence of a resonant cone (see the following section) and thus the absence of one of the principal waves. The symbols Lft, Rt, and Ch refer to the sense of rotation of the waves around $\underset{\sim}{B}_0$, and will be described in greater detail in subsequent chapters.

3.4 Propagation in Arbitrary Directions

The index of refraction for an arbitrary direction of propagation is given implicitly by the tangent formula (2.27) and in the following figures will be represented by polar plots of the phase velocity $\underset{\sim}{u} = c/\underset{\sim}{n}$. In these plots the magnetic field is directed upward, and the phase velocity surface is obtained by rotating the polar diagram of $\underset{\sim}{u}$ about the vertical axis passing through the origin of $\underset{\sim}{u}$. It should be noted again that these are not ray surfaces; that is, the planes of constant phase are perpendicular to the radius vector and are not tangent to the surface.

As one of the principal waves appears or disappears whenever a cutoff or a resonance line is crossed, the velocity surface is topologically different in each of the thirteen areas of the (α^2, β^2) plane. It is significant, therefore, to plot a sample velocity surface in each of the areas, and this has been done in Figures 3.7 and 3.8 for the lower and upper halves of the (α^2, β^2) plane, and in Figure 3.9 for the entire plane. As the thirteen areas are very different in size, it was necessary to draw each figure to a different scale, and this is shown in each case by the dotted circle that represents the velocity of light.

The velocity surface is in general double, corresponding to the two solutions for n^2. The two surfaces intersect only when the discriminant D (Equation 2.26) vanishes, which does not happen in general because D^2 is the sum of the two squares. However, the two surfaces do touch at the poles when $K_{\parallel} K_{\times} = 0$. This happens (a) when $\beta = 0$, $K_{\times} = 0$, but the medium is then isotropic, and the two surfaces are identical spheres; (b) when $\alpha = 1$, $K_{\parallel} = 0$, but this situation is quite singular, and will be discussed in the next paragraph; (c) when $\beta \to \infty$ and $K_{\times} \to 0$ as β^{-3}, and will be discussed in Section 3.7. When $K_r K_{\ell} = K_{\parallel} K_{\perp}$, which defines the straight line (3.15) along which $n_{\times} = n_o$, the two surfaces are in contact along the equator.

In general, one velocity surface contains the other, and we shall let u_1 denote the outer surface, u_2 the inner one. Eventually we shall find other velocity surfaces u_3 and u_4 inside u_2, and there may be further surfaces inside those when there are more than two kinds of particles. The principal waves u_{ℓ}, u_r occur at the poles, and u_o, $u_{\times}$ at the equator, of u_1 and u_2, and the proper correspondence is obtained by reference to Table 3.3. The principal waves u_{ℓ} and u_r are joined separately to u_o and $u_{\times}$ in such a way that the surfaces do not cross. If only three principal waves are real, the outer two are joined and the third joins with the origin, as shown on the diagrams. The distinction between u_1 and u_2 is not clear when there is only one surface as in regions 3, 4, 8, 9, and 10.

The resonance condition is obtained by setting $n = \infty$ in Equation 2.27. This yields

$$\tan^2 \theta_{res} = -\frac{K_{\parallel}}{K_{\perp}} = \frac{(\alpha^2 - 1)(1 - \beta_-^2)(1 - \beta_+^2)}{(1 - \beta_-^2)(1 - \beta_+^2) - \alpha^2(1 - \beta_+\beta_-)} \qquad (3.16)$$

or

$$\sin^2 \theta_{res} = \frac{\alpha^2 - 1}{\alpha^2} \frac{(1 - \beta_-^2)(1 - \beta_+^2)}{(1 - \beta_-^2)(1 - \beta_+^2) - 1 + \beta_+\beta_-}$$

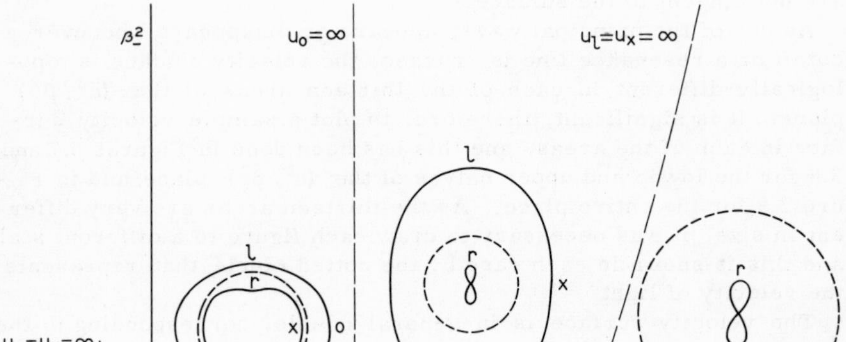

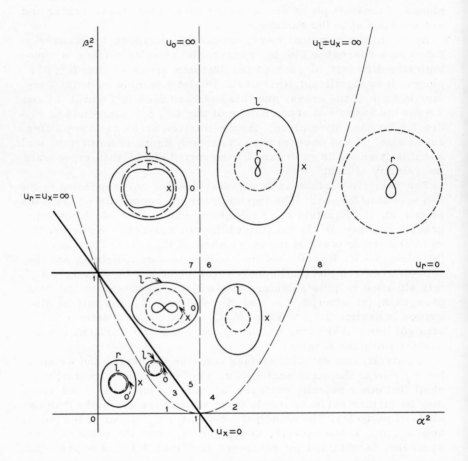

Figure 3.7. Phase velocity surfaces for electromagnetic waves at low magnetic fields or high frequencies, $m_+ \to \infty$. The symbols ℓ and r denote the left and right principal waves at the poles of the wave surface, and o and × the ordinary and extraordinary waves at the equator of the wave surface. The plots are not to scale, but the speed of light in relation to the velocities is shown by the dashed circle.

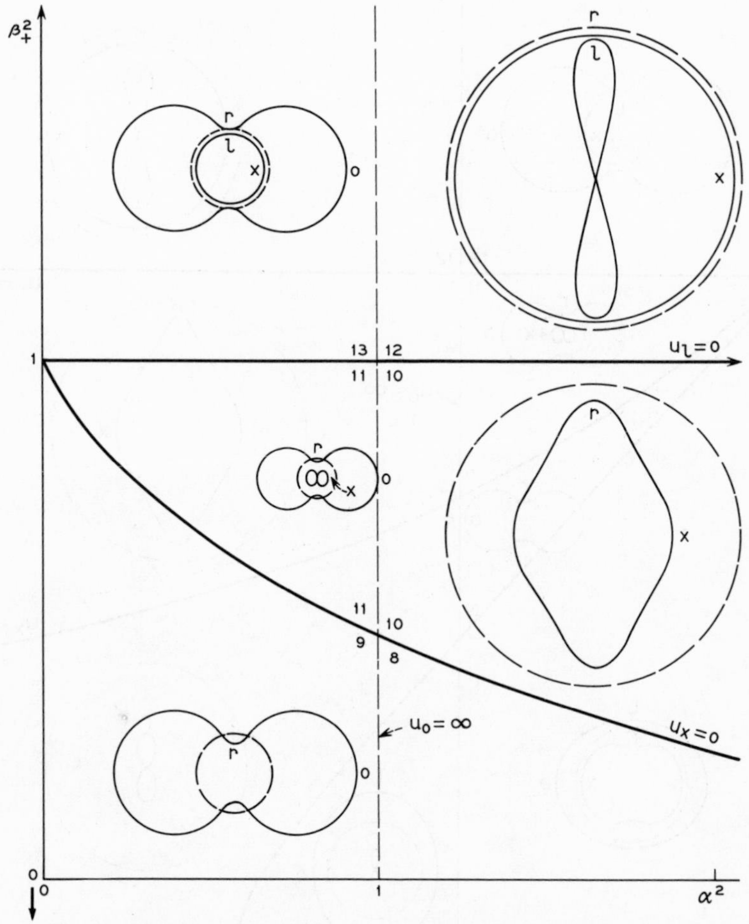

Figure 3.8. Phase velocity surfaces for electromagnetic waves at high magnetic fields or low frequencies, $\beta_- \gg 1$. The symbols ℓ and r denote the left and right principal waves at the poles of the wave surface and o and $\times$ the ordinary and extraordinary waves at the equator of the wave surface. The plots are not to scale, but the speed of light in relation to the velocities is shown by the dashed circle.

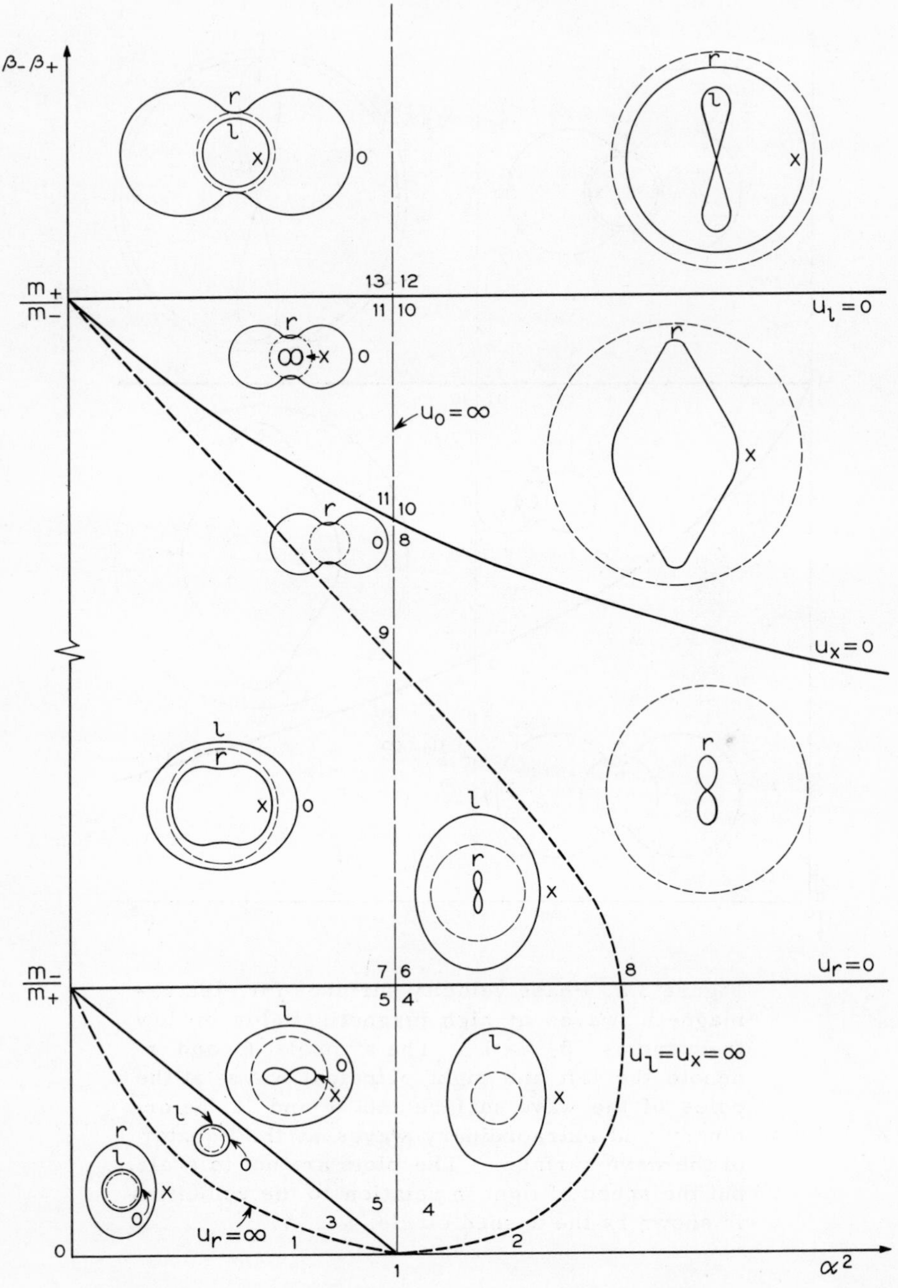

Figure 3.9. General (α^2, β^2) plot and phase velocity surfaces for electromagnetic waves.

which is, of course, the same as setting $A = 0$ in Equation 2.25. Equations 3.16 show that a resonance cone exists whenever $K_{\parallel}$ and $K_{\perp}$ are of opposite sign. This occurs only in regions 5, 6, 8, 11, and 12.

The resonance cone is very thin ($\theta_{res} \to 0$) near the cyclotron resonances and very flat ($\theta_{res} \to \pi/2$) near plasma resonance, and thus provides a continuous transition between these resonances in regions 5, 8, and 11. It is also very thin near plasma cutoff. Loci of constant cone angle $\theta_{res} = 0$ (cyclotron resonance), $\pi/12$, $\pi/6$, $\pi/4$, $\pi/3$, and $\pi/2$ (plasma resonance) are shown in Figure 3.10. Note that the resonance loci for small cone angles fall very close not only to the cyclotron resonances but also to the plasma cutoff. This is why the plasma cutoff condition $\alpha^2 = 1$ frequently appears as a resonance condition. This happens in particular for guided waves (see Part II). Some corresponding values of α^2 and β^2 are given in Table 3.4. Slight approximations have been made for the asymptotes at $\alpha^2 = \infty$, and the symbol $\mu = m_-/m_+$ has been

Table 3.4. Critical Values for the Resonance-Cone Loci

Region	β_+^2	α^2	Region	β_-^2	α^2
12	∞	$\sec^2 \theta_{res}$	8	$1 - \frac{1}{\mu} + \frac{1}{\mu^2}$	1
12	$1 + \mu \tan^2 \theta_{res}$	∞	8	$\frac{\sec^2 \theta_{res}}{1 + \mu \tan^2 \theta_{res}}$	∞
11	1	0	5	1	0
11	$1 - \mu + \mu^2$	1	5	0	1

used. In regions 6, 8, and 12, the directions of allowed propagation are inside the resonant cone, giving a somewhat dumbbell-shaped velocity surface; in regions 5 and 11, they are outside the cone, giving somewhat torus-shaped wave surfaces.

Studies of ray velocity show that for the dumbbell-shaped modes the group velocity is confined even more narrowly to directions close to the magnetic field. Thus radio waves in region 8 may be guided by the earth's magnetic field lines from one hemisphere to the other and even be reflected and come back quite close to their source. This gives rise to the phenomenon of "whistlers."[7] These are radio waves of audible frequencies which are produced by lightning flashes and travel along the earth's magnetic field lines from one hemisphere to the other with very little angular dispersion, and may then be reflected back again. The frequency dispersion produced by this long trajectory causes the whistling tone of steadily falling pitch when the wave train is received.

Equations 3.16 may be solved for the plasma density at which

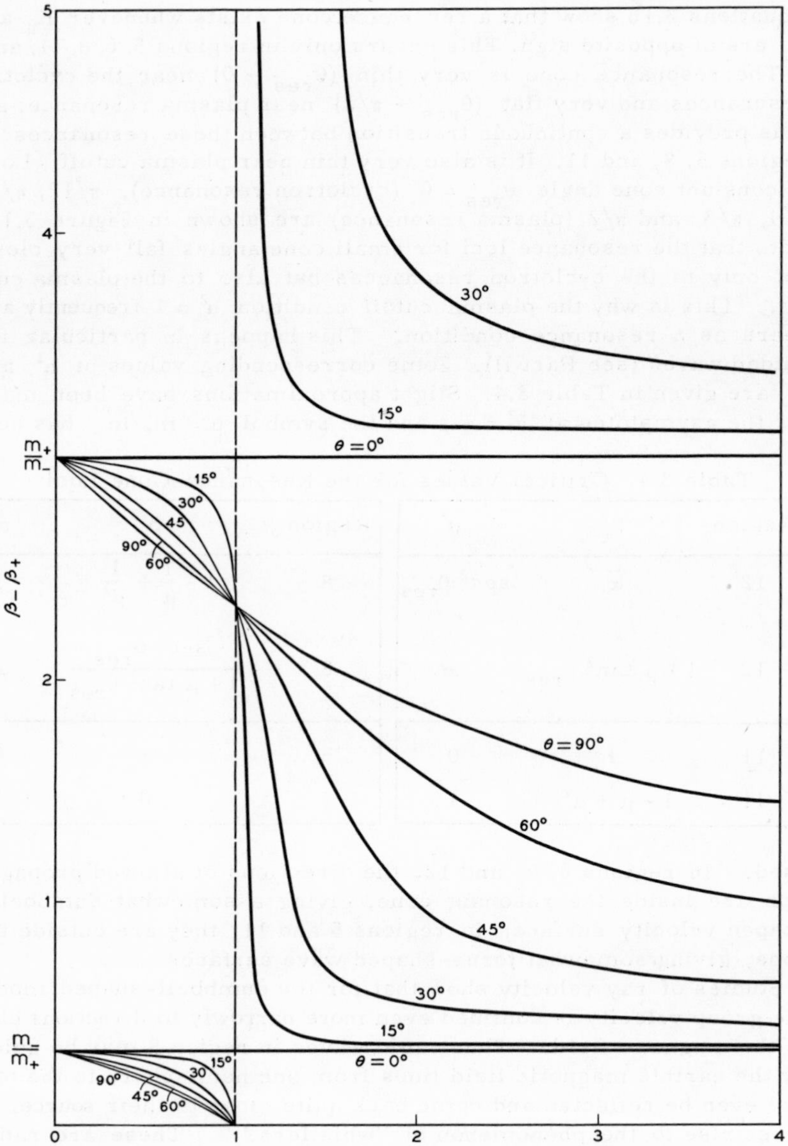

Figure 3.10. Loci of constant resonant angle θ_{res}. The 90° limit is the plasma resonance hyperbola. The 0° limit includes the cyclotron resonance and tends to plasma cutoff.

resonance occurs. This can be expanded to

$$\frac{\omega_p^2}{\omega^2} = \frac{(\omega^2 - \omega_{b-}^2)(\omega^2 - \omega_{b+}^2)}{(\omega^2 - \omega_{b-}^2)(\omega^2 - \omega_{b+}^2)\cos^2\theta_{res} + (\omega^2 - \omega_{b-}\omega_{b+})\sin^2\theta_{res}} \tag{3.17}$$

This equation is linear in ω_p^2, quadratic in B_0^2, and, provided θ is neither zero nor $\pi/2$, cubic in ω^2. This implies that for a given angle θ, an experiment in which B_0 and ω are held constant and the plasma density varied will yield one resonance, an experiment in which ω^2 and ω_p^2 are held constant but the magnetic field B_0 varied will yield two resonances, and an experiment in which B_0 and ω_p are held constant but the frequency ω varied will yield three resonances, provided propagation is neither along nor across $\underset{\sim}{B}_0$.

For a plasma consisting of g distinct species (different charge-to-mass ratios), the equation for the resonant cone is linear in ω_p^2, is of order g in B_0^2, and provided $\theta \neq 0$ or $\pi/2$, is of order $(g + 1)$ in ω^2, all ω^2 roots being real and positive definite. For $\theta = 0$ or $\pi/2$, one of the ω^2 roots is zero, so that there are only g resonant frequencies.

3.5 Waves with Stationary Ions

At sufficiently low magnetic fields or high frequencies (regions 1 to 5 and lower parts of regions 6 to 8), the ion motions can be neglected ($\beta_+ << 1$). Considerable simplification then results in the expression for n^2, which now is

$$n^2 = 1 - \frac{2\alpha^2(1 - \alpha^2)}{2(1 - \alpha^2) - \beta_-^2\sin^2\theta \pm \sqrt{\beta_-^4\sin^4\theta + 4\beta_-^2(1 - \alpha^2)^2\cos^2\theta}} \tag{3.18}$$

The corresponding velocity surfaces have been shown in Figure 3.7. Equation 3.18 is in a form first written down by Appleton and Hartree and is used extensively in ionospheric research. The plus sign gives the "ordinary" wave as defined by the students of the ionosphere. It possesses no resonance but has a cutoff at $\alpha^2 = 1$ for $\theta \neq 0$. The minus sign gives the "extraordinary" wave, which has a resonance at

$$\alpha^2 = \frac{1 - \beta_-^2}{1 - \beta_-^2\cos^2\theta} \tag{3.19}$$

for $\theta \neq 0$. Its cutoff is given by $\alpha^2 = 1 + \beta_-$ and $\alpha^2 = 1 - \beta_-$ for $\theta \neq 0$. At $\theta = 0$, Equation 3.18 becomes

$$n^2 = 1 - \frac{\alpha^2(1 - \alpha^2)}{(1 - \alpha^2) \pm \beta_-|1 - \alpha^2|} \tag{3.20}$$

where the absolute value of $1 - \alpha^2$ must be used after taking a square root. Thus

$$n^2 = \begin{cases} 1 - \dfrac{\alpha^2}{1 \pm \beta} & \text{for } \alpha^2 < 1 \\[2ex] 1 - \dfrac{\alpha^2}{1 \mp \beta} & \text{for } \alpha^2 > 1 \end{cases} \tag{3.21}$$

The + sign in Equation 3.20 gives n_ℓ^2 for $\alpha^2 < 1$ and n_r for $\alpha^2 > 1$. Thus by taking the absolute value of $1 - \alpha^2$, Equation 3.18 correctly represents the discontinuity at the poles of the velocity surface at $\alpha^2 = 1$, as described in Section 3.7. As these waves have been thoroughly discussed in the literature,[8] and also by Ratcliffe,[9] they will not be emphasized here.

3.6 Quasi-Circular and Quasi-Plane Waves

As was noted in Section 2.4, the discriminant

$$D^2 = (K_r K_\ell - K_\parallel K_\perp)^2 \sin^4\theta - 4K_\parallel^2 K_\times^2 \cos^2\theta \tag{3.22}$$

is a sum of two squares. When either of these is negligibly small, we obtain formulas without radicals. These approximations also have a physical meaning as it will be seen that the corresponding waves are circularly or plane-polarized, and hence we shall name them the quasi-circular and quasi-plane approximations[10]† and denote them QC or QP. A more detailed definition of "circular" and "plane polarization" will emerge in Chapter 4.

The QC approximation is obtained by setting

$$D = 2jK_\times K_\parallel \cos\theta \tag{3.23}$$

and yields

$$n_{R,L}^2 = K_\parallel \frac{K_\perp \pm jK_\times \cos\theta}{K_\parallel \cos^2\theta + K_\perp \sin^2\theta} \tag{3.24}$$

which represents two surfaces tangent at the equator $(\cos\theta = 0)$ and has the correct values for n_r^2, n_ℓ^2, and n_O^2, but not for n_X^2.

The QP approximation is obtained by setting

$$D = (K_r K_\ell - K_\parallel K_\perp)\sin^2\theta \tag{3.25}$$

and yields

$$n_O^2 = \frac{K_\parallel K_\perp}{K_\perp \sin^2\theta + K_\parallel \cos^2\theta} \tag{3.26}$$

$$n_X^2 = \frac{K_r K_\ell \sin^2\theta + K_\parallel K_\perp \cos^2\theta}{K_\perp \sin^2\theta + K_\parallel \cos^2\theta} \tag{3.27}$$

---

† Booker calls these approximations quasi-longitudinal and quasi-transverse, which would translate into our terminology as quasi-along and quasi-across.

This approximation yields two surfaces tangent at the poles and gives the correct values for n_o^2 and $n_\times^2$ and the average of n_r^2 and n_ℓ^2 at the poles.

The QC expression is exact for waves propagating along $\underset{\sim}{B}_0$, which are circularly polarized, and waves propagating nearly along $\underset{\sim}{B}_0$ are quasi-circular and given by Equation 3.24. Similarly, waves propagating across $\underset{\sim}{B}_0$ are plane-polarized; that is, the electric vector remains in a plane containing $\underset{\sim}{k}$ (Section 4.2). Those waves propagating nearly across $\underset{\sim}{B}_0$ are quasi-plane and given by Equation 3.26 or 3.27. The velocity surface is therefore divided into polar caps that are quasi-circular and an equatorial zone that is quasi-plane, and these are separated by a cone of angle θ_q defined by

$$q = \frac{2\cos\theta_q}{\sin^2\theta_q} = \frac{K_r K_\ell - K_\parallel K_\perp}{jK_\parallel K_\times} = \frac{\beta_-\beta_+}{\beta_- - \beta_+} + \frac{\beta_- - \beta_+}{1 - \alpha^2} - \frac{\beta_-^2\beta_+^2}{(1-\alpha^2)(\beta_- - \beta_+)} \tag{3.28}$$

The notation "right-" and "left-polarized" may appropriately be extended over the polar caps where the QC approximation is valid, and similarly "ordinary" and "extraordinary" may be extended over an equatorial zone, but these notations should never be extended beyond the cone θ_q.

There are considerable areas of the (α^2, β^2) plane in which the angle θ_q is either very small or is close to $\pi/2$, so that almost the entire wave surface is quasi-plane or quasi-circular. At plasma cutoff, $K_\parallel = 0$, or $n_o = 0$, the entire surfaces, excluding the poles, are quasi-plane. Along the line $K_\parallel K_\perp = K_r K_\ell$, or $n_o = n_\times$, the entire surfaces, excluding the equator, are quasi-circular. At the equator, as has been seen (Equation 3.15), the two velocity surfaces are in contact, and the polarization is arbitrary.

On either side of the line $K_\parallel K_\perp = K_r K_\ell$ (Equation 3.15), the two wave surfaces are distinct, but on one side of the line (indicated in Figure 3.6) the ordinary wave is on the outer surface and has the left circular polarization; on the other side of Equation 3.15, the ordinary wave is on the inner surface and has right circular polarization. As the line (3.15) is crossed, the two wave surfaces touch at the equator and exchange the ordinary and extraordinary waves.

The QP and QC approximations will be valid for some distance in the neighborhood of the lines $n_o = 0$ and $n_o = n_\times$, and we can obtain an estimate of the extent over which they are valid by considering the lines $q = \pm 1$.

This yields the two lines

$$q = \pm 1 \quad \begin{cases} \beta_+ = \dfrac{1 - \beta_+}{\beta_-} \\ \alpha^2 = 1 - \beta_+\beta_- \mp (\beta_- - \beta_+) \end{cases} \tag{3.29}$$

The first is a horizontal line on the (α^2, β^2) plane slightly below ion cyclotron resonance. The second is precisely the cyclotron cutoff (3.9b). Thus we see that the validity of the approximations in the thirteen regions is as follows:

13 QP	12 QP
11 QP	10 QP and QC
9 Neither	8 QC
7 QP and QC	6 QP
5 QP	4 QP
3 Neither	2 QC
1 QC	

and this is indicated in Figure 3.11. Note that the "whistler" modes that propagate only within a small cone angle along the magnetic field are quasi-circular.

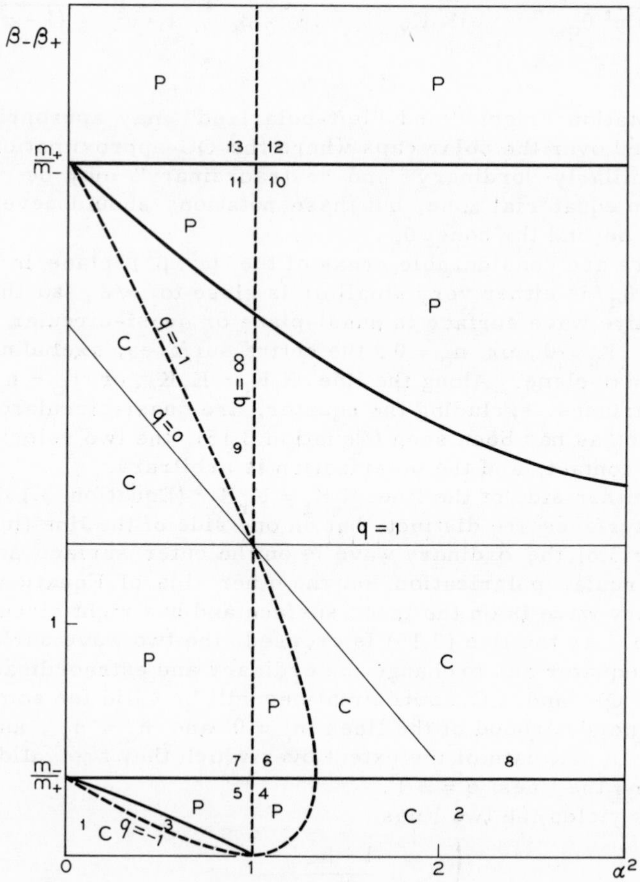

Figure 3.11. Division of the (α^2, β^2) plane into quasi-circular and quasi-plane regions.

3.7 The Quasi-Plane Approximation

The quasi-plane approximation is very good near plasma cutoff, and we see from Equations 3.26 and 3.27 that the velocity surfaces at $K_{\parallel} \approx 0$ are an "extraordinary" sphere

$$u = u_X \tag{3.30}$$

and a very large "ordinary" torus-shaped figure

$$u = u_o \sin\theta \tag{3.31}$$

where $u_o \to \infty$ as plasma cutoff is approached from $\alpha < 1$. However, the "torus" does not come in to the origin, as indicated by Equation 3.31, but to u_r or u_ℓ, whichever is larger. The extraordinary sphere is missing its poles, which are replaced by "dimples" extending into the smaller of u_r and u_ℓ, or to the origin if one of these is imaginary. Similarly, on the $\alpha > 1$ side of cutoff, the sphere has a "pimple" extending to the larger of u_r or u_ℓ, and there may or may not be an "ordinary" "figure-8" surface closing the resonant cone. This transition is shown in Figure 3.12.

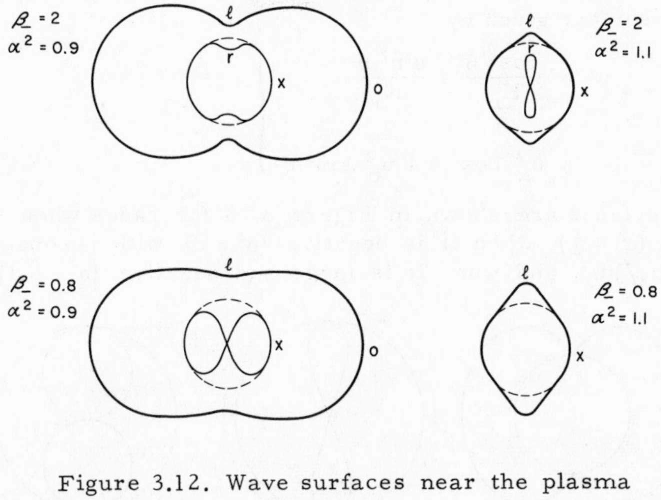

Figure 3.12. Wave surfaces near the plasma cutoff when $m_+ >> m_-$.

Thus the right- and left-handed waves jump discontinuously between the ordinary and extraordinary surfaces as plasma cutoff is crossed. The ordinary and extraordinary waves make similar jumps as the line $K_r K_\ell = K_\parallel K_\perp$ is crossed, and this is why it is often difficult to follow the "ordinary" and "extraordinary" waves by continuity arguments on propagation vector plots (k versus ω).

The QP approximation is particularly good at very large magnetic fields ($\beta_+ >> 1$) (upper parts of regions 12 and 13), because here $n_r \approx n_\ell$, since for $\beta_+ >> 1$

$$\left.\begin{aligned} K_r \approx K_\ell \approx K_\perp &\approx 1 + \frac{\omega_{p_-}^2}{\omega_{b_-}^2} + \frac{\omega_{p_+}^2}{\omega_{b_+}^2} = 1 + \frac{\omega_p^2}{\omega_{b_-}\omega_{b_+}} \\ K_\times &\approx j\left(\frac{\omega}{\omega_{b+}}\right)\left(\frac{\omega_p^2}{\omega_{b-}\omega_{b+}}\right) \to 0 \end{aligned}\right\} \quad (3.32)$$

Thus, here the QP approximation gives all four principal waves correctly (see the discussion after Equation 3.27). This arises because at the poles the left- and right-hand circularly polarized waves have the same phase velocity, and a linearly polarized plane wave can be constructed from them.

At large magnetic fields the QP approximation always yields a spherical extraordinary surface

$$u^2 = u_\times^2 = \frac{c^2}{K_\perp} \quad (3.33)$$

and as this wave obeys Snell's law, it is called the "ordinary" wave in texts that start from the magnetohydrodynamic approximations. The other wave, which we prefer to call the "ordinary," has a phase velocity surface given by

$$\left.\begin{aligned} u^2 &= \frac{\cos^2\theta}{K_\perp} + \frac{\sin^2\theta}{K_\parallel} \\ &= u_\times^2\cos^2\theta + u_o^2\sin^2\theta \end{aligned}\right\} \quad (3.34)$$

These surfaces are shown in Figure 3.13 for cases when $K_\parallel$ is positive ($\alpha^2 = \frac{1}{2}$), when it is negative ($\alpha^2 = 2$) with resonances at $\tan\theta = u_X/ju_o$, and when it is large and negative ($\alpha^2 >> 1$).

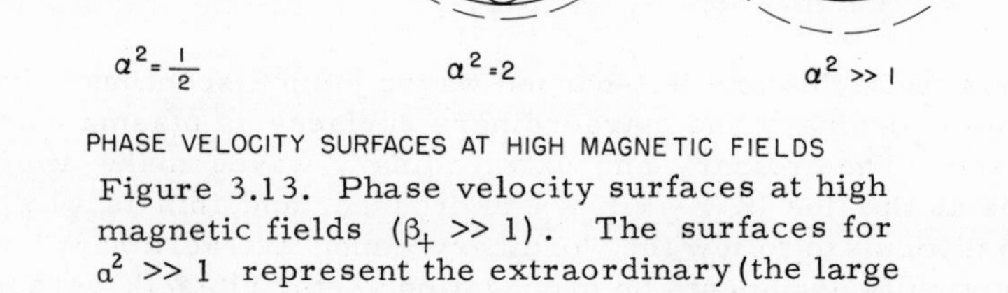

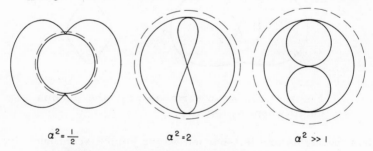

Figure 3.13. Phase velocity surfaces at high magnetic fields ($\beta_+ >> 1$). The surfaces for $\alpha^2 >> 1$ represent the extraordinary (the large sphere) and the ordinary (two small spheres tangent to each other at the origin) Alfvén waves.

This last condition, in which the displacement current is negligible, coupled with the condition $\beta_+ >> 1$, defines the so-called magnetohydrodynamic region (the upper right-hand corner of region 12). Here, the phase velocity surface of the ordinary wave is a double-sphere, that is, the surface consists of two spheres tangent to each other at the origin and tangent internally to the extraordinary sphere at the poles (Figure 3.13). It is the spherical extraordinary wave that is the fast wave. Its phase velocity is very nearly the Alfvén speed u_α, because in this limit

$$K_r \to K_\ell \to K_\perp \to 1 + \frac{c^2}{u_\alpha^2} \tag{3.35}$$

where

$$u_\alpha^2 = \frac{c^2\beta_+\beta_-}{\alpha^2} = \frac{B_0H_0}{N(m_+ + m_-)} \tag{3.36}$$

is the square of the Alfvén speed. Then the speed of the fast wave is given by

$$\frac{1}{u^2} = \frac{1}{u_\alpha^2} + \frac{1}{c^2} \tag{3.37}$$

and is generally close to the Alfvén speed.

The QP approximation is also valid in regions 6 and 7 provided the location specified is far enough away from cyclotron cutoff and cyclotron resonance. This is easier to accomplish than would appear from Figure 3.11 because the cyclotron cutoff curve extends much farther to the right for large ion-to-electron mass ratios. Formulas 3.33 and 3.34 hold in this region, but Condition 3.35 must be replaced by

$$\left.\begin{aligned} K_r &\to 1 + \frac{\alpha^2}{\beta_-} + \frac{\alpha^2}{\beta_-^2} \\ K_\ell &\to 1 - \frac{\alpha^2}{\beta_-} + \frac{\alpha^2}{\beta_-^2} \\ K_\perp &\to 1 + \frac{\alpha^2}{\beta_-^2} \end{aligned}\right\} \tag{3.38}$$

Thus, for $\alpha^2/\beta_- << 1$, the sequence of Figure 3.13 is repeated, but with the extraordinary wave propagating as though it were in free space. The reason for this behavior is that, within the appropriate limits, the ions effectively have infinite mass ($\beta_+ << 1$) and the electrons have zero mass ($\beta_- >> 1$). The restoring force arises from Coulomb interactions, but the entire inertia is caused by the magnetic field, so that no particle masses enter the effective dielectric coefficients.

3.8 Representation of the Index of Refraction on a $(n_\parallel^2, n_\perp^2)$ Diagram

One of the useful ways of representing the propagation properties of a plasma is a plot of $(k_\parallel c/\omega)^2 = n^2 \cos^2\theta = n_\parallel^2$ against $(k_\perp c/\omega)^2 = n^2 \sin^2\theta = n_\perp^2$.[11] In the case of a temperate plasma, this is at most a quadratic curve (hyperbola, parabola or straight lines). This fact makes plotting easy and has the advantage of showing imaginary components of $\underset{\sim}{k}$ as well as real ones.

For such representation the dispersion equation (2.24) is rewritten as

$$(K_\perp n_\perp^2 + K_\parallel^2 n_\parallel^2)(n_\perp^2 + n_\parallel^2) - (K_\parallel K_\perp + K_r K_\ell)n_\perp^2 - 2K_\parallel K_\perp n_\parallel^2 + K_\parallel K_r K_\ell = 0 \tag{3.39}$$

Equation 3.39 in general represents a hyperbola in the $(n_\parallel^2, n_\perp^2)$ plane, but for $K_\parallel = 0$, $K_\parallel = K_\perp$, or $K_\times = 0$, it gives simpler curves.

When $K_\parallel = 0$ (plasma cutoff), Equation 3.39 reduces to

$$n_\perp^2\left(n_\perp^2 + n_\parallel^2 - \frac{K_r K_\ell}{K_\perp}\right) = 0 \tag{3.40}$$

which gives the two straight lines $n_\perp^2 = 0$ and $n_\perp^2 + n_\parallel^2 = K_r K_\ell / K_\perp = n_\times^2$ (Figure 3.14).

When $K_\perp = K_\parallel$, which happens for $\beta_+^2 = 1 - (m_-/m_+) + (m_-/m_+)^2$, that is, for magnetic fields slightly lower than those required for ion cyclotron resonance, Equation 3.39 reduces to

$$(n_\perp^2 + n_\parallel^2 - K_\perp^2)^2 - \frac{K_\times^2}{K_\perp}(n_\perp^2 - K_\perp) = 0 \tag{3.41}$$

which is a parabola (Figure 3.14).

Finally, when $\beta_+ >> 1$, so that $K_\times \to 0$ (see Section 3.7), Equation 3.39 gives the two straight lines

$$n_\perp^2 + n_\parallel^2 = K_\perp \tag{3.42a}$$

and

$$K_\perp n_\perp^2 + K_\parallel n_\parallel^2 = K_\perp K_\parallel \tag{3.42b}$$

Otherwise, Equation 3.39 can be rewritten as

$$\left(n_\perp^2 + n_\parallel^2 - K_\perp + \frac{K_\times^2}{K_\parallel - K_\perp}\right)\left(\frac{K_\perp}{K_\parallel}n_\perp^2 + n_\parallel^2 - K_\perp - \frac{K_\times^2}{K_\parallel - K_\perp}\right) + K_\times^2 \frac{(K_\parallel - K_r)(K_\parallel - K_\ell)}{(K_\parallel - K_\perp)^2} = 0 \tag{3.43}$$

which is hyperbola with the two asymptotes

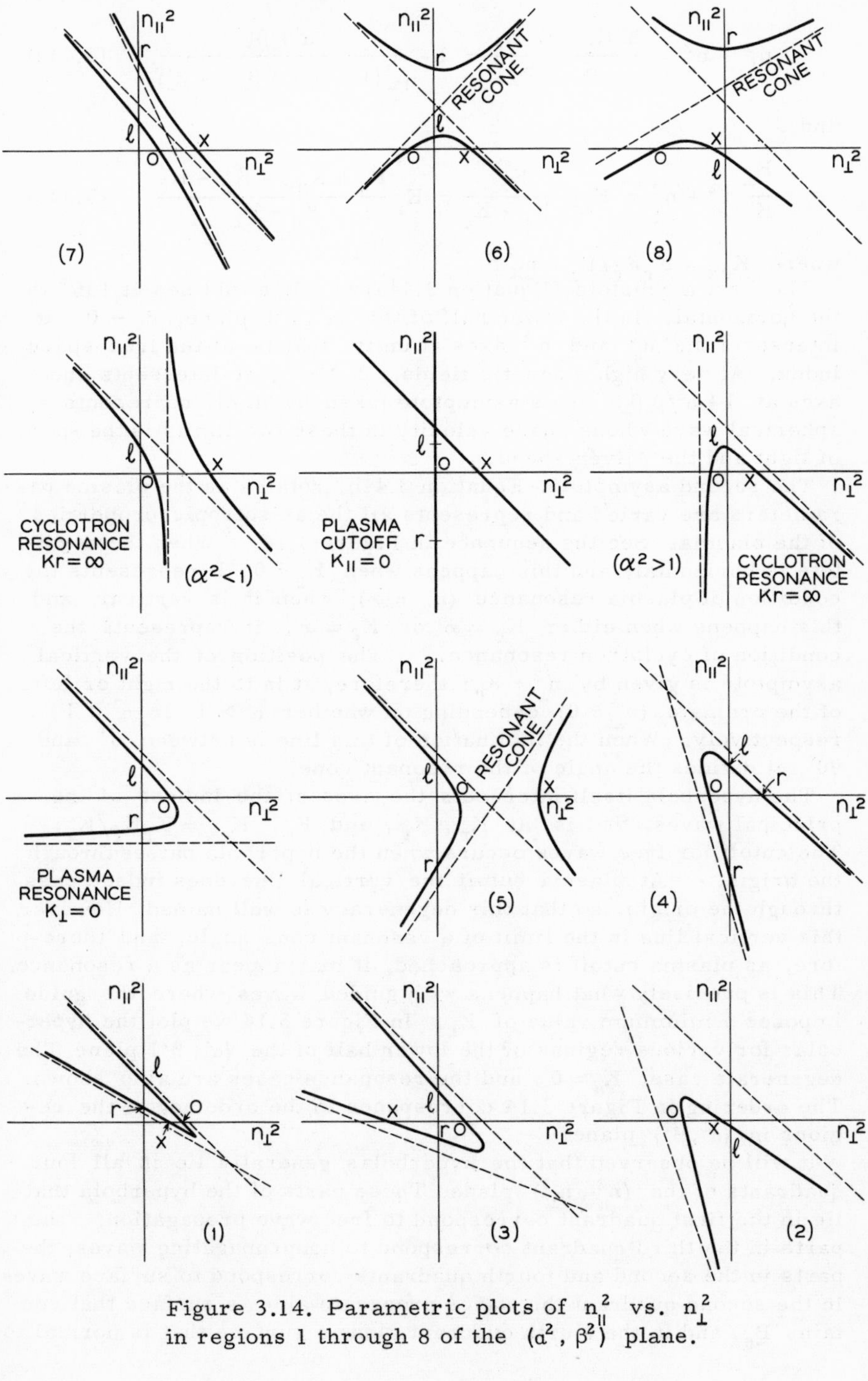

Figure 3.14. Parametric plots of $n_\parallel^2$ vs. $n_\perp^2$ in regions 1 through 8 of the (α^2, β^2) plane.

$$n_\perp^2 + n_\parallel^2 = \frac{K_\parallel K_\perp - K_r K_\ell}{K_\parallel - K_\perp} = 1 - \frac{\alpha^2 \beta_- \beta_+}{\beta_-^2(1 - \beta_+^2) + \beta_+(1 - \beta_-)} \quad (3.44a)$$

and

$$\frac{K_\perp}{K_\parallel} n_\perp^2 + n_\parallel^2 = K_\perp + \frac{K_\times^2}{K_\parallel - K_\perp} = K_\perp \frac{K_\parallel + K_{ex} - K_r - K_\ell}{K_\parallel - K_\perp} \quad (3.44b)$$

where $K_{ex} = K_r K_\ell / K_\perp = n_\times^2$.

The first asymptote (Equation 3.44a) is a line inclined at 135° to the horizontal. In the lower half of the (α^2, β^2) plane, $\beta_+ \to 0$, it intersects the $n_\parallel^2$ and $n_\perp^2$ axes at unity, that is, at the free-space index. At very high magnetic fields, $\beta_+ >> 1$, it intersects the axes at $1 + \alpha^2/\beta_- \beta_+$. This asymptote taken by itself represents a spherical wave whose phase velocity in these two limits is the speed of light and the Alfvén speed.

The second asymptote, Equation 3.44b, rotates as the plasma parameters are varied and represents all the anisotropic properties of the plasma. See the sequence in Figure 3.14. When the asymptote is horizontal, and this happens when $K_\perp = 0$, it represents the condition of plasma resonance ($n_\times = \infty$); when it is vertical, and this happens when either $K_r = \infty$ or $K_\ell = \infty$, it represents the condition of cyclotron resonance. The position of the vertical asymptote is given by $n_\perp^2 = K_\parallel$; therefore, it is to the right or left of the ordinate ($n_\perp^2 = 0$) depending on whether $\alpha^2 > 1$ or $\alpha^2 < 1$, respectively. When the inclination of this line is between 0° and 90°, it defines the angle of the resonant cone.

The hyperbola itself intersects the axes at the indices of the principal waves, that is, at K_r, K_ℓ, and $K_\parallel$, $K_{ex} = K_r K_\ell / K_\perp$. The cutoff for free waves occurs when the hyperbola passes through the origin. At plasma cutoff the vertical line does indeed pass through the origin, so that this degeneracy is well named. However, this vertical line is the limit of a resonant cone angle, and therefore, as plasma cutoff is approached, it may appear as a resonance. This is precisely what happens with guided waves where the guide imposes a minimum value of $K_\perp$. In Figure 3.14 we plot the hyperbolas for various regions of the lower half of the (α^2, β^2) plane. The degenerate case, $K_\parallel = 0$, and the resonance cases are also shown. The ordering in Figure 3.14 corresponds to the ordering of the regions in (α^2, β^2) plane.

It will be observed that the hyperbolas generally lie in all four quadrants of the $(n_\parallel^2, n_\perp^2)$ plane. Those parts of the hyperbola that lie in the first quadrant correspond to free wave propagation; the parts in the third quadrant correspond to nonpropagating waves; the parts in the second and fourth quadrants correspond to surface waves. In the second quadrant the waves propagate along a surface that contains $\underset{\sim}{B}_0$, and in the fourth quadrant along a surface that is normal to $\underset{\sim}{B}$

Such plots are also useful for studying bounded waves. For some special conditions, discussed in Part II, it is possible to fix the values of either $n_{\parallel}$ or $n_{\perp}$ and to investigate the other index as a function of various plasma parameters. It is worth while then to investigate the properties of the dispersion relation in two cases: (a) when $n_{\parallel}$ is assumed to be held constant and $n_{\perp}$ is a function of the various plasma parameters, and (b) when $n_{\perp}$ is assumed to be constant and $n_{\parallel}$ is the variable.

(a) $n_{\parallel}$ = constant
The dispersion relation for $n_{\perp}$ is obtained from Equation 2.24 and has the form

$$A_{\perp}n_{\perp}^4 + B_{\perp}n_{\perp}^2 + C_{\perp} = 0 \tag{3.45}$$

where

$$\left.\begin{aligned} A_{\perp} &= K_{\perp} \\ B_{\perp} &= (K_{\perp} + K_{\parallel})n_{\parallel}^2 - (K_r K_{\ell} + K_{\parallel}K_{\perp}) \\ C_{\perp} &= K_{\parallel}(n_{\parallel}^2 - K_r)(n_{\parallel}^2 - K_{\ell}) \end{aligned}\right\} \tag{3.46}$$

with $K_{\perp}$, K_r, K_{ℓ}, and $K_{\parallel}$ given by Equations 2.20 and 2.21. We note that the resonance condition for $n_{\perp}^2$ $(n_{\perp}^2 \to \infty)$ is given by $K_{\perp} = 0$, which is the plasma resonant condition for the extraordinary wave (Section 3.2).

The cutoff condition for $n_{\perp}^2$ $(n_{\perp}^2 = 0)$ is given by

$$C_{\perp} = 0 = K_{\parallel}(K_r - n_{\parallel}^2)(K_{\ell} - n_{\parallel}^2) \tag{3.47}$$

and depends strongly on the value of $n_{\parallel}$. The plasma cutoff is given, as in the case of free propagation, by $\alpha^2 = 1$; the two cyclotron cutoffs, however, depend on $n_{\parallel}$ and are given by

$$\alpha^2 = (1 - n_{\parallel}^2)(1 - \beta_-)(1 + \beta_+) \tag{3.48a}$$

and

$$\alpha^2 = (1 - n_{\parallel}^2)(1 + \beta_-)(1 - \beta_+) \tag{3.48b}$$

When $n_{\parallel}^2 = 0$, these reduce to the former cutoff condition (Equations 3.3a and 3.3b). Let us consider the right-hand cyclotron cutoff first (Equation 3.48a and Figure 3.15). When $n_{\parallel}^2 < 1$, α^2_{cutoff} is positive only when $\beta_- < 1$, as in the case of free propagation; however, when $n_{\parallel}^2 > 1$, there no longer exists a cutoff condition for $\beta_- < 1$, but only for $\beta_- > 1$, $[\alpha^2 = (n_{\parallel}^2 - 1)(\beta_- - 1)(\beta_+ + 1)]$. Similarly, the left-hand cyclotron cutoff (Equation 3.48b and Figure 3.16) is real only when $n_{\parallel}^2 < 1$

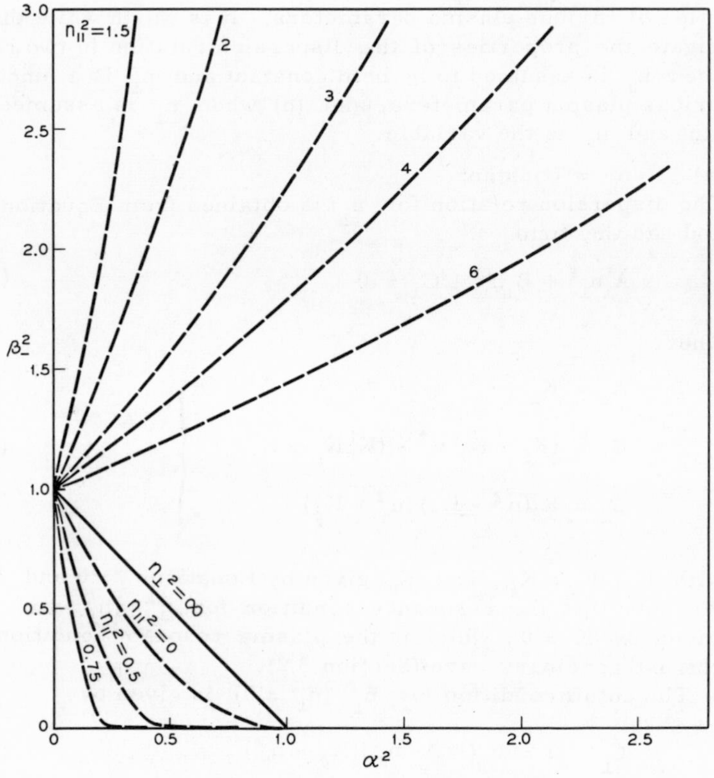

Figure 3.15. Loci of the cutoff and the resonance conditions for guided propagation across $\underset{\sim}{B}_0$ in plasma with ions of infinite mass. The dashed lines are the cutoff condition $\alpha^2 = (1 - n_{\|}^2)(1 - \beta_-)$ for various values of $n_{\|}^2$. The solid line is the first branch $(\alpha^2 + \beta_-^2 = 1)$ of the resonance condition. Note that for $n_{\|}^2 \geq 1$ and $\beta_- < 1$ no cutoff condition exists for positive α, and thus the resonance region can be reached by a wave which propagates across $\underset{\sim}{B}_0$ in a plasma whose density is increasing from zero along the direction of propagation.

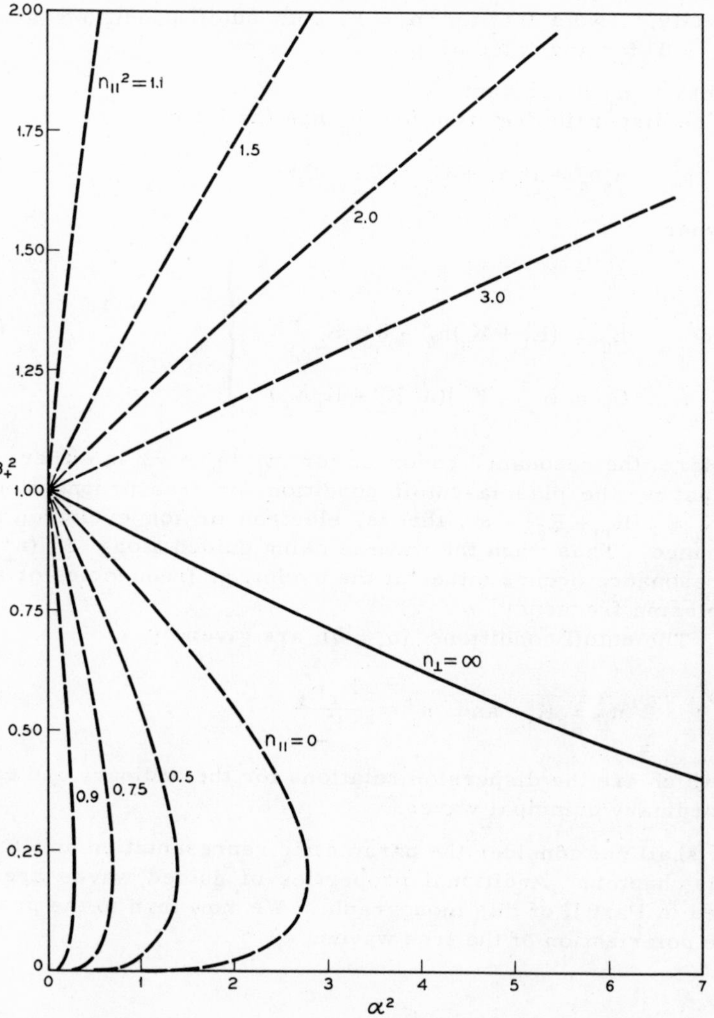

Figure 3.16. Loci of the cutoff and resonance conditions for guided propagation across $\underset{\sim}{B}_0$ in plasma with ion-to-electron mass ratio of 9 to 1. The dashed lines are the cutoff condition $\alpha^2 = (1 - n_{\parallel}^2)(1 + \beta_-)(1 - \beta_+)$ for various values of $n_{\parallel}^2$. The solid line is the second branch of the resonance condition $K_{\perp} = 0$. The plots of Figure 3.15 must be superimposed on those of this figure to obtain all the cutoff and resonance conditions for guided propagation across $\underset{\sim}{B}_0$.

and $\beta_+ < 1$, or $n_\parallel^2 > 1$ and $\beta_+ > 1$, are satisfied simultaneously. Note that for $n_\parallel = 1$, both cutoff conditions are at $\alpha^2 = 0$ for any value of β.

(b) $n_\perp$ = constant

The dispersion relation for $n_\parallel$ has the form

$$A_\parallel n_\parallel^4 + B_\parallel n_\parallel^2 + C_\parallel = 0 \tag{3.49}$$

where

$$\left.\begin{aligned} A_\parallel &= K_\parallel \\ B_\parallel &= (K_\perp + K_\parallel)n_\perp^2 - 2K_\parallel K_\perp \\ C_\parallel &= (n_\perp^2 - K_\parallel)(n_\perp^2 K_\perp - K_r K_\ell) \end{aligned}\right\} \tag{3.50}$$

Here, the resonance condition for $n_\parallel$ $(n_\parallel \to \infty)$ is either $K_\parallel = 0$, that is, the plasma cutoff condition for free propagation, or $K_\perp = \frac{1}{2}(K_r + K_\ell) = \infty$, that is, electron or ion cyclotron resonance. Thus when the wave is being guided along $\underset{\sim}{B}_0$ $(n_\perp \neq 0)$, resonance occurs either at the cyclotron frequencies or at the plasma frequency.

The cutoff conditions $(n_\parallel = 0)$ are given by

$$n_\perp^2 = K_\parallel \quad \text{and} \quad n_\perp^2 = \frac{K_r K_\ell}{K_\perp} \tag{3.51}$$

which are the dispersion relations for the ordinary and extraordinary principal waves.

We shall not consider the parametric representation any further in this chapter. Additional properties of guided waves are discussed in Part II of this monograph. We now turn to the problem of the polarization of the free waves.

Chapter 4

POLARIZATION

4.1 General Relations

The polarizations of the waves are determined by the components of the electric vector that have been given in Equations 1.34 and 1.36. When $K_{yz} = K_{zx} = 0$, they reduce, in Cartesian coordinates, to

$$\left.\begin{aligned} E_x &\approx (K_\perp - n^2)(K_\parallel - n^2 \sin^2\theta) \\ E_y &\approx K_\times (n^2 \sin^2\theta - K_\parallel) \\ E_z &\approx (n^2 - K_\perp) n^2 \sin\theta \cos\theta \end{aligned}\right\} \tag{4.1}$$

in spherical coordinates to

$$\left.\begin{aligned} E_\theta &\approx (K_\perp - n^2) K_\parallel \cos\theta \\ E_\phi &\approx K_\times (n^2 \sin^2\theta - K_\parallel) \\ E_k &\approx (K_\perp - n^2)(K_\parallel - n^2) \sin\theta \end{aligned}\right\} \tag{4.2}$$

and in rotating coordinates to

$$\left.\begin{aligned} \sqrt{2}\, E_r &= (n^2 \sin^2\theta - K_\parallel)(n^2 - K_\ell) \\ \sqrt{2}\, E_\ell &= (n^2 \sin^2\theta - K_\parallel)(n^2 - K_r) \end{aligned}\right\} \tag{4.3}$$

Thus, circular polarization around z occurs only when $n^2 = K_\ell$ or $n^2 = K_r$.

We note that only $E_y = E_\phi$ is imaginary (because $K_\times$ is imaginary) and the other electric field components are real. Unless $E_\phi = 0$, the electric vector therefore rotates in a plane containing the y direction and the resultant of E_θ and E_k. This plane (see Figure 4.1), which can be called the $\underset{\sim}{E}$ plane, forms an angle χ with the wave-normal plane, which contains the vector $\underset{\sim}{H}$, such that

$$\tan\chi = \frac{E_k}{E_\theta} = \frac{K_\parallel - n^2}{K_\parallel} \tan\theta = \frac{(n^2 - K_r)(n^2 - K_\ell)}{K_\perp n^2 - K_r K_\ell} \cot\theta \tag{4.4}$$

The $\underset{\sim}{H}$ plane, as has been seen directly from Maxwell's equations,

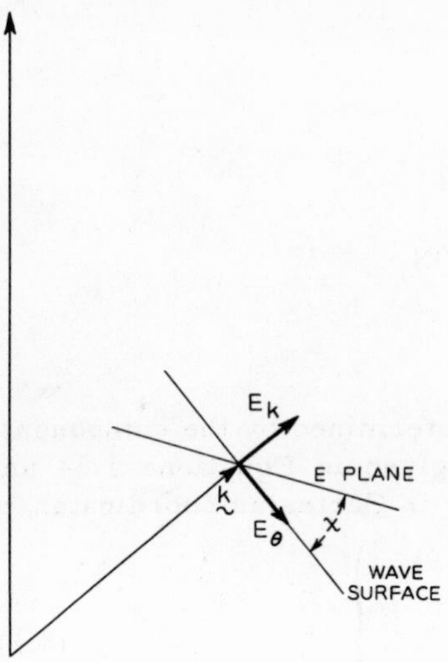

Figure 4.1. Orientation of the polarization plane with respect to the wave surface

is transverse to $\underset{\sim}{k}$. When $n^2 = K_\parallel$, $n^2 = K_r$, or $n^2 = K_\ell$, the angle χ is zero and the $\underset{\sim}{E}$ plane is also transverse. However, when $n^2 = K_r K_\ell / K_\perp$, as for the extraordinary wave, $\chi = \pi/2$ and the $\underset{\sim}{E}$ plane contains $\underset{\sim}{k}$.

An interesting situation occurs when $n^2 = K_\perp$. It is obvious from Equations 4.1 that $E_x/E_y = 0$, so that the $\underset{\sim}{E}$ plane is parallel to the z axis. We find that $K_\parallel - n^2 \sin^2 \theta = K_\parallel (K_\perp - n^2)/K_\perp$, so that $E_z/E_y \neq 0$, and the wave is not linearly polarized, as might be inferred from a casual inspection of Equations 4.1. As the wave parameters (θ or α or β) vary so that n^2 goes through the value $K_\perp$, the $\underset{\sim}{E}$ plane rotates, passing through the vertical position. The direction of rotation of the $\underset{\sim}{E}$ vector in this plane does not change, but the projection of $\underset{\sim}{E}$ on the (x, y) plane does change the sense of rotation. This phenomenon occurs only in regions 7 and 13 of the (α^2, β^2) plane and is indicated by the notation Ch in Table 3.3.

If we form the product

$$E_{\theta_1} E_{\theta_2} = K_\parallel^2 \cos^2 \theta (K_\perp - n_1^2)(K_\perp - n_2^2) \tag{4.5}$$

where n_1^2 and n_2^2 are the two solutions of the dispersion relation (2.24), and we note that

$$\left.\begin{aligned} n_1^2 + n_2^2 &= \frac{B}{A} \\ n_1 n_2 &= \frac{C}{A} \end{aligned}\right\} \tag{4.6}$$

we find that

$$AE_{\theta_1} E_{\theta_2} = CK_\times^2 \sin^2 \theta \cos^2 \theta = AE_{\phi_1} E_{\phi_2} \tag{4.7}$$

whence

$$\frac{E_{\theta_1}}{E_{\phi_1}} = \frac{E_{\phi_2}}{E_{\theta_2}} \tag{4.8}$$

Thus if we consider the transverse electric vectors (not including E_k) of the two possible waves traveling in a given direction, they describe similar ellipses but at right angles to each other. If one wave has a major axis along θ, the other has it along ϕ. Furthermore, these ellipses are described in the opposite directions, because if E_{ϕ_1}/E_{θ_1} is positive imaginary, E_{ϕ_2}/E_{θ_2} must be negative imaginary. Thus the transverse components of the electric vectors of the two waves are in a sense orthogonal to each other.

It has already been noted that E_k contains a higher power of n than either E_θ or E_ϕ, and therefore, the electric vector becomes longitudinal near a resonance $n^2 \to \infty$. Cyclotron resonance is an exception because E_k is also proportional to $\sin\theta$.

At a cutoff, $n^2 \to 0$, and we see that

$$\left.\begin{aligned} E_x &\approx K_\perp K_\parallel \\ E_y &\approx -K_\times K_\parallel \\ E_z &\approx 0 \end{aligned}\right\} \qquad (4.9)$$

and the $\underset{\sim}{E}$ plane is perpendicular to $\underset{\sim}{B}_0$. Plasma cutoff is an exception because here $K_\parallel = 0$. At cyclotron cutoff, either $K_r = 0$ or $K_\ell = 0$, and hence $K_\times = \pm jK_\perp$, so that the wave is circularly polarized around z over its entire surface. It follows that the Poynting vector is in this case directed along $\underset{\sim}{B}_0$.

4.2 The Quasi-Circular and Quasi-Plane Approximations

In the QC approximation,

$$(K_r K_\ell - K_\parallel K_\perp)\sin^2\theta \to 0 \qquad (4.10)$$

we find the following components for the electric vector:

$$\left.\begin{aligned} E_\theta &\approx K_\parallel \cos^2\theta + K_\perp \sin^2\theta \\ E_\phi &\approx \pm j(K_\parallel \cos^2\theta + K_\perp \sin^2\theta) \\ E_k &\approx \left[(K_\parallel - K_\perp)\cos\theta \mp jK_\times\right]\sin\theta \\ E_x &\approx K_\parallel \cos\theta \mp jK_\times \sin^2\theta \\ E_z &\approx -(K_\perp \pm jK_\times \cos\theta)\sin\theta \end{aligned}\right\} \qquad (4.11)$$

As

$$E_\phi = \pm jE_\theta \qquad (4.12)$$

this wave is circularly polarized around $\underset{\sim}{k}$, and hence its name. In general there is a longitudinal component as well, so that the actual trace of $\underset{\sim}{E}$ is the intersection of a circular cylinder parallel to $\underset{\sim}{k}$ with the $\underset{\sim}{E}$ plane. For propagation along $\underset{\sim}{B}_0$, however, $E_k = 0$, and the polarization is circular and transverse.

In the QP approximation

$$2K_{\parallel}K_{\times} \cos\theta \to 0 \tag{4.13}$$

There are separate formulas, Equations 3.26 and 3.27, for the ordinary and extraordinary indices, and the corresponding formulas for the electric vector in the "ordinary" wave are

$$\left.\begin{aligned} E_x &\approx K_{\parallel} \cos\theta \\ E_y &\approx 0 \\ E_z &\approx -K_{\perp} \sin\theta \end{aligned}\right\} \tag{4.14}$$

and for the "extraordinary" wave

$$\left.\begin{aligned} E_\theta &\approx 0 \\ E_\phi &\approx -K_{\perp} \sin^2\theta - K_{\parallel} \cos^2\theta \\ E_k &\approx K_{\times} \sin\theta \end{aligned}\right\} \tag{4.15}$$

In each case they have been given in the coordinate system in which the expressions are simplest. Both waves are plane-polarized; that is, the electric vector remains in a plane containing $\underset{\sim}{k}$. We shall use the term "linearly polarized" when the electric vector remains parallel to itself. The "ordinary" wave is indeed linearly polarized parallel to $\underset{\sim}{B}_0$, for propagation across $\underset{\sim}{B}_0$ ($\cos\theta \to 0$) and near plasma cutoff ($K_{\parallel} \to 0$). In general it is polarized in the plane of $\underset{\sim}{B}_0 + \underset{\sim}{k}$, but in the magnetohydrodynamic limit ($\alpha^2 >> 1$, $|K_{\parallel}| >> K_{\perp}$) it becomes linearly polarized along χ.

Note that in the magnetohydrodynamic limit the Poynting vector $\underset{\sim}{E} \times \underset{\sim}{H}$ is directed along the magnetic field over the entire velocity surface. No matter what the orientation of a plane wave, the energy travels along $\underset{\sim}{B}_0$ in this limit. As pointed out earlier, this applies to region 6 of the diagram as well as to region 12, and in this region, is responsible for "whistlers."

Of course the QP approximation degenerates as propagation along $\underset{\sim}{B}_0$ is approached. Then, E_z decreases with $\sin\theta$, E_x increases with $\cos\theta$, and as the pole of the velocity surface is approached, E_y becomes equal to E_x.

The "extraordinary" wave (Equations 4.15 and 3.27) is polarized in the plane of $\underset{\sim}{k}$ and $\underset{\sim}{B}_0 \times \underset{\sim}{k}$: The $\underset{\sim}{E}$ plane follows $\underset{\sim}{k}$ as θ is decreased from $\pi/2$ toward zero, and this wave nearly always has a longitudinal component, hence the name "extraordinary." Only in the high magnetic field limit ($K_{\times} \to 0$) does it become linearly polarized perpendicular to $\underset{\sim}{B}_0$ and $\underset{\sim}{k}$. As the pole is approached, the longitudinal component decreases as $\sin\theta$, and E_θ builds up to equal E_ϕ to give the circular polarization at the pole. Plasma cutoff

($K_{\parallel} = 0$) is, of course, peculiar. In that case, E_{ϕ} decreases relative to E_k, and the wave becomes more longitudinal near the poles. Only in the regions of the "dimples" or "pimples" of Figure 3.12 does the longitudinal component rapidly decay to give circular polarization exactly at the poles.

It is clear that the polar points do not belong to the same type of wave as the neighboring parts of the velocity surfaces. It will be seen later that the nearly longitudinal parts of the velocity surfaces belong to another family of waves which we shall name plasma electron waves and which couple to the electromagnetic waves near the plasma frequency.

As $K_{\perp}$ and $K_{\times}$ go through infinity together at both cyclotron resonances, and $K_{\times}$ is never zero, the ratio $-K_{\perp}/K_{\times}$ changes sign only when $K_{\perp} = 0$, that is, at plasma resonance. It follows that between the branches of the resonance hyperbola, $K_{\perp} = 0$, the extraordinary wave rotates left-handed around $\underset{\sim}{B}_0$, and right-handed above or below that hyperbola.

Comparison with the velocity surfaces sketched in Figure 3.9 shows that the extraordinary wave generally is polarized in the same direction as the wave at the polar point on the same velocity

Table 4.1. Conditions for Plane and Linear Polarizations

Plane Polarizations			
E Plane	Situation	Plasma Parameters	Angle
$\perp B_0$	Always	All	$\theta = 0$
$\perp B_0$	Extraordinary wave	All	$\theta = \pi/2$
$\perp B_0$	Cyclotron cutoff	$K_r K_\ell = 0$	All
$\perp x$	Change of rotation	$n^2 = K_{\perp}$	$\sin^2\theta = K_{\parallel}/K_{\perp}$
$\perp \theta$	Sphere at plasma cutoff	$K_{\parallel} = 0$	All
$\perp \phi$	Nonspherical wave	$\beta_+ >> 1$	All
Linear Polarizations			
Direction	Situation	Angle	Equation
$\underset{\sim}{B}_0$	Ordinary	$\theta = \pi/2$	Always
$\underset{\sim}{B}_0$	Plasma cutoff	All	$K_{\parallel} = 0$
$\underset{\sim}{x}$	Alfvén wave	All	$\alpha, \beta_+ >> 1$
$\underset{\sim}{k}$	Plasma resonance	$\theta = \pi/2$	$K_{\perp} = 0$
$\underset{\sim}{k}$	Conical resonance	$\tan\theta = -K_{\parallel}/K_{\perp}$	—
$\underset{\sim}{\phi}$	Spherical wave	All	$\beta_+ >> 1$

surface. However, as the polar points pass discontinuously from one surface to the other as plasma cutoff is crossed, the correspondence does not hold everywhere. In regions 7 and 13, the extraordinary wave rotates oppositely from the wave at the polar point on the same velocity surface. In general, the $\underset{\sim}{E}$ plane, which is horizontal for the extraordinary wave, follows the $\underset{\sim}{k}$ vector as θ is decreased from $\pi/2$ as long as the QP approximation is valid, and then drops back to the horizontal position at the poles, leaving the direction of rotation unchanged. When the direction of rotation at the pole is opposite to that at the equator, it is because the $\underset{\sim}{E}$ plane has turned over, as described in Section 4.1. The directions of rotation for both waves are indicated in Table 3.3. Note that in areas 5 and 11 both waves rotate in the same direction. Some of the special polarizations are summarized in Table 4.1.

4.3 Resonance Currents

At a resonance, conditions are such that a longitudinal electric field produces a transverse current without a transverse electric field. As has been seen, this happens for wave normals directed along a cone defined by $A = 0$, and the polar and equatorial limits of this cone have been rather arbitrarily termed "cyclotron" and "plasma" resonance.

In order to determine the physical nature of these resonances, we shall study the electron and ion orbits at these limits. As $E_z = 0$ at the two limits, we can write, from Equation 2.7,

$$\left.\begin{aligned} \underset{\sim}{v}_- &= \beta_- \frac{-j - \underset{\sim}{\beta}_- \times}{\beta_-^2 - 1} \frac{\underset{\sim}{E}}{B_0} \\ \underset{\sim}{v}_+ &= \beta_+ \frac{j + \underset{\sim}{\beta}_+ \times}{\beta_+^2 - 1} \frac{\underset{\sim}{E}}{B_0} \end{aligned}\right\} \qquad (4.16)$$

where the directions of $\underset{\sim}{\beta}_-$ and $\underset{\sim}{\beta}_+$ are defined by Equation 2.5. The two orbits are thus ellipses, described in the right- and left-handed directions, and whose ratios of transverse to longitudinal axes are β_- and β_+, respectively.

The two corresponding energies for ion (+) and electrons (-) are

$$\frac{m_\pm v_\pm^2}{2e} = \frac{\beta_\pm}{\omega} \frac{1 + \beta_\pm^2}{(\beta_\pm^2 - 1)^2} \frac{E^2}{2B_0} \qquad (4.17)$$

It is seen that these two energies are equal when

$$\beta_+\beta_- = 1 \qquad (4.18)$$

At frequencies above that defined by Equation 4.18, the electrons have most of the energy. At frequencies below it, the ions do.

The conduction current is

$$\underset{\sim}{J} = Ne(\underset{\sim}{v}_+ - \underset{\sim}{v}_-) = \epsilon_0 \omega a^2 \frac{j(\beta_+\beta_- - 1) - (\underset{\sim}{\beta}_+ + \underset{\sim}{\beta}_-)\times}{(\beta_+^2 - 1)(\beta_-^2 - 1)} \underset{\sim}{E} \qquad (4.19)$$

At cyclotron resonance, $\beta_- = 0$ or $\beta_+ = 0$, there is no steady-state solution of the equations. The particle orbits are ever-increasing spirals, as can be seen from the equations of motion when the existence of a steady-state solution is not assumed.

At plasma resonance,

$$a^2 = \frac{(\beta_-^2 - 1)(1 - \beta_+^2)}{\beta_+\beta_- - 1} \qquad (4.20)$$

and Equation 4.19 reduces to

$$\underset{\sim}{J} = -\epsilon_0\omega\left(j + \frac{\underset{\sim}{\beta}_- + \underset{\sim}{\beta}_+}{\beta_+\beta_- - 1}\times\right)\underset{\sim}{E} \qquad (4.21)$$

The first term shows the proper cancellation of conduction and displacement currents in the longitudinal direction:

$$J_k = -j\omega\epsilon_0 E_k \qquad (4.22)$$

The transverse current

$$J_t = -j\frac{\beta_- - \beta_+}{\beta_-\beta_+ - 1} J_k \qquad (4.23)$$

is finite, and therefore, $\underset{\sim}{H} = 0$ at this resonance. The transverse current J_t is smaller than J_k ($J_t \approx \beta_- J_k$) on the lower branch ($\beta_+ << 1$)

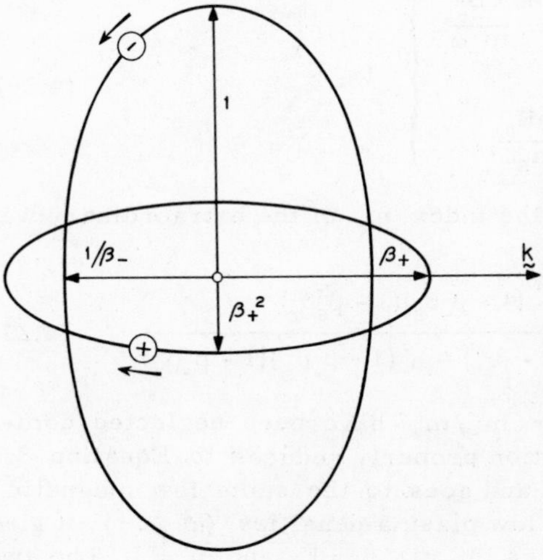

Figure 4.2. The electron and ion orbits induced by the wave at the electron-ion hybrid resonance $\beta_-\beta_+ = 1$ and $a^2 >> 1$. The eccentricities of the orbits are in the ratios shown.

of the resonance hyperbola, but it is always larger than J_k on the upper branch, tending to infinity at the hybrid resonance. Thus there are very large currents near this resonance, although the magnetic fields are small because the wavelength is small.

It is readily verified that the transverse current is mainly carried by the electrons except near ion cyclotron resonance. The longitudinal electron current is smaller than the ion current by a factor β_-, but may nevertheless be quite large, because the electron and ion currents tend to cancel longitudinally and add transversely. The orbits at the electron-ion hybrid resonance are sketched in Figure 4.2.

4.4 The Ion-Ion Hybrid Resonance

An interesting resonance condition occurs when a plasma possesses two (or more) ion species of different charge-to-mass ratios. Let x_1 and x_2 be the relative concentrations of the two ions $(x_1 + x_2 = 1)$, m_1 and m_2 their masses, ω_{b1} and ω_{b2} their cyclotron frequencies $(\omega_{bi} = e_i B/m_i)$, and ω_{p1} and ω_{p2} their plasma frequencies $\omega_{pi}^2 = (x_i N e^2)/(m_i \epsilon_0)$. Then

$$K_{r,\ell} = 1 - \frac{\alpha_-^2}{1 \mp \beta_-} - \frac{\alpha_1^2}{1 \pm \beta_1} - \frac{\alpha_2^2}{1 \pm \beta_2} \tag{4.24}$$

It is convenient to define two averages denoted by r and s :

$$\left.\begin{aligned} m_s &= x_1 m_1 + x_2 m_2 \\ \frac{1}{m_r} &= \frac{x_1}{m_1} + \frac{x_2}{m_2} \end{aligned}\right\} \tag{4.25}$$

Also

$$\left.\begin{aligned} \beta_r &= x_1\beta_1 + x_2\beta_2 = \frac{eB_0}{m_r\omega} \\ \frac{1}{\beta_s} &= \frac{x_1}{\beta_1} + \frac{x_2}{\beta_2} = \frac{eB_0}{m_s\omega} \end{aligned}\right\} \tag{4.26}$$

In terms of these quantities the index $n_\times$ of the extraordinary wave is given by

$$n_\times^2 = 1 + \alpha^2 \frac{\alpha^2(1 - \beta_r^2) - (1 - \beta_-\beta_s)(1 - \beta_s\beta_r)}{(1 - \beta_-^2)(1 - \beta_1^2)(1 - \beta_2^2) - \alpha^2(1 - \beta_-\beta_s)(1 - \beta_s\beta_r)} \tag{4.27}$$

where quantities of the order m_-/m_+ have been neglected compared with unity. This equation properly reduces to Equation 3.7 if either x_1 or x_2 is zero, and goes to the same low magnetic field limit in each case. At low plasma densities $(\alpha^2 << 1)$, it gives the three cyclotron resonances $\beta_- = 1$, $\beta_1 = 1$, and $\beta_2 = 1$. The two ion resonances head two "hyperbolas" whose high plasma density limits are

$$\beta_-\beta_s = 1 \tag{4.28}$$

and

$$\beta_s\beta_r = 1 \tag{4.29}$$

The first corresponds to the usual electron-ion "hybrid" resonance. The second involves only the ions. At this resonance the electron and ion velocities are given by

$$\left.\begin{aligned}
\mathbf{v}_- &= -\frac{jm_-\sqrt{m_r m_s}}{m_1 m_2}\frac{\mathbf{E}}{B_0} + \frac{\mathbf{E}\times\mathbf{B}_0}{B_0^2} \\
\mathbf{v}_1 &= \frac{j\sqrt{m_r m_s}\,\dfrac{\mathbf{E}}{B_0} - m_2\mathbf{E}\times\dfrac{\mathbf{B}_0}{B_0^2}}{x_1(m_2^2 - m_1^2)}\,\frac{m_1 m_2}{m_s} \\
\mathbf{v}_2 &= \frac{j\sqrt{m_r m_s}\,\dfrac{\mathbf{E}}{B_0} - m_1\mathbf{E}\times\dfrac{\mathbf{B}_0}{B_0^2}}{x_2(m_1^2 - m_2^2)}\,\frac{m_1 m_2}{m_s}
\end{aligned}\right\} \tag{4.30}$$

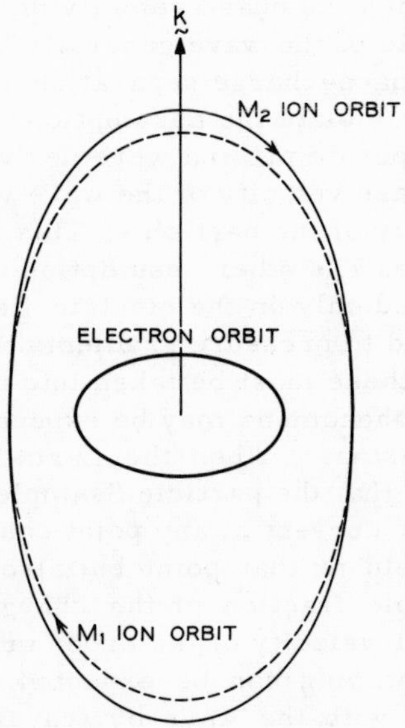

Figure 4.3. Not-to-scale sketch of induced orbits at the ion-ion hybrid resonance. The relative size of the electron orbit is exaggerated. At a given instant the ions are at the two positions indicated by the arrows.

The three orbits are shown in Figure 4.3. Here the ion velocities along $\mathbf{E}$ are large compared with any component of electron velocity and are out of phase with each other. At the "ion-ion hybrid resonance" the electrons remain relatively motionless, and it is the oscillation of the two ion clouds out of phase with each other that constitutes the resonance condition; while at the "electron-ion hybrid resonance" it is the <u>joint</u> oscillation of the whole plasma along the direction of $\mathbf{E}$, with only a small slippage between the negatively and positively charged plasma constituents, that constitutes the resonance condition. The ion drift energies are much larger than the electron energy at the ion-ion hybrid resonance, u_+/u_- being of the order of m_+/m_-.

Chapter 5

TRANSPORT THEORY

5.1 Transport Equations

In the previous sections we have discussed the properties of waves in temperate plasmas. Among others, two facts emerged: (1) Resonance conditions existed at which the phase velocity of the wave was zero, and (2) the electric field of the wave generally had a longitudinal component so that ac space-charge separation existed. In a hot plasma these phenomena violate the assumptions on which the properties of waves in a temperate plasma were derived. One of the assumptions was that the phase velocity of the wave was large compared with the random velocity of the particles. This assumption breaks down near a resonance. The other assumption was that the current in the plasma depended only on the electric field. When a temperature exists, it will lead to pressure gradients that will also drive electric currents, and these must be taken into account. In a warm plasma additional phenomena may be expected that are not possible in a temperate plasma. When the Larmor orbit of an ion (or electron) is so large that the particle "samples" a large fraction of the wavelength, the current at any point can be expected to depend not only on the field at that point but also on the field gradient. When an appreciable fraction of the charged particles in the plasma have a thermal velocity of the order of the phase velocity of the wave, Landau damping can be expected. These particles will interact strongly with the wave by traveling with it, they may slow it down (or speed it up!), and they may damp it.

In order to account properly for some of these expected phenomena, we shall in the next chapter derive in a self-consistent manner from the Boltzmann equation (1.32) the current $\underset{\sim}{J}$ and the space charge ρ. The mathematical details of this procedure are sufficiently involved, however, that it is instructive to consider first in this section an approximation to the Boltzmann equation which is the Boltzmann transport equation. The Boltzmann transport equation for any quantity $X(v)$ is obtained by multiplying the Boltzmann equation by $X(v)$ and integrating over all velocities.[2] In this manner some information is obtained about the transport of X without detailed knowledge of the distribution function $f(r, v, t)$. We shall take $X = 1$ and $X = \underset{\sim}{v}$ to obtain the first two moments:

$$\frac{\partial N}{\partial t} + \nabla \cdot \underset{\sim}{\Gamma} = 0 \tag{5.1}$$

$$\frac{\partial \underset{\sim}{\Gamma}}{\partial t} + \frac{\nabla \cdot \underset{\approx}{P}}{m} \pm \frac{e}{m} (N\underset{\sim}{E} + \underset{\sim}{\Gamma} \times \underset{\sim}{B}_0) = 0 \tag{5.2}$$

Here the upper sign is for electrons and the lower sign for ions. Collisions have been neglected, and the subscripts distinguishing different kinds of particles are suppressed until needed. The particle density is

$$N = \int f \, d^3v \tag{5.3}$$

the particle current is

$$\underset{\sim}{\Gamma} = \int \underset{\sim}{v} f \, d^3v \tag{5.4}$$

and the pressure tensor is

$$\underset{\approx}{P} = \int m \underset{\sim}{v}\underset{\sim}{v} f \, d^3v \tag{5.5}$$

The transport method always leads to more unknowns than equations, and some assumption must be made concerning the highest moment, in this case $\underset{\approx}{P}$. We assume that $\underset{\approx}{P}$ is diagonal, that the diagonal terms P are related to the density N by an equation of state†

$$P \approx N^\gamma \tag{5.6}$$

By truncating the set of moment equations, we lose those wave phenomena that depend for their existence on particles with particular velocities (such as Landau damping). By assuming $\underset{\approx}{P}$ to be a scalar, we lose sight of the "shear waves" that depend for their existence on an anisotropic pressure tensor. These are outside the scope of this monograph. By neglecting collisions, we also lose sight of all collisional damping phenomena.

With these assumptions the second transport equation becomes

$$\frac{\partial \underset{\sim}{\Gamma}}{\partial t} + \nabla \frac{P}{m} \pm \frac{e}{m} (N\underset{\sim}{E} + \underset{\sim}{\Gamma} \times \underset{\sim}{B}_0) = 0 \tag{5.7}$$

5.2 The Dielectric Tensor

We now expand the density N into an equilibrium part N_0 and a small ac part N_1:

$$N = N_0 + N_1 \exp[j(\omega t - \underset{\sim}{k} \cdot \underset{\sim}{r})] \tag{5.8}$$

The term ∇P is also linearized

$$\nabla P \approx \gamma \frac{P_0}{N_0} \nabla N_1 = \gamma eT \nabla N_1 \tag{5.9}$$

---

† Note that in this chapter γ denotes the ratio of specific heats and not the normalized collision frequency.

where we set $P_0/N_0 = eT$, with T in electron volts. The continuity equation

$$\frac{\partial N_1}{\partial t} + \nabla \cdot \Gamma = j\omega N_1 - j\mathbf{k} \cdot \Gamma = 0 \tag{5.10}$$

provides a relation by which we can express ∇N in terms of Γ in Equation 5.7. Using Equations 5.1 and 5.10 in Equation 5.7 and linearizing by neglecting the product EN_1, we obtain

$$(\omega + j\omega_b \times)\Gamma - \frac{\gamma eT}{m\omega} \mathbf{k}\mathbf{k} \cdot \Gamma = \pm j \frac{N_0 e}{m} \mathbf{E} \tag{5.11}$$

Define

$$\delta = \frac{\gamma eT}{mc^2} = \frac{\gamma}{3} \frac{\overline{v^2}}{c^2} \tag{5.12}$$

and let ξ, $\eta = 0$, and ζ be the direction cosines of $\mathbf{k}$. Then the solution of Equation 5.11 is

$$\Gamma = \pm \frac{jN_0 e}{m\omega} \tau \cdot \mathbf{E} \tag{5.13}$$

where

$$[1 - \delta n^2 - \beta^2(1 - \delta n^2 \zeta^2)]\tau = \begin{bmatrix} 1 - \delta n^2\zeta^2 & \pm j\beta(1 - \delta n^2\zeta^2) & \delta n^2 \xi\zeta \\ \mp j\beta(1 - \delta n^2\zeta^2) & 1 - \delta n^2 & \mp j\beta\delta n^2 \xi\zeta \\ \delta n^2 \xi\zeta & \pm j\beta\delta n^2 \xi\zeta & 1 - \delta n^2 \xi^2 - \beta^2 \end{bmatrix}$$

We can now form the current $\mathbf{J} = (\Gamma_+ - \Gamma_-)e$ and from it the conductivity

$$\sigma = \frac{N_0 e^2}{j\omega}\left(\frac{\tau_+}{m_+} + \frac{\tau_-}{m_-}\right) \tag{5.14}$$

and hence the dielectric coefficient

$$\mathbf{K} = \mathbf{1} + \frac{\sigma}{j\omega\epsilon_0} = \mathbf{1} - \alpha_-^2 \tau_- - \alpha_+^2 \tau_+ \tag{5.15}$$

from which everything follows as in Chapters 2 to 4. The result is quite unwieldy, so that we shall first restrict ourselves to high frequencies.

5.3 Electron Plasma Waves along $\mathbf{B}_0$

At high frequencies the ion motion can be neglected. The dispersion relation may be obtained by using Equation 5.15 with $m_+ = \infty$ in the determinant (1.24). It is not convenient to use the tangent formula, Equation 1.27, because of the explicit appearance of the angles in the expression for $\underset{\approx}{K}$. With some manipulation, we obtain

$$\tan^2\theta = -\frac{(n^2 - K_r)(n^2 - K_\ell)[n^2\epsilon(K_\parallel - 1) + K_\parallel]}{(n^2 - K_\parallel)\left\{\epsilon(K_\perp - 1)n^4 + [K_\perp - \epsilon(K_r K_\ell - K_\perp)]n^2 - K_r K_\ell\right\}} \tag{5.16}$$

where $\epsilon = \delta/\alpha^2$. Equation 5.16 indicates that along the magnetic field ($\theta = 0$) three distinct waves can propagate, the right- and left-handed circularly polarized waves whose phase velocity is of the order of the velocity of light, and a third wave, called the "electron plasma wave," whose velocity is of the order of the electron thermal velocity and whose dispersion equation is

$$n_{po}^2 = \frac{K_\parallel}{\delta_-} \tag{5.17a}$$

The subscript po indicates that it is a plasma wave with the "ordinary" polarization. The three waves are distinct because the two electromagnetic waves are transverse, while the plasma wave is longitudinal, so that there is no coupling between them. The plasma wave has the usual Vlasov[12] or Bohm and Gross[13] dispersion formula (when $\gamma = 1$)

$$\omega^2 = \omega_{p-}^2 + \tfrac{1}{3}k^2\overline{v_-^2} \tag{5.18}$$

and reduces to the singular solution $\omega = \omega_{p-}$ described in Section 3.1, when the temperature is neglected. Note that

$$n_{po}^2 = \frac{n_o^2}{\delta_-} = 3\frac{c^2}{\overline{v^2}}n_o^2 \tag{5.17b}$$

so that the phase velocity u_p of the plasma wave along $\underset{\sim}{B}_0$ remains at a fixed fraction $\sqrt{\overline{v^2}/3c^2}$ of the velocity u_o of the ordinary wave. This relationship holds in spite of their different directions of propagation, along and across $\underset{\sim}{B}_0$, because they are both polarized parallel to $\underset{\sim}{B}_0$. Neither wave has a resonance, and they both cut off at $\alpha_-^2 = 1$.

As plasma cutoff is approached, the plasma wave is no longer "slow." In fact, u_{po} passes u_r and then u_ℓ, and this resolves the singularity that we have observed at plasma cutoff. The extraordinary velocity surface is carried from u_r to u_ℓ by u_{po} according to the sequence of sketches in Figure 5.1.

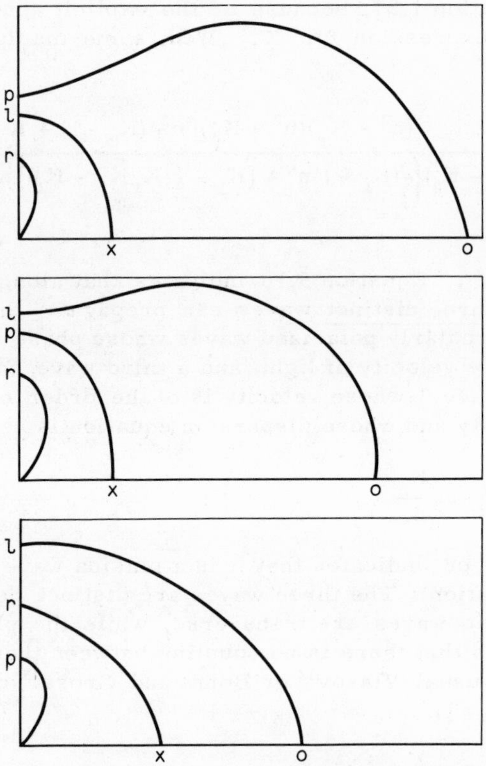

Figure 5.1. Phase velocity surfaces as plasma cutoff is approached from region 7. Only one quarter of the surface is shown.

5.4 Electron Plasma Waves across $\underset{\sim}{B}_0$

For propagation across the magnetic field only the ordinary wave is distinct. As the extraordinary wave is partially longitudinal, it couples with the plasma wave, which is now partially transverse. The indices of the two waves are given by the quadratic equation

$$\left\{\epsilon(1 - K_\perp)n^4 - \left[K_\perp + \epsilon(K_\perp - K_r K_\ell)\right] n^2 + K_r K_\ell\right\} = 0 \qquad (5.19)$$

or

$$\delta_- n^4 + \left[(\alpha_-^2 - 1)(1 + \delta_-) + \beta_-^2\right] n^2 + (\alpha_-^2 - 1)^2 - \beta_-^2 = 0 \qquad (5.20)$$

The discriminant of Equation 5.20 is positive definite:

$$D^2 \equiv \left[(\alpha_-^2 - 1)(1 + \delta_-) + \beta_-^2\right]^2 - 4\delta_-\left[(\alpha_-^2 - 1)^2 - \beta_-^2\right]$$
$$\equiv \left[(\alpha_-^2 - 1)(1 - \delta_-) + \beta_-^2\right]^2 + 4\delta_-\alpha_-^2\beta_-^2 \qquad (5.21)$$

so that the waves either do or do not propagate, but there is no attenuation.

Equation 5.20 is of the form

$$\epsilon' n^4 + an^2 + b = 0; \qquad \epsilon' = \epsilon(1 - K_\perp) \qquad (5.22)$$

where the highest power of n has a very small coefficient. Provided $|4\epsilon' b| < a^2$, it has two solutions given by the expansions

$$n_\times^2 = -\frac{b}{a}\left(1 + \frac{\epsilon' b}{a^2} \cdots\right) \approx \frac{K_r K_\ell}{K_\perp}$$

and

$$n_{p\times}^2 = -\frac{a}{\epsilon'} + \frac{b}{a} + \cdots \approx \frac{K_\perp}{\epsilon'(1 - K_\perp)} = \frac{1 - \alpha_-^2 - \beta_-^2}{\delta_-} \qquad (5.23)$$

The subscript $p\times$ indicates that it is the electron plasma wave coupled with the extraordinary wave. The extraordinary wave cuts off as previously at $K_r K_\ell = 0$, and the plasma wave has no resonance because $K_\perp \neq 1$ in the high-frequency range. The resonance of the extraordinary wave is, however, confused with the cutoff of the plasma wave because the condition $K_\perp = 0$ makes $a \approx 0$ and violates the requirement for the approximate solutions. We must look more closely into the solutions of the biquadratic equation when the preceding inequality is not satisfied. A plot of n^2 against a, taken as independent variable, is a hyperbola whose asymptotes are $n_\times^2 = 0$ and $n_{p\times}^2 = -a/\epsilon'$, and the quadrants in which the curve lies depend on the sign of ϵ'/b, as shown in Figure 5.2.

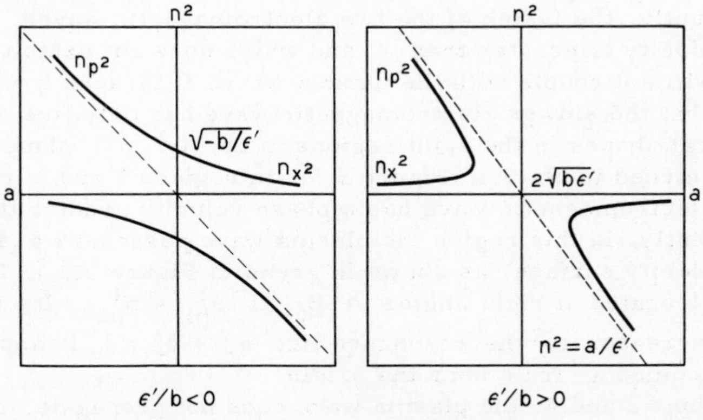

Figure 5.2. Diagram showing the coupling of the extraordinary with the electron plasma waves.

In either case only one branch of the hyperbola gives a positive n^2, but if $\epsilon'/b > 0$, the electromagnetic wave joins the plasma wave at the "resonance" line $a^2 = 4\epsilon' b$, and there is no solution on the other side. This is not the case here, however, where

$$\frac{\epsilon'}{b} = \epsilon \frac{1 - K_\perp}{K_r K_\ell} = \frac{\delta_-}{(\alpha_-^2 - 1)^2 - \beta_-^2} < 0 \tag{5.24}$$

near $K_\perp = 0$. The extraordinary wave is real above the plasma resonance line on the (α^2, β^2) plot, and the plasma wave is real below it. They merge into each other as that line is crossed. Thus $K_\perp = 0$ represents a resonance between the extraordinary and electron plasma waves, and for this reason deserves the name "plasma resonance."

5.5 Waves in Arbitrary Directions

Except near plasma cutoff, n_p^2 is large compared with $K_\parallel$, K_r, or K_ℓ, so that the tangent formula for the plasma wave reduces to

$$\tan^2 \theta = - \frac{n_p^2 \epsilon (K_\parallel - 1) + K_\parallel}{n_p^2 \epsilon (K_\perp - 1) + K_\perp} \tag{5.25}$$

This shows a "cut off" at $\tan^2 \theta = - K_\parallel / K_\perp$, where the plasma wave is speeded up to join with the slowed-down electromagnetic wave, and a resonance angle at

$$\tan^2 \theta_{rp} = \frac{1 - K_\parallel}{K_\perp - 1} = \beta_-^2 - 1 \tag{5.26}$$

This angle exists everywhere above the cyclotron resonance line.

Coupling between the plasma waves and the electromagnetic waves will be strong only when their phase velocities are nearly equal. Consequently, the faster of the two electromagnetic waves, whose phase velocity is greater than c and which does not exhibit a resonance, will not couple with the plasma wave. It is seen from Figure 3.7 that the slower electromagnetic wave has only five different topological shapes in the eight regions in the (α^2, β^2) plane. These are resketched in black in Figure 5.3. In regions 1 and 3, even the slower electromagnetic wave has a phase velocity greater than c. Consequently, in this region the plasma wave possesses a distinct phase velocity surface, as shown in green in Figure 5.3. It is an ellipse elongated at right angles to $\underset{\sim}{B}_0$ as $n_{px}^2 < n_{po}^2$. Its eccentricity increases as the resonance line $\alpha_-^2 + \beta_-^2 = 1$ is approached, and it becomes a circle near the origin.

In regions 2 and 4, the plasma wave does not propagate. In region 5, the electromagnetic wave possesses a resonance angle that is a cutoff for the plasma wave. Thus the plasma wave is real inside

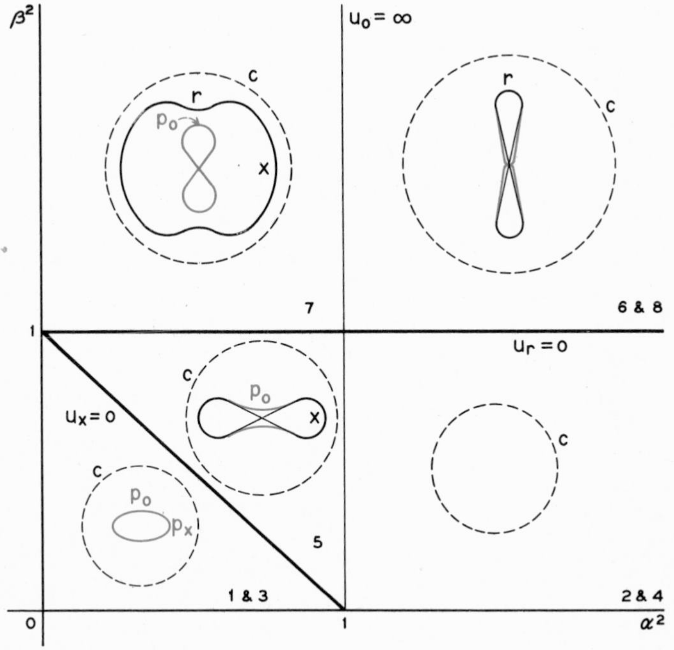

Figure 5.3. Phase velocity surfaces at low magnetic fields or high frequencies for the electron plasma wave and the slower of the two electromagnetic waves.

the resonance cone and joins the electromagnetic wave, as shown in Figure 5.3. Note that only two, and not three, distinct velocity surfaces exist in this region, the fast electromagnetic wave not being shown.

In region 7, the plasma wave possesses a resonance angle at $\tan^2 \theta = \beta_-^2 - 1$, but it has no cutoff because the electromagnetic wave has no resonance. Consequently, the waves do not couple, the plasma wave being spindle-shaped as shown.

In regions 6 and 8, the electromagnetic wave has a resonance at

$$\tan^2 \theta_{res} = -\frac{K_\parallel}{K_\perp} = (\beta_-^2 - 1)\left(1 - \frac{\beta_-^2}{\beta_-^2 + \alpha_-^2 - 1}\right) \qquad (5.27)$$

The angle given by Equation 5.27 is thus always smaller than the resonant angle of the plasma wave, Equation 5.26. Therefore, in this region the effect of the plasma wave is to increase the resonant angle, as shown in Figure 5.3.

5.6 Dispersion Equation for Coupled Electron and Ion Plasma Waves

For waves of sufficiently low frequency, the ion mass can no longer be set infinite, and Equation 5.15 for the dielectric coefficient tensor $\underset{\approx}{K}$ must be used in its general form in the dispersion relation (1.24). Note that $\underset{\approx}{K}$ is of first-order in n^2 if only electrons are considered, and of second-order in n^2 if one type of ions is included. It can be easily generalized to a plasma containing g species of ions of different charge-to-mass ratios, in which case $\underset{\approx}{K}$ is of order $(g+1)$ in n^2. The dispersion relation (1.24) is itself of second-order in n^2. Consequently, in a plasma of g different ion species (and one electron species), there will in general exist $(g+3)$ waves. Of these, two can be labeled electromagnetic, one electron plasma wave, and the remainder ion plasma waves. The various ion plasma waves involve the oscillation of the various ion species out of phase with each other (and with the electrons) in a manner not unlike that found in the case of two-ion hybrid resonance in Section 4.4. For simplicity we shall consider only a plasma with a single-ion species.

The dispersion relation (1.24) with electron and ion thermal motion included is sufficiently involved that no simple form of the type of Equation 5.16 has yet been found for it. Consequently, we shall proceed as follows. At first, we shall neglect the coupling between the electromagnetic and the plasma waves and study only the coupled electron and ion plasma waves. This will yield a correct description of the plasma waves everywhere except near, and at, cutoff of the plasma waves where, as we have seen in Section 5.5, the speeded-up plasma wave may couple strongly with the slowed-down electromagnetic waves. We shall describe this coupling intuitively on the basis of what was learned in Sections 5.3 to 5.5 for electron plasma waves.

The dispersion equation for the coupled electron and ion plasma waves may be obtained by noting that the coupling between the electromagnetic waves and the plasma waves is weakened by allowing the velocity of light c to tend to infinity. This procedure is correct in all regions of (α^2, β^2) plane (Figure 3.6) except in the upper right-hand corner of region 12 (the Alfvén region), where, as we saw in Section 3.7, the phase velocity of the wave u is related not to c but to the Alfvén velocity u_a. We shall discuss this region separately.

As $n = c/u$, in the limit $c = \infty$, the equation

$$An^4 - Bn^2 + C = 0 \tag{5.28}$$

yields

$$A = 0 \tag{5.29}$$

as the dispersion relation for the coupled electron and ion plasma waves. Note that when $\underset{\approx}{K}$ was independent of n^2 (temperate plasma), Equation 5.29 represented the resonant condition for the electromagnetic waves. It can be easily verified that the expression for $\underset{\approx}{K}$, as given by Equation 5.15, tends to its value in temperate plasmas either in the limit of $n^2 = 0$ (cutoff of plasma waves) or the limit of $\delta_\pm = 0$ (temperate plasma). Thus, the cutoff condition for the plasma waves is synonymous with the resonance condition for the electromagnetic waves <u>in a temperate plasma</u>, indicating that these waves couple there, and that the velocity surfaces join each other where their respective cutoff and resonance conditions are satisfied.

We now recall from Equation 1.26a that

$$A = K_{xx} \sin^2\theta + 2K_{xz} \sin\theta \cos\theta + K_{zz} \cos^2\theta \tag{5.30}$$

and from Equation 5.15 that

$$K_{xx} = 1 - \frac{\alpha_-^2(1 - \delta_- n^2 \cos^2\theta)}{[1 - \beta_-^2 - \delta_- n^2(1 - \beta_-^2 \cos^2\theta)]} - \frac{\alpha_+^2(1 - \delta_+ n^2 \cos^2\theta)}{[1 - \beta_+^2 - \delta_+ n^2(1 - \beta_+^2 \cos^2\theta)]} \tag{5.31a}$$

$$K_{xz} = - \frac{\alpha_-^2 \delta_- n^2 \cos\theta \sin\theta}{[1 - \beta_-^2 - \delta_- n^2(1 - \beta_-^2 \cos^2\theta)]} - \frac{\alpha_+^2 \delta_+ n^2 \cos\theta \sin\theta}{[1 - \beta_+^2 - \delta_+ n^2(1 - \beta_+^2 \cos^2\theta)]} \tag{5.31b}$$

$$K_{zz} = 1 - \frac{\alpha_-^2(1 - \beta_-^2 - \delta_- n^2 \sin^2\theta)}{[1 - \beta_-^2 - \delta_- n^2(1 - \beta_-^2 \cos^2\theta)]} - \frac{\alpha_+^2(1 - \beta_+^2 - \delta_+ n^2 \sin^2\theta)}{[1 - \beta_+^2 - \delta_+ n^2(1 - \beta_+^2 \cos^2\theta)]} \tag{5.31c}$$

Using Equations 5.30 and 5.31 in Equation 5.29, we find

$$1 = \frac{a_-^2}{\dfrac{1-\beta_-^2}{1-\beta_-^2\cos^2\theta} - \delta_- n^2} + \frac{a_+^2}{\dfrac{1-\beta_+^2}{1-\beta_+^2\cos^2\theta} - \delta_+ n^2} \tag{5.32}$$

as the dispersion relation for the coupled electron and ion plasma waves. Equation 5.32 can be easily generalized to a plasma with many ion species

$$1 = \sum_i \frac{a_i^2}{\dfrac{1-\beta_i^2}{1-\beta_i^2\cos^2\theta} - \delta_i n^2} \tag{5.33}$$

5.7 Electron and Ion Plasma Waves along $\underset{\sim}{B}_0$

Setting $\cos\theta = 1$ in Equation 5.32 gives the dispersion relation for propagation along $\underset{\sim}{B}_0$:

$$1 = \frac{a_-^2}{1-\delta_- n^2} + \frac{a_+^2}{1-\delta_+ n^2} \tag{5.34}$$

As $\delta_- \gg \delta_+$, the two solutions of this biquadratic are well separated.

For the electron wave,

$$\delta_+ n^2 \ll 1 \tag{5.35}$$

which yields

$$\delta_- n^2 = \frac{1-a_-^2}{1-a_+^2} \tag{5.36}$$

or

$$\omega^2 = \omega_p^2 + \frac{\gamma_-}{3} k^2 \overline{v_-^2}(1-a_+^2) \tag{5.37}$$

which is the same dispersion relation as was obtained in Section 5.3, with a small correction for the shielding by the ions. Note that $\omega_p^2 = \omega_{p-}^2 + \omega_{p+}^2$ involves only the reduced mass.

For the ion wave,

$$\delta_- n^2 \gg 1 \tag{5.38}$$

This leads to the relation

$$\frac{1}{n^2} = \delta_+ + \frac{a_+^2}{n^2 + \dfrac{a_-^2}{\delta_-}} \tag{5.39}$$

and for the phase velocity

$$u^2 = \frac{e}{m_+}\left(\gamma_+ T_+ + \frac{\gamma_- T_-}{\gamma_- k^2 \ell_0^2 + 1}\right) \tag{5.40}$$

where

$$\ell_0^2 = \frac{\epsilon_0 T_-}{Ne^2} = \frac{\delta_- c^2}{\gamma_- \omega_{p-}^2} \tag{5.41}$$

is the Debye length squared. When $k^2\ell_0^2 << 1$, this is an ordinary acoustic wave with the "sound" speed

$$u_s^2 = \frac{e}{m_+}(\gamma_+ T_+ + \gamma_- T_-) \tag{5.42}$$

and with little dispersion. Note that "collisions" are not required for the propagation of this wave. Amplitudes of electron and ion oscillations are not quite the same, and the Coulomb repulsion provides the potential energy to drive the wave. On the other hand, a near equality between phase and particle velocity is the condition for strong Landau damping; therefore, the ion wave is strongly damped[14] unless $T_- >> T_+$, and the phase velocity is between the electron and ion random velocities.

Equation 5.39 can be formally solved for the frequency

$$\omega^2 = \frac{\omega_{p+}^2}{1 + \dfrac{\alpha_-^2}{\delta_- n^2}} + \frac{\gamma_+}{3} k^2 \overline{v_+^2} \tag{5.43}$$

This shows the possibility of ion plasma oscillations $\omega = \omega_{p+}$, but only when $\alpha_-^2 << \delta_- n^2$, that is, for wavelengths much <u>shorter</u> than a Debye length.

5.8 Electron and Ion Plasma Waves across $\underset{\sim}{B}_0$

For propagation across $\underset{\sim}{B}_0$ we set $\cos\theta = 0$ in Equation 5.32 to obtain

$$1 = \frac{\alpha_-^2}{1 - \beta_-^2 - \delta_- n^2} + \frac{\alpha_+^2}{1 - \beta_+^2 - \delta_+ n^2} \tag{5.44}$$

As before, we obtain the electron wave by neglecting $\delta_+ n^2$ and find

$$\delta_- n^2 = 1 - \beta_-^2 - \alpha_-^2 \frac{1 - \beta_+^2}{1 - \beta_+^2 - \alpha_+^2}$$

$$= \frac{(1 - \beta_-^2)(1 - \beta_+^2) - \alpha^2(1 - \beta_+\beta_-) + 2\alpha_+^2\beta_+\beta_-}{1 - \beta_+^2 - \alpha_+^2} \tag{5.45}$$

Note that the last term in the numerator is small and the first two terms are precisely the ones which give cyclotron cutoff, so that the electron wave cuts off at a frequency ω_x' very near the cyclotron cutoff frequency ω_x. Equation 5.45 can be rewritten

$$\omega^2 = \omega_x'^2 + \frac{\alpha_-\gamma_-}{3} k^2 \overline{v_-^2} (1 - \alpha_+^2 - \beta_+^2) \qquad (5.46)$$

which is similar to the dispersion relation (5.37) for the electron wave propagating along $\underset{\sim}{B}_0$.

As the approximation (5.38) does not lead to a simple expression for the ion wave, we shall make a more drastic approximation. For a sufficiently dense plasma, the left-hand side of Equation 5.44 may be neglected. Noting that

$$\frac{\alpha_+^2}{\alpha_-^2} = \frac{\beta_+}{\beta_-} = \frac{m_-}{m_+} \qquad (5.47)$$

we find that

$$n^2 = \frac{m_+ + m_-}{m_+\delta_+ + m_-\delta_-} (1 - \beta_+\beta_-) \qquad (5.48)$$

or

$$u^2 = \frac{\gamma_+ T_+ + \gamma_- T_-}{m_+ + m_-} \frac{e}{1 - \beta_+\beta_-} \qquad (5.49)$$

At low magnetic fields this is essentially the same formula as Equation 5.40 for propagation along $\underset{\sim}{B}_0$. But this wave has a cutoff at the hybrid resonance $\beta_+\beta_- = 1$.

For wavelengths much shorter than the Debye length, that is, $\delta_- n^2 >> \alpha_-^2$, we obtain from Equation 5.44 the relation

$$\omega^2 = \omega_{p+}^2 + \omega_{b+}^2 + \frac{\gamma}{3} k^2 \overline{v_+^2} \qquad (5.50)$$

5.9 Plasma Waves in Arbitrary Directions

As was stated in Section 5.5, the cutoff cone of the plasma waves ($n_p^2 = 0$) coincides with the resonance cone (θ_{res}) of the electromagnetic waves ($n^2 = \infty$):

$$\tan^2 \theta_{res} = -\frac{K_\parallel}{K_\perp} = \frac{(1 - \alpha^2)(1 - \beta_-^2)(1 - \beta_+^2)}{\alpha^2(1 - \beta_-\beta_+) - (1 - \beta_-^2)(1 - \beta_+^2)} \qquad (5.51)$$

so that a plasma wave joins smoothly to an electromagnetic wave along the cone θ_{res}. The loci of constant θ_{res} are shown in Figure 3.10.

The plasma waves also have resonance cones of their own, which we denote by θ_{rp-} and θ_{rp+}. These resonances occur when the longitudinal component of the magnetic field is at electron or ion cyclotron resonance. This follows from Equation 5.32 as $n_p^2 \to \infty$ when

$$\beta_-^2 \cos^2 \theta_{rp-} = 1 \quad \text{and} \quad \beta_+^2 \cos^2 \theta_{rp+} = 1 \qquad (5.52a)$$

or

$$\tan^2 \theta_{rp-} = \beta_-^2 - 1 \quad \text{and} \quad \tan^2 \theta_{rp+} = \beta_+^2 - 1 \qquad (5.52b)$$

These may be called the longitudinal cyclotron resonance cones. They are independent of plasma density. They coincide with the cyclotron resonances of the electromagnetic waves when $\beta_\pm = 1$, $\theta_{rp} = 0$. The resonance cones open out as the magnetic field is increased. There is only one longitudinal resonance cone when $\beta_- \geq 1$ and $\beta_+ < 1$, and there are two cones when $\beta_+ \geq 1$. However, when $\beta_+ > 1$, the electron longitudinal resonance cone (θ_{rp-}) is practically $\pi/2$ because $m_+/m_- >> 1$.

Note that the electromagnetic resonance cone (θ_{res}) also decreases as cyclotron resonance is approached, but at a rate which depends on α^2, so that the two cone angles may cross, and in fact they do. We find that

$$\theta_{res} \leq \theta_{rp-}; \qquad \alpha^2 \leq \left(\frac{m_+}{m_-}\right) \frac{(1 - \beta_+^2)}{\left(1 - \dfrac{m_-}{m_+}\right)} \qquad (5.53a)$$

and

$$\theta_{res} \leq \theta_{rp+}; \qquad \alpha^2 \leq \left(\frac{m_-}{m_+}\right)^2 \frac{(\beta_-^2 - 1)}{\left(1 - \dfrac{m_-}{m_+}\right)} \qquad (5.53b)$$

These equalities plot as straight lines on the (α^2, β^2) plane. There is, of course, no contradiction when a cutoff of a plasma wave (θ_{res}) coincides with a resonance of a plasma wave (θ_{rp}). There are two plasma waves, and the one with a cutoff is not the one with the resonance.

The two solutions of Equation 5.32 may be called the fast and the slow plasma wave, but we prefer to follow our precedent with the electromagnetic waves and make a distinction in the names based on polarizations. From the particle current, Equation 5.11, it follows that in the limit $c \to \infty$, $E_t \to 0$; comparing the left-hand sides of Equation 5.10 for the two currents, we obtain the relationship between the longitudinal components of Γ_- and Γ_+:

$$\frac{\Gamma_{-k}}{\alpha_-^2}\left(\frac{1 - \beta_-^2}{1 - \beta_-^2 \cos^2 \theta} - \delta_- n^2\right) = \frac{\Gamma_{+k}}{\alpha_+^2}\left(\frac{1 - \beta_+^2}{1 - \beta_+^2 \cos^2 \theta} - \delta_+ n^2\right) \qquad (5.54)$$

One solution of Equation 5.32 makes the two brackets of Equation 5.54 of the same sign while the other solution makes them of opposite sign. Thus for one solution the electrons and ions move in phase, for the other they move out of phase. We shall call the solu-

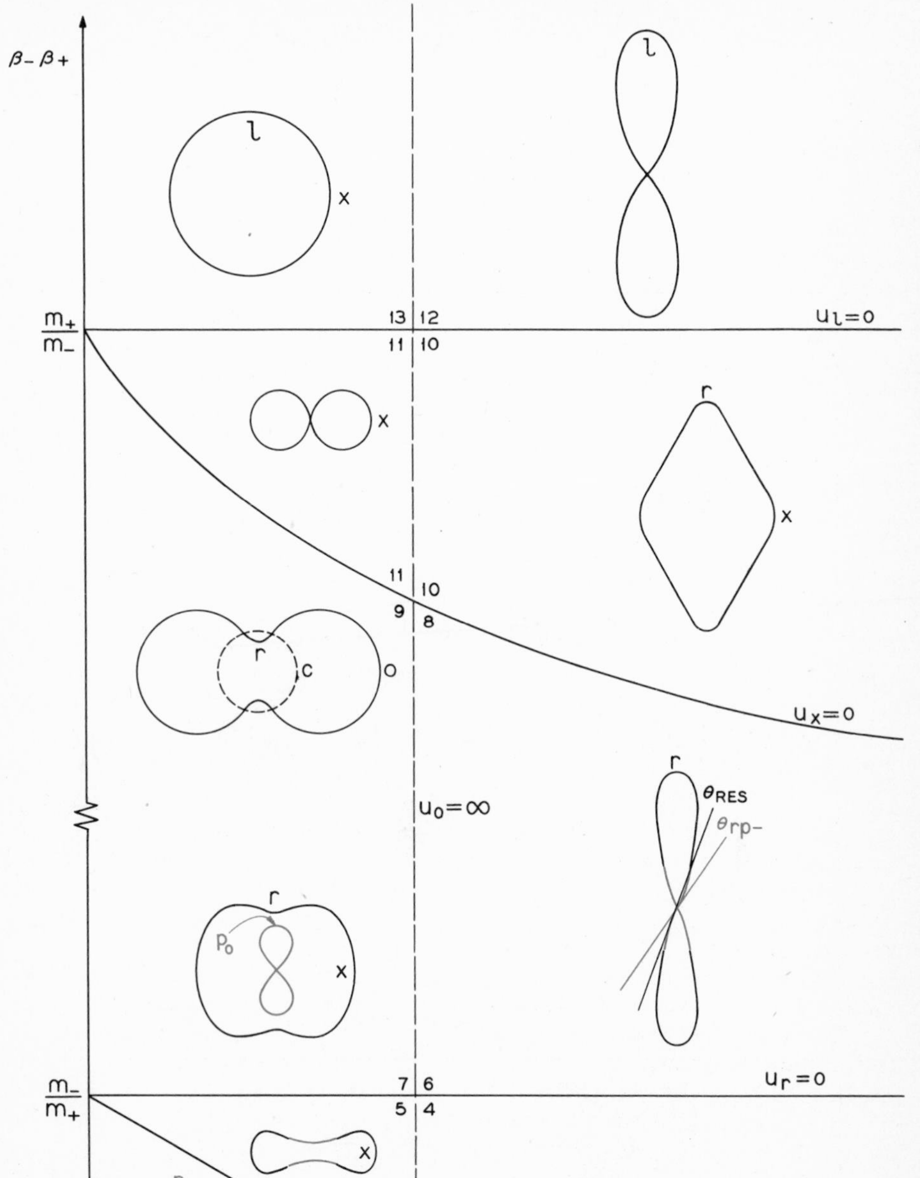

Figure 5.4. Phase velocity surfaces for electron plasma wave and slow electromagnetic wave.

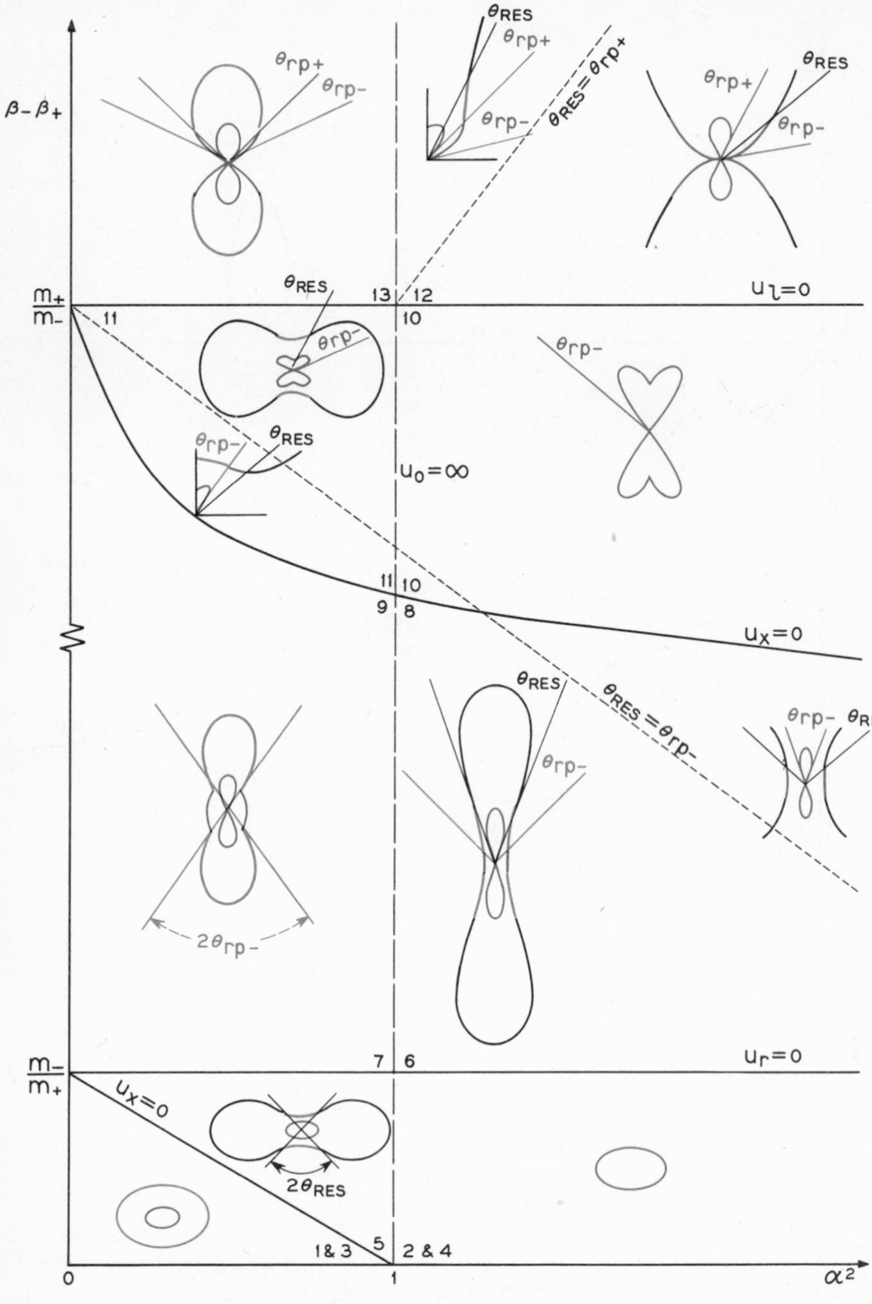

Figure 5.5. Phase velocity surfaces for electron and ion plasma waves and the relevant slow magnetic wave.

tion for which the particles move in phase the ion plasma wave. It is a wave driven by total pressure gradients. It is similar to an acoustic wave and may properly be given that name also. We shall call the wave in which electrons and ions move out of phase, being pulled back toward each other at least in part by their Coulomb attractions, the electron plasma wave. When both plasma waves can propagate in a given direction, the electron plasma wave is always the faster of the two because the restoring force is greater. When only one plasma wave can propagate, it is always of the ion or acoustic type.

We shall now illustrate these waves on the (α^2, β^2) plane. As was seen in Section 5.5, only the slow electromagnetic wave, which has a resonance cone, couples strongly with the plasma waves, and only this wave will be shown on the phase velocity surfaces. The phase velocity surfaces shown in Figure 5.3 will be modified to account for the ion plasma wave, but to do this the range of this figure must be extended to higher β's $(\beta_+ > 1)$. However, the thirteen regions of Figure 3.9 reduce to only nine because the cyclotron cutoff line is irrelevant to the plasma waves and to the slow electromagnetic wave. The nine regions are shown in Figure 5.4 with the ion plasma wave not yet included. Figure 5.5 shows the (α^2, β^2) plane with the ion plasma phase velocity surface shown in red, the electron plasma surface in green, and the electromagnetic wave, where it is relevant, shown in black.† We shall now discuss the nine regions in Figure 5.5 separately.

In regions (1 and 3), (2 and 4), and 5, the combined electromagnetic and electron plasma waves exhibit no resonance. In these regions the inclusion of the ion plasma wave is a simple matter. As its phase velocity is generally much smaller than that of the other three waves, it does not couple to them, and it plots as an ellipse flattened at the poles well inside the other waves. In region (2 and 4) it is the only wave present aside from the fast electromagnetic wave present in region 4. Its properties at the poles and at the equator of the velocity surface were discussed in Sections 5.7 and 5.8.

In region (7 and 9) the ion plasma ellipse intersects the electron plasma ellipse. The two waves couple strongly at the crossing points, and the resultant phase velocity surfaces are shown in Figure 5.5. As no strong coupling to the electromagnetic wave exists in this region, the combined electron and ion plasma waves are obtainable to an excellent approximation from Equation 5.32. Note that the inner wave surface is all red; that is, it is the ion plasma wave that has a resonance cone at θ_{rp-}. This is a feature found in all regions of (α^2, β^2) plane in which θ_{rp-} is real. The outer wave surface belongs partly to the electron and partly to the ion plasma wave, the crossover occurring at θ_{rp-}. At this angle Γ_{-k}/Γ_{+k}

† We are indebted to H. R. Radoski for these plots.

passes through zero, being negative for $\theta < \theta_{rp-}$ and positive for $\theta > \theta_{rp-}$. In region (7 and 9) then, the ion plasma wave exists for all angles θ.

In region (6 and 8) the electron plasma wave is tightly coupled with the electromagnetic wave when β_+ is neglected (see Figure 5.4. Since the phase velocity of the ion plasma is sufficiently small that the ion-plasma-wave ellipse intersects the combined electromagnetic-electron plasma wave very close to the origin, that is, far from the cutoff of the electron plasma wave, Equation 5.32 suffices to study the coupling of the two plasma waves provided that the more exact formula, Equation 5.16, is subsequently used to obtain the coupling to the electromagnetic wave. It is on this basis that the wave normal plot in region (6 and 8) of Figure 5.5 was obtained. As in region (7 and 9), the inner surface is all red; therefore, it is the ion wave that possesses the resonance cone at θ_{rp-}. The composition of the outer surface depends on whether θ_{res} is smaller or larger than θ_{rp-} (see Equation 5.53a). When $\theta_{res} < \theta_{rp-}$, the outer surface is red (ion wave) for $\theta > \theta_{rp-}$, becomes green (electron wave) for $\theta_{res} \leq \theta \leq \theta_{rp-}$, and then goes over into the electromagnetic wave for $\theta < \theta_{res}$. But when $\theta_{res} > \theta_{rp-}$, the green part disappears, and it is the ion plasma wave that couples directly to the electromagnetic wave at $\theta = \theta_{res}$. This situation exists to the right of the dotted line, that is, for $\alpha^2 > (m_+/m_-)(1 - \beta_+^2)/(1 - m_-/m_+)$.

In regions 10 and 13 the electromagnetic waves do not possess a resonance, so that Equation 5.32 is adequate for the description of the plasma waves. Note that in region 10 there is only one plasma wave and it is all red. In region 13 the inner surface is all red, but the outer surface again consists partly of an electron and partly of an ion plasma wave, the crossing occurring at θ_{rp+}.

Region 11 is similar to region 5 in that the slow electromagnetic wave has the same "torus-like" topology. But the plasma waves and their coupling to the electromagnetic wave are different for two reasons: (a) the resonance angle θ_{rp-} is real in region 11 (and imaginary in region 5), and (b) the line $\theta_{res} = \theta_{rp-}$ passes through region 11. As a result, the inner plasma wave is all red (ion wave). The slow electromagnetic wave couples to electron plasma wave (green) at θ_{res} when $\theta_{res} < \theta_{rp-}$. When $\theta_{res} > \theta_{rp-}$, the outer phase velocity surface shown in Figure 5.5 possesses all three waves: for $\theta < \theta_{rp-}$, the wave has the polarization of the electron plasma wave; at $\theta = \theta_{rp-}$, it crosses over to the polarization of the ion plasma waves; and it is the ion plasma wave that couples with electromagnetic wave at $\theta = \theta_{res}$.

In the remaining region, 12, both electromagnetic waves can have a velocity that is less than the velocity of light, although only one of them exhibits a resonance. In the upper right-hand corner of this region, the wave velocity is intimately related to the Alfvén speed rather than to the velocity of light. There the approximation that led to Equation 5.32 breaks down when the Alfvén speed is smaller than the thermal velocity of electrons (and of ions). When this is so, the Alfvén waves must be treated as the limit of plasma waves rather than as the limit of electromagnetic waves, and the exact dispersion relation must be analyzed. However, in the left-hand side of region 12 the phase velocity of the electromagnetic waves is near c, and the topology of the slow electromagnetic wave is similar to that of region (6 and 8). In the left-hand side of region 12 the plasma waves and their coupling to the electromagnetic wave can thus be analyzed in the same way as in the rest of the (α^2, β^2) plane. Here both θ_{rp-} and θ_{rp+} are real; it also follows from Equation 5.53 that here θ_{rp-} is everywhere larger than θ_{res}, but that θ_{rp+} can be either smaller or larger than θ_{res}, the equality between the two occurring along the dotted line shown in Figure 5.5. Again the inner wave is all red with a resonance cone at θ_{rp+}. The effect of the plasma waves on the electromagnetic wave is to increase its resonance cone from θ_{res} to θ_{rp-}. Note, however, the manner in which this is achieved on the two sides of the $\theta_{res} = \theta_{rp+}$ line.

We now return to the upper right-hand corner of region 12, that is, the magnetohydrodynamic region, to discuss the coupling of Alfvén waves with the ion plasma wave when the Alfvén speed u_a is of the same order of magnitude or smaller than the velocity of the plasma wave. This problem has received considerable attention in the literature.[15]

Of the three phase velocity surfaces that exist in this region, the ordinary Alfvén wave

$$u_o = u_a \cos\theta \tag{5.55}$$

remains unaffected by the ion plasma wave. The remaining two waves, which are generally called magnetoacoustic waves, are given by the equation

$$u^4 - (u_a^2 + u_s^2)u^2 + u_a^2 u_s^2 \cos^2\theta = 0 \tag{5.56}$$

where u_a is the Alfvén speed and u_s is the phase velocity of the ion plasma wave in the high plasma density limit ($k^2 \ell_0^2 << 1$), where the ion plasma wave that propagates along $\underset{\sim}{B}_0$ becomes an acoustic wave (see Equation 5.42) with the speed

$$u_s^2 = \frac{e}{m_+}(\gamma_+ T_+ + \gamma_- T_-) \tag{5.57}$$

The two solutions of Equation 5.56 are

$$u^2 = \frac{1}{2}(u_a^2 + u_s^2)\left[1 \pm \sqrt{1 - \frac{4u_a^2 u_s^2 \cos^2\theta}{(u_a^2 + u_s^2)^2}}\right] \tag{5.58}$$

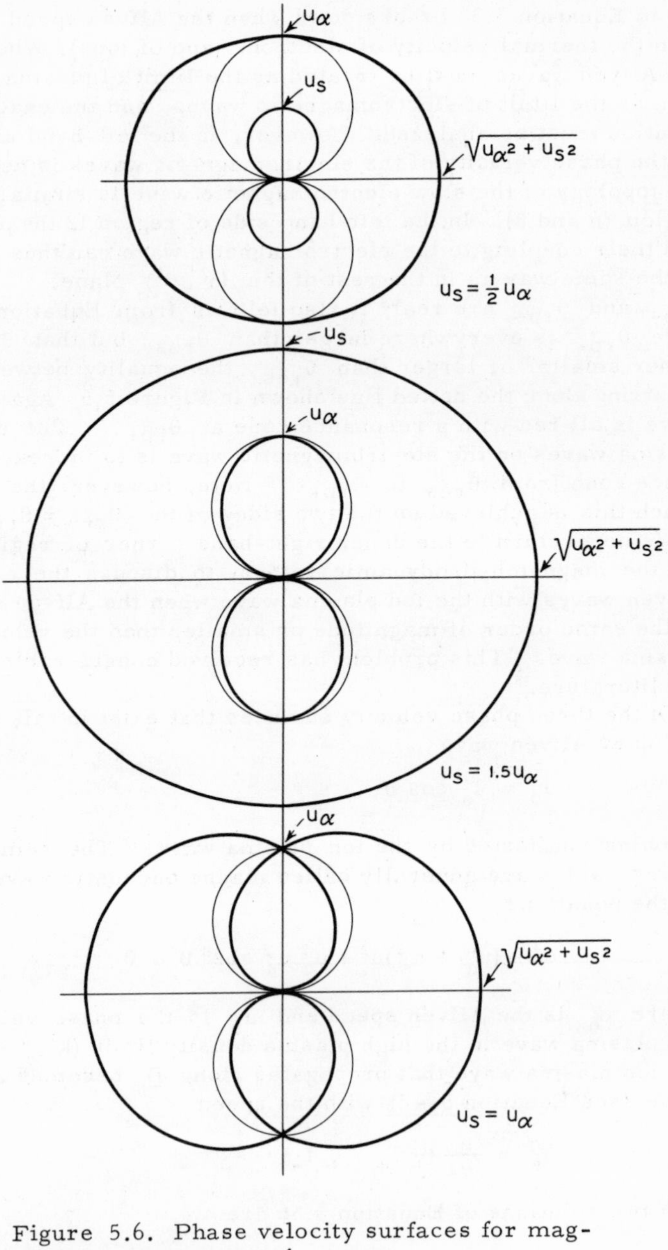

Figure 5.6. Phase velocity surfaces for magnetoacoustic and Alfvén waves.

At the poles of the velocity surface $(\theta = 0)$, the two solutions are $u = u_\alpha$ and $u = u_s$. At the equator $(\theta = \pi/2)$, the two solutions are $u = \sqrt{u_\alpha^2 + u_s^2}$ and $u = 0$. That velocity surface which at the equator has the speed $\sqrt{u_\alpha^2 + u_s^2}$ joins at the poles with the larger of u_α and u_s. Of the three surfaces, it is outermost; that is, it represents the fastest of the three waves. The remaining wave has at the poles the lesser of speeds u_α and u_s, which then decreases to zero as θ varies from zero to $\pi/2$. In Figure 5.6 are plotted the three phase velocity surfaces when $u_s/u_\alpha = \frac{1}{2}$, $u_s/u_\alpha = \frac{3}{2}$, and $u_s = u_\alpha$.

In this chapter we have discussed the properties of plasma waves and their coupling to the electromagnetic waves in all regions of the (α^2, β^2) plane without regard to any condition of validity of the chosen model. As was stated in the Section 5.1, the truncation of the infinite set of the Boltzmann transport equations after the first two and the neglect of collisions removed all damping mechanisms in the plasma. However, there is an additional factor that limits the validity of the results even further. As was noted earlier, it is mainly the pressure gradient that drives the plasma waves. It is transmitted, as a force on the particles, through the long-range collective interaction. As such force requires a distance of a Debye length to become effective, waves whose wavelength is less than a Debye length do not remain coherent and are dispersed.[16] This means that electron plasma waves propagating along the magnetic field are generally evanescent for $\alpha^2 < 1$ and ion plasma waves similarly for $\alpha^2 < m_+/m_-$. An important exception occurs when the phase velocity of the wave is larger than the corresponding thermal velocity $\sqrt{eT/m}$, as happens near plasma cutoff. There may be other exceptions, as the limits of collective interactions in the presence of a magnetic field have not been fully worked out. For this, and other, reasons we have not attempted to eliminate evanescent waves from our diagrams.

Chapter 6

BOLTZMANN THEORY

In previous chapters we obtained the tensor dielectric coefficient of a plasma, first from the Lorentz force equation, then from a truncated set of the Boltzmann transport equations. Both procedures, in addition to linearization, involved certain physical assumptions that restricted the range of validity of the results obtained. In the so-called temperate plasma, which was based on the Lorentz force equation, nonlocal effects were excluded, and currents due to pressure gradients were neglected. This led to a reasonably proper description of the set of the electromagnetic waves in a plasma in which the thermal velocity of the carriers was much smaller than the phase velocity of the wave. The model was then improved through the inclusion of the effects of pressure gradients via the Boltzmann transport equation. This brought out the existence of electron and ion plasma waves and the nature of their coupling to the electromagnetic waves. However, important physical processes were neglected even in this model. In a hot plasma whose electrons (or ions) are distributed in velocity space according to some distribution and in which a wave propagates with a phase velocity smaller than the speed of light, there will exist electrons with velocity vectors nearly equal to the phase velocity of the wave. Those traveling in the same direction as the waves can be expected to interact very strongly with the field and thus have a relatively stronger effect on the properties of the wave than the electrons in the remainder of the distribution. There are also resonant effects when the electron's Larmor orbit is of the order of the wavelength measured across the magnetic field. In that case resonant interactions with the wave can be expected not only at cyclotron resonance but also at harmonics of the cyclotron frequency.

These velocity-sensitive effects are not brought out in the transport equations because various averages over the velocity distribution are performed before a detailed interaction of the carriers with the electromagnetic field is considered. These effects would enter the high moments of the distribution function, but the truncation of the infinite set of transport equations after the first two and the assumption of a diagonal pressure tensor lose sight of them.

Any nonlocal effects in a plasma can be described rigorously by an expression for any Fourier component of the conduction current $\underset{\sim}{J}$ at $\underset{\sim}{r}$ in terms of the Fourier component of the electric field $\underset{\sim}{E}$ at all other points $\underset{\sim}{r}'$ in the plasma:

$$\underset{\sim}{J}(\underset{\sim}{r}, \omega) = \int \underset{\approx}{\sigma}(\underset{\sim}{r} - \underset{\sim}{r}', \omega) \cdot \underset{\sim}{E}(\underset{\sim}{r}', \omega)\, d\underset{\sim}{r}' \tag{6.1}$$

where the integration is over all real space r'. This integral relation between current and field expresses the fact that $\underset{\sim}{J}(\underset{\sim}{r})$, the current at the point $\underset{\sim}{r}$, depends on the fields $\underset{\sim}{E}(\underset{\sim}{r}')$ throughout the whole medium. The tensor conductivity $\underset{\approx}{\sigma}(\underset{\sim}{r} - \underset{\sim}{r}')$ is a kernel that properly weights the contribution to $\underset{\sim}{J}$ at $\underset{\sim}{r}$ from points at $\underset{\sim}{r}'$. The Fourier-space transform of Equation 6.1 is given by

$$\underset{\sim}{J}(\omega, \underset{\sim}{k}) = \underset{\approx}{\sigma}(\omega, \underset{\sim}{k}) \cdot \underset{\sim}{E}(\omega, \underset{\sim}{k}) + \text{terms arising from boundary conditions} \tag{6.2}$$

Provided these conditions are known, we can therefore obtain a complete solution to a problem, at least in principle, <u>if</u> the conductivity $\underset{\approx}{\sigma}(\omega, \underset{\sim}{k})$ is known. It is the quantity $\underset{\approx}{\sigma}(\omega, \underset{\sim}{k})$ that determines the response of the medium to an applied electromagnetic field and thus describes the physics of the situation.

In this chapter we shall calculate the conductivity tensor of the plasma $\underset{\approx}{\sigma}$ resulting from electrons only, and through it the dielectric coefficient $\underset{\approx}{K}$, from the Boltzmann equation

$$\frac{\partial f}{\partial t} + \underset{\sim}{v} \cdot \underset{\sim}{\nabla} f - \frac{e}{m}(\underset{\sim}{E} + \underset{\sim}{v} \times \underset{\sim}{B}) \cdot \frac{\partial f}{\partial \underset{\sim}{v}} = \left(\frac{\partial f}{\partial t}\right)_{\text{coll}} \tag{6.3}$$

where the fields $\underset{\sim}{E}$ and $\underset{\sim}{B}$ satisfy Maxwell's equations. We shall show that in general $\underset{\approx}{\sigma}$ is a tensor whose components are integral operators involving the propagation vector $\underset{\sim}{k}$.

A detailed discussion of the properties of the collision integral $(\partial f/\partial t)_{\text{coll}}$ is outside the scope of this monograph. For our purposes it is sufficient to replace the integral by a relaxation term of the form

$$\left(\frac{\partial f}{\partial t}\right)_{\text{coll}} = -\nu(f - f_0) \tag{6.4}$$

where f_0 is the equilibrium distribution function and f is the distribution function in the presence of the wave. We now linearize Equation 6.3 by <u>assuming</u> that the electromagnetic field is given by one of its Fourier components, so that $\underset{\sim}{E} = \underset{\sim}{E}_1 \exp j(\omega t - \underset{\sim}{k} \cdot \underset{\sim}{r})$ (we neglect any dc electric fields in the plasma), $\underset{\sim}{B} = \underset{\sim}{B}_0 + \underset{\sim}{B}_1 \exp j(\omega t - \underset{\sim}{k} \cdot \underset{\sim}{r})$, and $f = f_0 + f_1 \exp j(\omega t - \underset{\sim}{k} \cdot \underset{\sim}{r})$, and by treating all quantities that vary as $\exp j(\omega t - \underset{\sim}{k} \cdot \underset{\sim}{r})$ as being small compared with the dc values. Then the equation for the distribution f_1 of electrons is

$$j(\omega - \underset{\sim}{v} \cdot \underset{\sim}{k}) f_1 - \frac{e}{m} \underset{\sim}{E}_1 \cdot \frac{\partial f_0}{\partial \underset{\sim}{v}} - \frac{e}{m}(\underset{\sim}{v} \times \underset{\sim}{B}_0) \cdot \frac{\partial f_1}{\partial \underset{\sim}{v}} - \frac{e}{m}(\underset{\sim}{v} \times \underset{\sim}{B}_1) \cdot \frac{\partial f_0}{\partial \underset{\sim}{v}} = -\nu f_1 \tag{6.5}$$

with a similar equation for ions. The last term on the left-hand side of Equation 6.5 involves the product $\underset{\sim}{v} \times (\partial \underset{\sim}{f}_0/\partial \underset{\sim}{v})$. This is identically zero provided f_0 is spherically symmetric so that f_0 is a function of v^2 only. The effects of nonisotropic equilibrium distributions have been considered by others.[17] Here we shall treat only isotropic distributions for which the term $(\underset{\sim}{v} \times \underset{\sim}{B}_1) \cdot (\partial f_0/\partial \underset{\sim}{v})$ vanishes.

The relaxation form (6.4) contains an error in principle: It does not conserve particles. The local density of particles

$$\int f\, d^3v = N_0 + N_1 \exp j(\omega t - \underset{\sim}{k} \cdot \underset{\sim}{r})$$

has an alternating component N_1, and Equation 6.4 causes collisions to destroy (or create) these particles. This error can be corrected by requiring f to relax to a local Maxwellian rather than to an absolute Maxwellian, but this is not simple when the collision frequency ν depends on velocity and Equation 6.4 does not lead to appreciable errors when $\nu \ll \omega$ and $\ell_0 \ll \lambda$.

To solve Equation 6.5, we follow essentially the procedure of Bernstein.[18] We choose a coordinate system in velocity space such that

$$\underset{\sim}{v} = \underset{\sim}{a}_x w \cos \phi + \underset{\sim}{a}_y w \sin \phi + \underset{\sim}{a}_z u \tag{6.6}$$

where $\underset{\sim}{a}_x$, $\underset{\sim}{a}_y$, and $\underset{\sim}{a}_z$ are unit vectors along the x, y, and z axes. With this choice of coordinates,

$$\frac{e}{m} \underset{\sim}{B}_0 \cdot \left(\underset{\sim}{v} \times \frac{\partial f_1}{\partial \underset{\sim}{v}} \right) = \omega_b \frac{\partial f_1}{\partial \phi} \tag{6.7}$$

so that Equation 6.5 becomes

$$\frac{\partial f_1}{\partial \phi} + j \frac{(\omega - \underset{\sim}{v} \cdot \underset{\sim}{k} - j\nu)}{\omega_b} f_1 = \frac{e}{m} \frac{\underset{\sim}{E}_1}{\omega_b} \cdot \frac{\partial f_0}{\partial \underset{\sim}{v}} \tag{6.8}$$

As in previous chapters, we assume without loss of generality that

$$\underset{\sim}{k} = \underset{\sim}{a}_x k_\perp + \underset{\sim}{a}_z k_\parallel = \underset{\sim}{a}_x k \sin \theta + \underset{\sim}{a}_z k \cos \theta \tag{6.9}$$

Thus Equation 6.8 becomes

$$\frac{\partial f_1}{\partial \phi} + j \frac{(\omega - j\nu - uk_\parallel - wk_\perp \cos \phi)}{\omega_b} f_1 = \frac{e}{m\omega_b} \underset{\sim}{E}_1 \cdot \frac{\partial f_0}{\partial \underset{\sim}{v}} \tag{6.10}$$

Equation 6.10, a first-order, linear, inhomogeneous equation, has the solution

$$f_1 = \frac{e}{m\omega_b} \underset{\sim}{E}_1 \cdot \int_C^{\phi} \frac{\partial f_0(v')}{\partial \underset{\sim}{v}'} \exp - j\Big[a(\phi - \phi') - b(\sin \phi - \sin \phi') \Big] d\phi' \tag{6.11}$$

where

$$a = \frac{\omega - j\nu - uk_{\parallel}}{\omega_b}$$

$$b = \frac{wk_{\perp}}{\omega_b}$$

and C is a constant of integration. From the condition that f_1 be bounded and be periodic in ϕ with a period 2π, C is determined to be $-\infty$.

The conduction current that flows in the plasma as a result of the wave is

$$\underset{\sim}{J} = -e\int \underset{\sim}{v} f_1\, d^3v$$

$$= -\frac{e^2}{m\omega_b}\int \underset{\sim}{v}\, d^3v \int_{-\infty}^{\phi} \exp\left\{-j[a(\phi - \phi') - b(\sin\phi - \sin\phi')]\right\} \frac{\partial f_0(v')}{\partial \underset{\sim}{v}'} \cdot \underset{\sim}{E}_1\, d\phi'$$

$$= \underset{\approx}{\sigma} \cdot \underset{\sim}{E}_1 \tag{6.12}$$

so that

$$\underset{\approx}{\sigma} = -\frac{e^2}{m\omega_b}\int \underset{\sim}{v}\, d^3v \int_{-\infty}^{\phi} \exp\left\{-j[a(\phi - \phi') - b(\sin\phi - \sin\phi')]\right\} \frac{\partial f_0(v')}{\partial \underset{\sim}{v}'}\, d\phi' \tag{6.13}$$

The tensor nature of $\underset{\approx}{\sigma}$ arises from the juxtaposition of the two vectors $\underset{\sim}{v}$ and $(\partial f_0/\partial \underset{\sim}{v})$.

The evaluation of the various integrals is simplified somewhat if $\underset{\approx}{\sigma}$ is expressed in rotating coordinates: $(x + jy)/\sqrt{2}$, $(x - jy)/\sqrt{2}$, z. Thus we find

$$\sigma' = U \cdot \sigma \cdot U^{-1}$$

$$= -\frac{e^2}{2m\omega_b}\int_{-\infty}^{\infty} du \int_0^{\infty} w\, dw \int_0^{2\pi} d\phi \int_{-\infty}^{\phi} \exp\{-j[(\phi - \phi') - b(\sin\phi - \sin\phi')]\}$$

$$\times \begin{bmatrix} w\frac{\partial f_0}{\partial w} e^{j(\phi-\phi')} & w\frac{\partial f_0}{\partial w} e^{j(\phi+\phi')} & \sqrt{2}w\frac{\partial f_0}{\partial u} e^{j\phi} \\ w\frac{\partial f_0}{\partial w} e^{-j(\phi+\phi')} & w\frac{\partial f_0}{\partial w} e^{-j(\phi-\phi')} & \sqrt{2}w\frac{\partial f_0}{\partial u} e^{-j\phi} \\ \sqrt{2}u\frac{\partial f_0}{\partial w} e^{-j\phi'} & \sqrt{2}u\frac{\partial f_0}{\partial w} e^{j\phi'} & 2u\frac{\partial f_0}{\partial u} \end{bmatrix} d\phi' \tag{6.14}$$

There exists a basic difficulty in evaluating the integrals in Equation 6.14. It arises from the fact that in this, the steady-state problem, where we take $\underset{\sim}{k}$ to be a complex quantity, the integrals are not analytic functions of $\underset{\sim}{k}$. In the corresponding initial-value problem, where k is taken as real and ω is complex, it was first shown by Landau[19] that this leads to a damping of the wave, since called Landau damping. In this monograph we shall not discuss in detail the question of Landau damping as it applies to the steady-state problem. Here, we shall give only the expansion of Equation 6.14 in the limit of low temperature (or strong magnetic field), where Landau damping can be neglected. For details of this expansion, the reader is referred to Appendices A and B.

As usual, we define the dielectric coefficient tensor $\underset{\approx}{K}$ by

$$\underset{\approx}{K} = \underset{\approx}{I} + \frac{\underset{\approx}{\sigma}}{j\omega\epsilon_0} \tag{6.15}$$

The dielectric tensor is given in Cartesian coordinates by

$$\left.\begin{aligned}
K_{xx} &= 1 - \frac{\alpha^2}{1-\beta^2}\left[1 + \frac{1+3\beta^2}{(1-\beta^2)^2}\delta n^2 \cos^2\theta + \frac{3}{1-4\beta^2}\delta n^2 \sin^2\theta\right] \\
K_{yy} &= 1 - \frac{\alpha^2}{1-\beta^2}\left[1 + \frac{1+3\beta^2}{(1-\beta^2)^2}\delta n^2 \cos^2\theta + \frac{1+8\beta^2}{1-4\beta^2}\delta n^2 \sin^2\theta\right] \\
K_{zz} &= 1 - \alpha^2\left(1 + 3\delta n^2 \cos^2\theta + \frac{\delta n^2 \sin^2\theta}{1-\beta^2}\right) \\
K_{xy} &= -K_{yx} = \frac{-j\alpha^2\beta}{1-\beta^2}\left[\frac{3+\beta^2}{(1-\beta^2)^2}\delta n^2 \cos^2\theta + \frac{6}{1-4\beta^2}\delta n^2 \sin^2\theta\right] \\
K_{xz} &= K_{zx} = \frac{-2\alpha^2\delta}{(1-\beta^2)^2} n^2 \sin\theta\cos\theta \\
K_{yz} &= -K_{zy} = j\alpha^2\beta\frac{3-\beta^2}{(1-\beta^2)^2}\delta n^2 \sin\theta\cos\theta
\end{aligned}\right\} \tag{6.16}$$

and in rotating coordinates by

$$\left.\begin{aligned}
K_r &= 1 - \frac{\alpha^2}{1-\beta}\left[1 + \frac{\delta n^2\cos^2\theta}{(1-\beta)^2} + \frac{\delta n^2 \sin^2\theta}{1-2\beta}\right] \\
K_\ell &= 1 - \frac{\alpha^2}{1+\beta}\left[1 + \frac{\delta n^2\cos^2\theta}{(1+\beta)^2} + \frac{\delta n^2 \sin^2\theta}{1+2\beta}\right] \\
K_{r\ell} &= -\frac{\alpha^2\delta n^2 \sin^2\theta}{1-\beta^2} \\
K_{rp} &= -\frac{2-\beta}{(1-\beta)^2}\alpha^2\delta n^2\frac{\sin\theta\cos\theta}{\sqrt{2}} \\
K_{\ell p} &= -\frac{2+\beta}{(1+\beta)^2}\alpha^2\delta n^2\frac{\sin\theta\cos\theta}{\sqrt{2}} \\
K_p &= K_{zz}
\end{aligned}\right\} \tag{6.17}$$

For easy reference, the symbols in these expressions are

$$\alpha = \frac{\omega_p}{\omega}, \quad \beta = \frac{\omega_{b-}}{\omega}, \quad \delta n^2 = \frac{eT_-}{m}\frac{k^2}{\omega^2} \qquad (6.18)$$

and if we desire to retain collisions, we replace

$$\alpha^2 \text{ by } \frac{\alpha^2}{1 - j\gamma}, \quad \beta \text{ by } \frac{\beta}{1 - j\gamma}, \quad \text{and } \gamma = \frac{\nu}{\omega} \qquad (6.19)$$

Note that these "dielectric coefficients" are not simply properties of the medium. We have become accustomed to having the dielectric coefficients depend on the frequency of the electric field and on its direction. Now they depend also on the wavelength and the direction of the wave normal. These coefficients include the effects of pressure gradients, and therefore, if they are introduced in the general dispersion equation (1.25), the two electromagnetic and the electron plasma waves will be obtained. The ions have been left out of this formulation, but they can be added in a perfectly straightforward manner. In fact, the electron waves turn out somewhat differently from the ones pictured in Figure 5.4 because the present treatment allows for anisotropic distortions to the pressure.

These coefficients show that at the first harmonic of the cyclotron frequency ($\beta_- = \frac{1}{2}$) there is a resonance ($K_{rr} = \infty$) for the electromagnetic wave which is right-rotating and a cutoff ($\delta n^2 = 0$) for the electron plasma wave, provided the wave is not propagating along $\underset{\sim}{B}_0$ ($\theta \neq 0$).

The expressions for the K's were obtained from the first terms of expansions in powers of δn^2. They contain, therefore, only terms linear in the temperature. Even these are not quite correct because one part of the expansion is in powers of $\delta n^2 \cos^2 \theta/(1 - \beta)^2$ and is therefore invalid at cyclotron resonance if propagation is partly along $\underset{\sim}{B}_0$. This is the situation in which the electron wave is coupling to the electromagnetic wave, and the mathematics shows that the interaction is strong. No resonance at harmonics of the cyclotron frequency is evident for propagation along the magnetic field. This is consistent with our understanding of the nature of these waves. Since they are purely transverse in the plane of the electron orbit, no mechanism exists for resonance to occur at the harmonics of the cyclotron frequency.

For propagation across $\underset{\sim}{B}_0$ ($\theta = \pi/2$), the series is in terms of $(\delta n^2 \sin^2 \theta)^{s-1}/(1 - s\beta)$ with s an integer. Successive terms bring in the higher harmonics. We note the presence of resonance terms at $\omega = 2\omega_b$. If terms were retained of the order δ^2 in the expansion, there would exist a resonance term at $\omega = 3\omega_b$, and so on. The origin of the resonance term at cyclotron harmonics lies in the fact that when the wave propagates at an angle to the magnetic field, the phase of the field varies across the electron orbit, and this leads to resonance at $\omega = s\omega_b$. Stated in another way, it is well known that an electron radiates not only at the cyclotron frequency but at all of its harmonics (albeit with decreasing amplitudes) when viewed at an angle to the normal to its orbit. The synchrotron emission is strongest at $\theta = \pi/2$, and the resonance at $\omega = s\omega_b$ is the manifestation of this radiation.

We also note that the off-diagonal terms K_{xz}, K_{zx}, K_{yz}, and K_{zy}, which were identically zero in a temperate plasma, still vanish for the principal directions of propagation.

We note finally that three principal transverse waves (the r, ℓ, and o waves), which according to the transport theory were not affected by a finite temperature, are in fact so affected, according to the more correct kinetic theory. For example, the dispersion relation for the ordinary wave now is

$$n_o^2 = K_{33}\Big|_{n_{\|}=0} = 1 - \alpha^2 - \frac{\alpha^2 \delta n_o^2}{1 - \beta^2}$$

or

$$n_o^2 = \frac{1 - \alpha^2}{1 + \dfrac{\delta \alpha^2}{1 - \beta^2}} \tag{6.20}$$

For finite temperatures $(\delta \neq 0)$, n_o^2 is a function, albeit a weak function, of the magnetic field. Again we must disregard the singularity at $\beta = 1$ because there the expansion is invalid. Exact calculations show, however, that the effect of finite temperature is indeed largest near $\beta = 1$. As mentioned earlier, Equation 6.20 is correct only to order δ. If terms in δ^2 were retained, they would contain terms in $1/(\omega^2 - 4\omega_b^2)$, that is, resonance at the second harmonic.

For circularly polarized waves propagating along $\underset{\sim}{B}_0$, we find similarly

$$n_{r,\ell}^2 = \frac{1 - \dfrac{\alpha^2}{1 \mp \beta}}{1 + \dfrac{\delta \alpha^2}{(1 \mp \beta)^2}} \tag{6.21}$$

where again the equation for n_r^2 is incorrect at $\beta = 1$.

Finally, we mention that Expression 6.18 is not complete even when the expansion is valid. This is because we have retained terms to order eT/mc^2 in what is essentially a nonrelativistic theory. Peskoff[20] has solved the relativistically invariant Boltzmann equation and has shown that in Equation 6.18 additional terms of the order (eT/mc^2) appear. We note, however, that the terms (eT/mc^2) which already appear in Equation 6.18 are multiplied by n^2. As Equation 6.18 is correct only for $(eT/mc^2) \ll 1$, temperature effects are appreciable only when $n^2(eT/mc^2) \approx 1$, that is, when n^2 is large. Consequently, the relativistic corrections are quite negligible in the low-temperature limit.

PART II

ENERGY-POWER THEOREMS AND GUIDED WAVES

Abraham Bers

Chapter 7

CONSERVATION PRINCIPLES FOR TEMPERATE PLASMAS

The theory of waves in plasmas as presented in this monograph is based entirely upon a linearized macroscopic description of the fields in a plasma medium. Thus every time-dependent macroscopic field quantity, such as electric field, magnetic field, velocity, current density, and charge density, is a small perturbation upon some assumed unperturbed state of the medium. These first-order fields obey a linearized set of field equations and equations of motion. In analogy to field analyses of conservative linear media, the linearized set of equations for the plasma in the absence of collisions leads to conservation principles involving products of the first-order field quantities. From these conservation principles it is convenient to identify terms analogous to energy and power in a linear conservative medium. Such terms, involving products of only first-order field quantities, are not in general the second-order energy and power obtainable from a nonlinear analysis. In fact, the conservation principles that we shall derive from the linearized equations of the plasma cannot usually be derived from the nonlinear equations.[21] From the conservation principles based on the linearized field equations and linearized equations of motion, we shall be able to derive some important properties of the field solutions and relate these to both free and guided wave propagation in plasmas. In Chapter 7 we restrict ourselves to temperate plasmas, and in Chapter 8 we generalize the treatment to warm and hot plasmas.

7.1 Energy and Power

The linearized equations of motion for a temperate plasma, in the absence of collisions, are

$$\underset{\sim}{\Gamma}_i = N_{0i}\underset{\sim}{v}_i \qquad (7.1)$$

$$0 = \nabla \cdot \underset{\sim}{\Gamma}_i + \frac{\partial N_i}{\partial t} \qquad (7.2)$$

$$0 = -m_i \frac{\partial \underset{\sim}{v}_i}{\partial t} + e_i \underset{\sim}{E} + e_i \underset{\sim}{v}_i \times \underset{\sim}{B}_0 \qquad (7.3)$$

The author of Part II is indebted to Professor H. A. Haus and Messrs. P. Serafim and R. J. Briggs for many useful comments on all aspects of Chapters 7 through 10.

$$\nabla \times \underset{\sim}{E} = -\mu_0 \frac{\partial \underset{\sim}{H}}{\partial t} \tag{7.4}$$

$$\nabla \times \underset{\sim}{H} = \epsilon_0 \frac{\partial \underset{\sim}{E}}{\partial t} + \sum_i e_i \underset{\sim}{\Gamma}_i \tag{7.5}$$

In these equations, N_{0i} is the unperturbed density of the ith species of charged particles, in general a function of position but time-independent. Similarly, $\underset{\sim}{B}_0$, the externally applied magnetic field will be assumed time-independent but in general may be a function of position. In the unperturbed state, we assume the plasma particles to be stationary and the electric field to be zero. All other field variables in Equations 7.1 through 7.5, $\underset{\sim}{\Gamma}_i$, N_i, $\underset{\sim}{v}_i$, $\underset{\sim}{E}$, and $\underset{\sim}{H}$, are first-order perturbations of small amplitudes. We assume that the continuity equation, Equation 7.2, and the force equation, Equation 7.3, hold for each species of charged particles of mass m_i and charge e_i making up the plasma.

A conservation equation for the linearized field quantities is obtained by dot-multiplying Equation 7.3 by $\underset{\sim}{\Gamma}_i$ and summing over all i, Equation 7.4 by $\underset{\sim}{H}$, Equation 7.5 by $-\underset{\sim}{E}$, and adding the resulting equations:

$$\nabla \cdot (\underset{\sim}{E} \times \underset{\sim}{H}) = -\frac{\partial}{\partial t}\left(\tfrac{1}{2}\mu_0 H^2 + \tfrac{1}{2}\epsilon_0 E^2 + \sum_i \tfrac{1}{2} N_{0i} m_i v_i^2\right) \tag{7.6}$$

The term in parentheses on the left-hand side of Equation 7.6 can be identified as the instantaneous electromagnetic power density, the Poynting vector; the terms in parentheses on the right-hand side of Equation 7.6 can be identified as the instantaneous magnetic, electric, and kinetic energy densities in space. We note that the three energy densities are positive definite quantities for all time and everywhere in space. For a volume V enclosed by a surface S, containing the plasma, the electromagnetic power flow out of V is

$$S = \oint_S \underset{\sim}{E} \times \underset{\sim}{H} \cdot d\underset{\sim}{S} \tag{7.7}$$

and the energy stored in V is

$$W = \int_V \left(\tfrac{1}{2}\mu_0 H^2 + \tfrac{1}{2}\epsilon_0 E^2 + \sum_i \tfrac{1}{2} N_{0i} m_i v_i^2\right) dV \tag{7.8}$$

7.2 Sinusoidal Time Variation

When the linearized field quantities are Fourier analyzed in time with components exp $(j\omega t)$, the equations of motion, Equations 7.2 through 7.5, become

$$0 = \nabla \cdot \underset{\sim}{\Gamma}_i + j\omega N_i \tag{7.9}$$

$$0 = -j\omega m_i \underset{\sim}{v}_i + e_i \underset{\sim}{E} + e_i \underset{\sim}{v}_i \times \underset{\sim}{B}_0 \tag{7.10}$$

$$\nabla \times \underset{\sim}{E} = -j\omega\mu_0 \underset{\sim}{H} \tag{7.11}$$

$$\nabla \times \underset{\sim}{H} = j\omega\epsilon_0 \underset{\sim}{E} + \sum_i e_i \underset{\sim}{\Gamma}_i \tag{7.12}$$

where all field quantities are now the corresponding complex Fourier amplitudes and are functions of spatial coordinates. Dot-multiplying Equation 7.10 by $\underset{\sim}{\Gamma}_i^*$, Equation 7.11 by $\underset{\sim}{H}^*$, the complex conjugate of Equation 7.12 by $-\underset{\sim}{E}$, and adding the resulting equations, we obtain

$$\tfrac{1}{2}\nabla \cdot (\underset{\sim}{E} \times \underset{\sim}{H}^*) = -j2\omega \left[\tfrac{1}{4}\mu_0 |\underset{\sim}{H}|^2 - \tfrac{1}{4}\epsilon_0 |\underset{\sim}{E}|^2 + \sum_i \tfrac{1}{4} N_{0i} m_i |\underset{\sim}{v}_i|^2 - \sum_i \tfrac{1}{2} N_{0i} e_i \frac{(\underset{\sim}{v}_R \times \underset{\sim}{v}_I)_i}{\omega} \cdot \underset{\sim}{B}_0 \right] \tag{7.13}$$

where in the last term $\underset{\sim}{v}_R$ and $\underset{\sim}{v}_I$ are, respectively, the real and the imaginary parts of the complex velocity vector

$$\underset{\sim}{v} = \underset{\sim}{v}_R + j\underset{\sim}{v}_I \tag{7.14}$$

All the terms within the brackets on the right-hand side of Equation 7.13 are real. The first two are, respectively, the time-averaged magnetic and the electric energy densities. The third term is the time-averaged kinetic energy density. The last term can be interpreted as arising from the interaction of the orbiting particle's dipole with the magnetic field $\underset{\sim}{B}_0$, Figure 7.1. This term vanishes everywhere in space if either (a) $\underset{\sim}{B}_0 = 0$, that is, the plasma is isotropic; or (b) $\underset{\sim}{v}_R \times \underset{\sim}{v}_I = 0$, that is, $\underset{\sim}{v}$ is linearly polarized; or (c) $\underset{\sim}{v} \times \underset{\sim}{B}_0 = 0$, that is, $\underset{\sim}{v}$ and $\underset{\sim}{B}_0$ are aligned, as may be the case for very strong magnetic fields. However, this term does not appear in the time-averaged energy density of the plasma. If we assume a sinusoidal time variation, the time average of Equation 7.8 gives

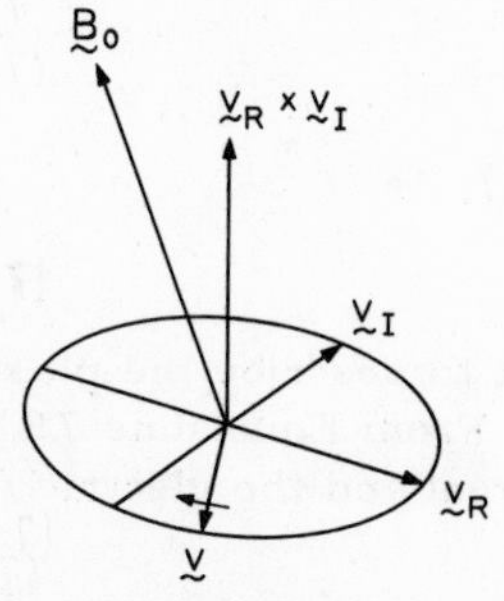

Figure 7.1. Elliptically polarized velocity vector and its relation to $\underset{\sim}{B}_0$.

$$\overline{W} = \int_V \left(\tfrac{1}{4}\mu_0 |\underset{\sim}{H}|^2 + \tfrac{1}{4}\epsilon_0 |\underset{\sim}{E}|^2 + \sum_i \tfrac{1}{4} N_{0i} m_i |\underset{\sim}{v}_i|^2 \right) dV \tag{7.15}$$

which is the time-averaged energy within the volume V enclosing the plasma. The left-hand side of Equation 7.13 when integrated over a closed volume is the complex power flow; its real part is the time average of Equation 7.7:

$$\overline{S} = \tfrac{1}{2}\mathrm{Re} \oint_S \underset{\sim}{E} \times \underset{\sim}{H}^* \cdot d\underset{\sim}{S} \tag{7.16}$$

which is the time-averaged electromagnetic power flow out of the volume. When the complex power flow is pure real, the volume is said to be in resonance. This resonance condition follows from Equation 7.13,

$$\int_V \left(\tfrac{1}{4}\mu_0 |\underset{\sim}{H}|^2 - \tfrac{1}{4}\epsilon_0 |\underset{\sim}{E}|^2 + \sum_i \tfrac{1}{4} N_{0i} m_i |\underset{\sim}{v}_i|^2 - \sum_i \tfrac{1}{2} N_{0i} e_i \frac{(\underset{\sim}{v}_R \times \underset{\sim}{v}_I)_i}{\omega} \cdot \underset{\sim}{B}_0 \right) dV = \tag{7.17}$$

It is convenient to introduce the following abbreviations:

$$\overline{W}_m = \int_V \tfrac{1}{4}\mu_0 |\underset{\sim}{H}|^2 \, dV \tag{7.18}$$

$$\overline{W}_e = \int_V \tfrac{1}{4}\epsilon_0 |\underset{\sim}{E}|^2 \, dV \tag{7.19}$$

$$\overline{W}_k = \int_V \sum_i \tfrac{1}{4} N_{0i} m_i |\underset{\sim}{v}_i|^2 \, dV \tag{7.20}$$

$$\overline{W}_d = \int_V \sum_i \tfrac{1}{2} N_{0i} e_i \frac{(\underset{\sim}{v}_R \times \underset{\sim}{v}_I)_i}{\omega} \cdot \underset{\sim}{B}_0 \, dV \tag{7.21}$$

Thus the time-averaged energy, Equation 7.15, is

$$\overline{W} = \overline{W}_m + \overline{W}_e + \overline{W}_k \tag{7.22}$$

while the resonance condition, Equation 7.17, is

$$\overline{W}_m - \overline{W}_e + \overline{W}_k - \overline{W}_d = 0 \tag{7.23}$$

It is often mathematically more convenient to describe the plasma in terms of the electromagnetic fields only. From Equations 7.9 and 7.10, we can obtain a relation between the current and the electric field

$$\sum_i e_i \underset{\sim}{\Gamma}_i = \underset{\approx}{\sigma} \cdot \underset{\sim}{E} \tag{7.24}$$

which defines the conductivity tensor $\underset{\approx}{\sigma}$. Equations 7.11 and 7.12 may be written as

$$\nabla \times \underset{\sim}{E} = -j\omega\mu_0 \underset{\sim}{H} \tag{7.25}$$

and

$$\nabla \times \underset{\sim}{H} = j\omega\epsilon_0 \underset{\approx}{K} \cdot \underset{\sim}{E} \tag{7.26}$$

where

$$\underset{\approx}{K} = \underset{\approx}{I} + \frac{\underset{\approx}{\sigma}}{j\omega\epsilon_0} \tag{7.27}$$

with $\underset{\approx}{I}$ the identity tensor. Since collisions were neglected in Equation 7.10, the dielectric tensor $\underset{\approx}{K}$ is Hermitian, $\underset{\approx}{K}^+ = \underset{\approx}{K}$; its general form may be written as

$$\underset{\approx}{K} = \left[\begin{array}{c|c} \underset{\approx}{K}_T & \underset{\approx}{K}_{Tz} \\ \hline \underset{\approx}{K}_{zT} & K_z \end{array}\right] \tag{7.28}$$

where $\underset{\approx}{K}_T^+ = \underset{\approx}{K}_T$ is a two-by-two tensor, $\underset{\approx}{K}_{Tz}^+ = \underset{\approx}{K}_{zT}$, K_z is real, and the subscript T denotes orthogonal coordinates in a plane transverse to the z direction. If the magnetic field $\underset{\sim}{B}_0$ is entirely in the z direction, then $\underset{\approx}{K}_{Tz} = 0$ and $\underset{\approx}{K}_{zT} = 0$.

The conservation equation equivalent to Equation 7.13 is obtained by dot-multiplying Equation 7.25 by $\underset{\sim}{H}^*$, the complex conjugate of Equation 7.26 by $-E^*$, and adding the resulting equations:

$$\tfrac{1}{2}\nabla \cdot (\underset{\sim}{E} \times \underset{\sim}{H}^*) = -j2\omega\left(\tfrac{1}{4}\mu_0 |\underset{\sim}{H}|^2 - \tfrac{1}{4}\epsilon_0 \underset{\sim}{E}^* \cdot \underset{\approx}{K} \cdot \underset{\sim}{E}\right) \tag{7.29}$$

Comparing with Equation 7.13, we see that

$$\tfrac{1}{4}\epsilon_0 \underset{\sim}{E}^* \cdot \underset{\approx}{K} \cdot \underset{\sim}{E} = \tfrac{1}{4}\epsilon_0 |\underset{\sim}{E}|^2 - \sum_i \tfrac{1}{4} N_{0i} m_i |\underset{\sim}{v}_i|^2 + \sum_i \tfrac{1}{2} N_{0i} e_i \frac{(\underset{\sim}{v}_R \times \underset{\sim}{v}_I)_i}{\omega} \cdot \underset{\sim}{B}_0 \tag{7.30}$$

It is clear that this term cannot be regarded as the time-averaged electric energy density, even though Equation 7.29 might invite such interpretation. In fact, the time-averaged energy density as interpreted from the integrand of Equation 7.15 is found to be

$$\tfrac{1}{4}\mu_0 |\underset{\sim}{H}|^2 + \tfrac{1}{4}\epsilon_0 |\underset{\sim}{E}|^2 + \sum_i \tfrac{1}{4} N_{0i} m_i |\underset{\sim}{v}_i|^2 = \tfrac{1}{4}\mu_0 |\underset{\sim}{H}|^2 + \tfrac{1}{4}\epsilon_0 \underset{\sim}{E}^* \cdot \frac{\partial(\omega \underset{\approx}{K})}{\partial\omega} \cdot \underset{\sim}{E} \tag{7.31}$$

Here we note also that the second term on the right-hand side of Equation 7.31 is made up of the time-averaged free-space electric energy density and the time-averaged kinetic energy density. This term is positive definite, but the left-hand side of Equation 7.30 although real may be either positive or negative. For the tensor formalism, it is convenient to introduce the abbreviations

$$\overline{W}_\epsilon = \int_V \tfrac{1}{4}\epsilon_0 \underset{\sim}{E}^* \cdot \underset{\approx}{K} \cdot \underset{\sim}{E}\, dV \tag{7.32}$$

$$\overline{W}_{ek} = \int_V \tfrac{1}{4}\epsilon_0 \underset{\sim}{E}^* \cdot \frac{\partial(\omega \underset{\sim}{K})}{\partial\omega} \cdot \underset{\sim}{E}\, dV \tag{7.33}$$

Comparing with Equations 7.18 through 7.21 and 7.30 through 7.31, we h

$$\overline{W}_{\epsilon} = \overline{W}_e - \overline{W}_k + \overline{W}_d \tag{7.34}$$

$$\overline{W}_{ek} = \overline{W}_e + \overline{W}_k \tag{7.35}$$

The time-averaged energy, Equation 7.22, is

$$\overline{W} = \overline{W}_m + \overline{W}_{ek} \tag{7.36}$$

while the resonance condition, Equation 7.23, is

$$\overline{W}_m - \overline{W}_{\epsilon} = 0 \tag{7.37}$$

The tensor formalism, although simpler mathematically, is harder to interpret. The physical interpretation of the various energy terms must come from the equations containing the mechanical variables explicitly.

7.3 Variation Theorem and Applications

Consider the variation of the first-order field variables in Equations 7.9 through 7.12 caused by variations in the unperturbed densities N_{0i}, the applied magnetic field $\underset{\sim}{B}_0$, and the frequency ω. By taking such a variation of Equations 7.9 through 7.12, we obtain a new set of equations for the varied field quantities $\delta\underset{\sim}{\Gamma}$, $\delta\underset{\sim}{v}_i$, δN_i, $\delta\underset{\sim}{E}$, and $\delta\underset{\sim}{H}$. These equations together with Equations 7.9 through 7.12 then give

$$\tfrac{1}{4}\nabla\cdot(\delta\underset{\sim}{E}\times\underset{\sim}{H}^* + \underset{\sim}{E}^*\times\delta\underset{\sim}{H}) = -j\omega\left[\overline{w}\,\frac{\delta\omega}{\omega} - \sum_i(\overline{w}_k - \overline{w}_d)\,\frac{\delta N_{0i}}{N_{0i}} + \sum_i \tfrac{1}{2}N_{0i}e_i\,\frac{(\underset{\sim}{v}_R\times\underset{\sim}{v}_I)_i}{\omega}\cdot\delta\underset{\sim}{B}_0\right] \tag{7.38}$$

where $\overline{w}_k$, $\overline{w}_d$, and $\overline{w}$ are the energy densities as defined by the integrands of Equations 7.20, 7.21, and 7.22.

7.3.1 Frequency Variation in a Plasma Resonator. Consider a plasma enclosed by perfectly conducting walls, as shown in Figure 7.2. By integrating Equation 7.38 over the volume enclosed by the surface of the perfect conductor, which is assumed unvaried, the left-hand side vanishes, and we obtain

$$\frac{\delta\omega}{\omega} = \frac{\sum_i \int_V \left[(\overline{w}_k - \overline{w}_d)\,\frac{\delta N_{0i}}{N_{0i}} - \tfrac{1}{2}N_{0i}e_i\,\frac{(\underset{\sim}{v}_R\times\underset{\sim}{v}_I)_i}{\omega}\cdot\delta\underset{\sim}{B}_0\right] dV}{\overline{W}} \tag{7.39}$$

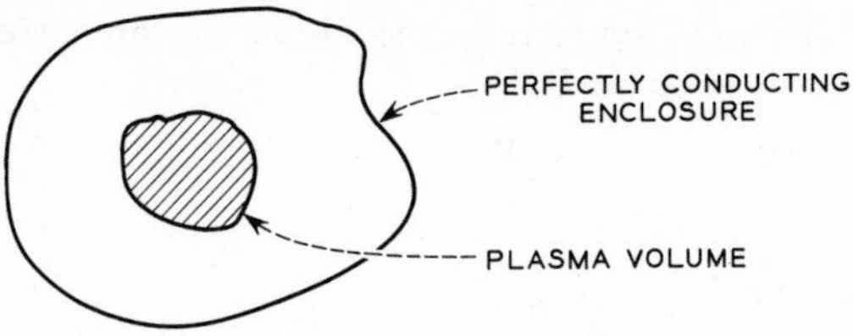

Figure 7.2. Plasma volume completely enclosed by perfectly conducting walls.

Equation 7.38 gives the change in frequency caused by variations in unperturbed density N_{0i} and magnetic field $\underset{\sim}{B}_0$. Near a resonance Equation 7.23 can be used to simplify further the evaluation of Equation 7.39.

7.3.2 Impedance Variation. We now consider the plasma system of Figure 7.2 to be coupled to a waveguide, as shown in Figure 7.3. At the reference plane in the waveguide cross section, we assume

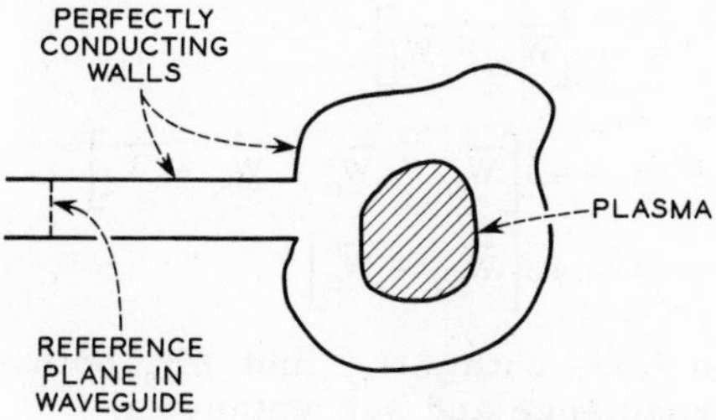

Figure 7.3. Plasma volume coupled to a waveguide.

that only the lowest waveguide mode is above cutoff. The fields at this cross section can be written as[22]

$$\underset{\sim}{E} = V\underset{\sim}{e}_T \tag{7.40}$$

$$\underset{\sim}{H} = I\underset{\sim}{h}_T \tag{7.41}$$

where $\underset{\sim}{e}_T$ and $\underset{\sim}{h}_T$ are functions of coordinates in the waveguide cross section and are normalized so that $\int \frac{1}{2} \underset{\sim}{E} \times \underset{\sim}{H}^* \cdot d\underset{\sim}{a} = -\frac{1}{2} VI^*$ is the complex power flow in the waveguide.

By integrating Equation 7.38 over the volume enclosed by the perfectly conducting walls and the reference plane in the waveguide, the left-hand side becomes

$$\frac{1}{4}\oint_S (\delta\underset{\sim}{E} \times \underset{\sim}{H}^* + \underset{\sim}{E}^* \times \delta\underset{\sim}{H}) \cdot d\underset{\sim}{S} = -\frac{1}{4}(\delta V\, I^* + V^*\, \delta I)$$

$$= -\frac{1}{4} j\, \delta B\, |V|^2 = -\frac{1}{4} j\, \delta X\, |I|^2 \tag{7.42}$$

where for the lossless system to the right of the reference plane in the guide we have written†

$$I = jBV \tag{7.43}$$

and

$$V = jXI \tag{7.44}$$

Hence,

$$\left.\begin{aligned} &\tfrac{1}{4}\,\delta B\,|V|^2 \\ &\text{or} \\ &\tfrac{1}{4}\,\delta X\,|I|^2 \end{aligned}\right\} = \omega\left\{\overline{W}\,\frac{\delta\omega}{\omega} - \sum_i \int_V \left[(\overline{w}_k - \overline{w}_d)\frac{\delta N_{0i}}{N_{0i}} - \tfrac{1}{2}N_{0i}e_i\frac{(\underset{\sim}{v}_R \times \underset{\sim}{v}_I)_i}{\omega}\cdot\delta\underset{\sim}{B}_0\right]dV\right\} \tag{7.45}$$

The reactance X or susceptance B at the waveguide reference plane can be evaluated from Equation 7.13, or its equivalent Equation 7.29, integrated over the same volume as in Figure 7.3,

$$\begin{aligned} X|I|^2 &= 4\omega\left[\overline{W}_m - \overline{W}_e + \overline{W}_k - \overline{W}_d\right] \\ &= 4\omega\left[\overline{W}_m - \overline{W}_\epsilon\right] \end{aligned} \tag{7.46}$$

$$\begin{aligned} B|V|^2 &= -4\omega\left[\overline{W}_m - \overline{W}_e + \overline{W}_k - \overline{W}_d\right] \\ &= -4\omega\left[\overline{W}_m - \overline{W}_\epsilon\right] \end{aligned} \tag{7.47}$$

From Equation 7.45, with δN_{0i} and $\delta\underset{\sim}{B}_0$ both zero, the frequency variation of the reactance and susceptance is

$$\left.\begin{aligned} &\frac{\partial X}{\partial\omega}\,|I|^2 \\ &\text{or} \\ &\frac{\partial B}{\partial\omega}\,|V|^2 \end{aligned}\right\} = 4\overline{W} = 4\left[\overline{W}_m + \overline{W}_e + \overline{W}_k\right] \tag{7.48}$$

Since $\overline{W} > 0$, both X and B are continuously increasing functions of frequency, similar to isotropic, lossless, waveguide or network systems that are passive. Thus for the temperate plasma,

$$\frac{\partial X}{\partial\omega} > 0 \tag{7.49}$$

and

$$\frac{\partial B}{\partial\omega} > 0 \tag{7.50}$$

† If the plasma system is coupled to more than one waveguide, Equations 7.43 and 7.44 become matrix relations, and Equation 7.42 is modified accordingly.

However, unlike the isotropic lossless systems where $(\partial X/\partial\omega) > |X|/\omega$ and $(\partial B/\partial\omega) > |B|/\omega$, we now have

$$\frac{\frac{\partial X}{\partial \omega}}{\frac{|X|}{\omega}} = \frac{\frac{\partial B}{\partial \omega}}{\frac{|B|}{\omega}} = \frac{\overline{W}_m + \overline{W}_k + \overline{W}_e}{|\overline{W}_m + \overline{W}_k - \overline{W}_e - \overline{W}_d|} \tag{7.51}$$

For $\overline{W}_d < 0$ and $|\overline{W}_d| > \overline{W}_e$, we then may have $\partial X/\partial\omega < |X|/\omega$ and $\partial B/\partial\omega < |B|/\omega$. For $\overline{W}_d = 0$, the right-hand side of Equation 7.51 is always greater than unity, as in the isotropic case.

Equation 7.45 also provides a convenient relation for evaluating variations in the reactance or susceptance that are caused by changes in density and magnetic field. It can thus be used for evaluating the density of a plasma in a magnetic field by measurements of perturbations in impedance.

7.4 Free Waves

In a homogeneous medium, the spatial dependence of the fields may be taken as $\exp(-j\underset{\sim}{k}\cdot\underset{\sim}{r})$. Maxwell's equations, Equations 7.25, 7.26, and their divergence, then take on the form

$$\underset{\sim}{k} \times \underset{\sim}{E} = \omega\mu_0 \underset{\sim}{H} \tag{7.52}$$

$$\underset{\sim}{k} \times \underset{\sim}{H} = -\omega\epsilon_0 \underset{\approx}{K}\cdot\underset{\sim}{E} \tag{7.53}$$

$$\underset{\sim}{k}\cdot\epsilon_0\underset{\approx}{K}\cdot\underset{\sim}{E} = 0 \tag{7.54}$$

$$\underset{\sim}{k}\cdot\mu_0\underset{\sim}{H} = 0 \tag{7.55}$$

We shall assume throughout this section that $\underset{\sim}{k}$ is real. The detailed solutions for this case are given in Part I. The complex power density vector in the direction of $\underset{\sim}{k}$ follows from Equations 7.52 and 7.53:

$$\begin{aligned}\underset{\sim}{k}\cdot(\tfrac{1}{2}\underset{\sim}{E}\times\underset{\sim}{H}^*) &= 2\omega(\tfrac{1}{4}\mu_0|\underset{\sim}{H}|^2)\\ &= 2\omega(\tfrac{1}{4}\epsilon_0\underset{\sim}{E}^*\cdot\underset{\approx}{K}\cdot\underset{\sim}{E})\end{aligned} \tag{7.56}$$

Hence, for the free propagating wave we have a balance similar to the resonance condition of Equations 7.37 and 7.23. Furthermore, Equation 7.56 shows that the power density in the direction of $\underset{\sim}{k}$ is pure real and independent of the magnitude of $\underset{\sim}{k}$. Similarly, we can find the complex power density vector in the direction perpendicular to $\underset{\sim}{k}$,

$$\underset{\sim}{k}\times(\tfrac{1}{2}\underset{\sim}{E}\times\underset{\sim}{H}^*) = -\tfrac{1}{2}(\underset{\sim}{k}\cdot\underset{\sim}{E})\underset{\sim}{H}^* \tag{7.57}$$

which is in general complex. The total complex power density vector can then be written as

$$\tfrac{1}{2}(\underset{\sim}{E} \times \underset{\sim}{H}^*) = \frac{2\omega \overline{w}_m}{k^2}\underset{\sim}{k} + \frac{(\underset{\sim}{k} \cdot \underset{\sim}{E})}{2k^2}(\underset{\sim}{k} \times \underset{\sim}{H}^*) \tag{7.58}$$

In particular, the time-averaged power density is

$$\tfrac{1}{2}\,\mathrm{Re}\,(\underset{\sim}{E} \times \underset{\sim}{H}^*) = \frac{2\omega \overline{w}_m}{k^2}\underset{\sim}{k} + \mathrm{Re}\left[\frac{(\underset{\sim}{k} \cdot \underset{\sim}{E})}{2k^2}(\underset{\sim}{k} \times \underset{\sim}{H}^*)\right] \tag{7.59}$$

and the reactive power density is

$$\tfrac{1}{2}\,\mathrm{Im}\,(\underset{\sim}{E} \times \underset{\sim}{H}^*) = \mathrm{Im}\left[\frac{(\underset{\sim}{k} \cdot \underset{\sim}{E})}{2k^2}(\underset{\sim}{k} \times \underset{\sim}{H}^*)\right] \tag{7.60}$$

In isotropic media the free propagating waves (uniform plane waves) have their complex power density entirely in the direction of $\underset{\sim}{k}$, because $\underset{\sim}{k} \cdot \underset{\sim}{E} = 0$. In the anisotropic plasma, Equations 7.59 and 7.60 show that this situation is in general different.

Consider the principal free waves in the anisotropic plasma. The right and left circularly polarized waves propagating along $\underset{\sim}{B}_0$ and the ordinary wave propagating across $\underset{\sim}{B}_0$ have their electric fields perpendicular to $\underset{\sim}{k}$ at all instants of time. Hence $\underset{\sim}{k} \cdot \underset{\sim}{E} = 0$, and

$$\tfrac{1}{2}\,\mathrm{Re}\,(\underset{\sim}{E} \times \underset{\sim}{H}^*) = \frac{2\omega \overline{w}_m}{k^2}\underset{\sim}{k} \tag{7.61}$$

$$\tfrac{1}{2}\,\mathrm{Im}\,(\underset{\sim}{E} \times \underset{\sim}{H}^*) = 0 \tag{7.62}$$

On the other hand, for the extraordinary wave propagating across $\underset{\sim}{B}_0$, $\underset{\sim}{k} \cdot \underset{\sim}{E}$ is not zero at all instants of time, and we find both real and reactive power density vectors. Using Equations 3.7 and 4.1, we obtain

$$\tfrac{1}{2}\,\mathrm{Re}\,(\underset{\sim}{E} \times \underset{\sim}{H}^*) = \frac{2\omega \overline{w}_m}{k^2}\underset{\sim}{k} \tag{7.63}$$

$$\tfrac{1}{2}\,\mathrm{Im}\,(\underset{\sim}{E} \times \underset{\sim}{H}^*) = -\underset{\sim}{i}_{\underset{\sim}{k}\times\underset{\sim}{B}_0}\,\tfrac{1}{2}\sqrt{\frac{\epsilon_0}{\mu_0}}\,n_\times\,|E_k|^2\,\mathrm{Im}\left(\frac{K_\perp}{K_\times}\right) \tag{7.64}$$

where $\underset{\sim}{i}_{\underset{\sim}{k}\times\underset{\sim}{B}_0}$ is a unit vector in the direction $\underset{\sim}{k} \times \underset{\sim}{B}_0$. Similar to the other principal waves, the time-averaged power density vector is in the direction of $\underset{\sim}{k}$. The reactive power density vector is in a direction perpendicular to both $\underset{\sim}{k}$ and $\underset{\sim}{B}_0$.

Equation 7.56 also provides a general relationship among the time-averaged power and magnetic energy density and the phase velocity of the wave

$$\frac{\underset{\sim}{k} \cdot \tfrac{1}{2}\,\mathrm{Re}\,(\underset{\sim}{E} \times \underset{\sim}{H}^*)}{k2\,\overline{w}_m} = \frac{\omega}{k} = u \tag{7.65}$$

where, by the balance condition for the wave (Equation 7.56), we also can write $\overline{w}_m = \overline{w}_\epsilon$.

The group velocity of free waves is defined as

$$\underset{\sim}{u}_g = \underset{\sim}{i}_x \frac{\partial \omega}{\partial k_x} + \underset{\sim}{i}_y \frac{\partial \omega}{\partial k_y} + \underset{\sim}{i}_x \frac{\partial \omega}{\partial k_z} = \frac{\partial \omega}{\partial \underset{\sim}{k}} \tag{7.66}$$

It can be evaluated with the aid of the dispersion relation for the waves. Thus from the wave equations for the E field,

$$\underset{\sim}{k} \times (\underset{\sim}{k} \times \underset{\sim}{E}) + k_0^2 \underset{\approx}{K} \cdot \underset{\sim}{E} = 0 \tag{7.67}$$

we obtain, for nontrivial solutions of $\underset{\sim}{E}$,

$$F(k, \omega) = 0 \tag{7.68}$$

where k_0 is the free-space phase constant and $F(k, \omega)$ is the dispersion relation, Equation 1.25. Equations 7.66 and 7.68 can be combined to give

$$\underset{\sim}{u}_g = \frac{-\nabla_k F}{\dfrac{\partial F}{\partial \omega}} \tag{7.69}$$

For the principal waves, it is clear that the group velocity has the same direction as $\underset{\sim}{k}$ and, hence, the direction of the time-averaged power density vector.

The general relationship between group velocity and energy flow becomes clearest when we consider the variation theorem, Equation 7.38, applied to free propagating waves. Assuming $\delta N_{0i} = 0$ and $\delta \underset{\sim}{B}_0 = 0$, we obtain

$$\tfrac{1}{4} \nabla \cdot (\delta \underset{\sim}{E} \times \underset{\sim}{H}^* + \underset{\sim}{E}^* \times \delta \underset{\sim}{H}) = -j\, \delta\omega\, \overline{w} \tag{7.70}$$

where $\overline{w}$ is the time-averaged energy density as given by the integrand of Equation 7.22 or 7.36. The fields of a free propagating wave can be written as

$$\underset{\sim}{E} = \underset{\sim}{E}_0 \exp(-j \underset{\sim}{k} \cdot \underset{\sim}{r}) \tag{7.71}$$

$$\underset{\sim}{H} = \underset{\sim}{H}_0 \exp(-j \underset{\sim}{k} \cdot \underset{\sim}{r}) \tag{7.72}$$

where $\underset{\sim}{E}_0$, $\underset{\sim}{H}_0$, and $\underset{\sim}{k}$ are independent of position, and $\underset{\sim}{k}$ is real. Using Equations 7.71 and 7.72 to evaluate the left-hand side of Equation 7.70, we obtain

$$\underset{\sim}{p} \cdot \delta \underset{\sim}{k} = \delta\omega\, \overline{w} \tag{7.73}$$

where we have written

$$\underset{\sim}{p} = \tfrac{1}{2} \operatorname{Re} \underset{\sim}{E} \times \underset{\sim}{H}^* \tag{7.74}$$

for the time-averaged electromagnetic power density vector. From Equation 7.73, we note the following: At a fixed frequency

$$\underset{\sim}{p} \cdot \delta \underset{\sim}{k} = 0 \tag{7.75}$$

which shows that the real part of the complex Poynting vector is perpendicular to the index surface[23] (see Figure 7.4). Also from Equation 7.73, we find

$$\frac{\partial \omega}{\partial \underset{\sim}{k}} = \frac{\underset{\sim}{p}}{\overline{\overline{w}}} = \frac{\underset{\sim}{p}}{\overline{\overline{w}}_m + \overline{\overline{w}}_e + \overline{\overline{w}}_k} = \frac{\underset{\sim}{p}}{\overline{\overline{w}}_m + \overline{\overline{w}}_{ek}} \tag{7.76}$$

which shows that the group velocity vector is also the energy velocity vector; each directional component of the group velocity vector is the time-averaged power flow density in that direction divided by the time-averaged energy density.

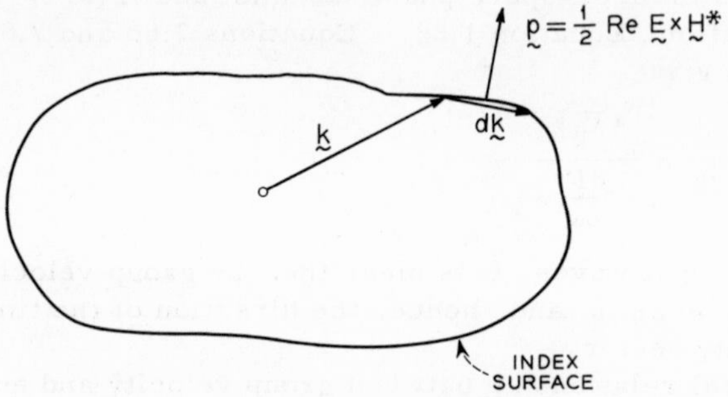

Figure 7.4. Time-averaged electromagnetic power density vector perpendicular to the index surface (Equation 7.75).

7.5 Complex Waves

We shall now generalize the wave solutions of Section 7.4. In a homogeneous medium, we consider the fields to have a spatial dependence $\exp(-\underset{\sim}{\gamma} \cdot \underset{\sim}{r})$, where†

$$\underset{\sim}{\gamma} = \underset{\sim}{\alpha} + j\underset{\sim}{\beta} \tag{7.77}$$

In homogeneous isotropic media, these solutions are the nonuniform plane waves.[24] In anisotropic media, these solutions have not yet been studied in detail. Here we shall give only some of their properties.

From Maxwell's equations, Equations 7.25 and 7.26, we have

$$\underset{\sim}{\gamma} \times \underset{\sim}{E} = j\omega\mu_0 \underset{\sim}{H} \tag{7.78}$$

$$\underset{\sim}{\gamma} \times \underset{\sim}{H} = -j\omega\epsilon_0 \underset{\approx}{K} \cdot \underset{\sim}{E} \tag{7.79}$$

$$\underset{\sim}{\gamma} \cdot \underset{\approx}{K} \cdot \underset{\sim}{E} = 0 \tag{7.80}$$

$$\underset{\sim}{\gamma} \cdot \underset{\sim}{H} = 0 \tag{7.81}$$

† In this part of the monograph β and γ will no longer be used for ω_b/ω and ν/ω.

If we denote the complex Poynting vector as

$$\tfrac{1}{2}\underset{\sim}{E} \times \underset{\sim}{H}^* = \underset{\sim}{p} + j\underset{\sim}{q} \tag{7.82}$$

and assume a lossless medium, $\underset{\approx}{K}^{+} = \underset{\approx}{K}$, Equations 7.78 and 7.79 give

$$\underset{\sim}{\gamma} \cdot (\underset{\sim}{p} + j\underset{\sim}{q}) = j2\omega\overline{w}_m \tag{7.83}$$

$$\underset{\sim}{\gamma}^* \cdot (\underset{\sim}{p} + j\underset{\sim}{q}) = -j2\omega\overline{w}_\epsilon \tag{7.84}$$

where $\overline{w}_m$ and $\overline{w}_\epsilon$ are real quantities, given by the integrands of Equations 7.18 and 7.32, respectively. Adding and subtracting Equations 7.83 and 7.84, we find

$$\underset{\sim}{\alpha} \cdot (\underset{\sim}{p} + j\underset{\sim}{q}) = j\omega(\overline{w}_m - \overline{w}_\epsilon) \tag{7.85}$$

$$\underset{\sim}{\beta} \cdot (\underset{\sim}{p} + j\underset{\sim}{q}) = \omega(\overline{w}_m + \overline{w}_\epsilon) \tag{7.86}$$

The real and imaginary parts of Equations 7.85 and 7.86 give the following four relations:

$$\underset{\sim}{\alpha} \cdot \underset{\sim}{p} = 0 \tag{7.87}$$

$$\underset{\sim}{\alpha} \cdot \underset{\sim}{q} = \omega(\overline{w}_m - \overline{w}_\epsilon) \tag{7.88}$$

$$\underset{\sim}{\beta} \cdot \underset{\sim}{p} = \omega(\overline{w}_m + \overline{w}_\epsilon) \tag{7.89}$$

$$\underset{\sim}{\beta} \cdot \underset{\sim}{q} = 0 \tag{7.90}$$

We note that the complex wave may carry both real and reactive power, with the direction of $\underset{\sim}{p}$ at right angles to $\underset{\sim}{\alpha}$, and with the direction of $\underset{\sim}{q}$ at right angles to $\underset{\sim}{\beta}$ (Equations 7.87 and 7.90).

When the wave is purely propagating, $\underset{\sim}{\alpha} = 0$ and $\underset{\sim}{\gamma} = j\underset{\sim}{\beta} = j\underset{\sim}{k}$, as in Section 7.4; Equation 7.88 then gives the balance condition $\overline{w}_m = \overline{w}_\epsilon$; Equations 7.89 and 7.90 become identical to Equation 7.56; and Equation 7.89 provides the relation between phase velocity, power, and energy, as in Equation 7.65.

For a pure cutoff wave, $\underset{\sim}{\beta} = 0$, $\underset{\sim}{\gamma} = \underset{\sim}{\alpha}$, and hence Equation 7.89 requires $\overline{w}_m = -\overline{w}_\epsilon$. Since $\overline{w}_m$ is positive, this requires that $\overline{w}_\epsilon$ be negative or equivalently $\overline{w}_e - \overline{w}_k + \overline{w}_d < 0$. When $\overline{w}_d$ vanishes, as for example, in the isotropic case, this cutoff condition leads to $\omega < \omega_p$. In the anisotropic plasma, the cutoff condition is determined by both the density and the magnetic field.

A variation theorem for the complex waves can also be derived. When the frequency, the propagation constant, and the fields are varied, Equations 7.78 and 7.79 become

$$\delta\underset{\sim}{\gamma} \times \underset{\sim}{E} = -\underset{\sim}{\gamma} \times \delta\underset{\sim}{E} + j\,\delta(\omega\mu_0)\,\underset{\sim}{H} + j\omega\mu_0\delta\underset{\sim}{H} \tag{7.91}$$

$$\delta \underset{\sim}{\gamma} \times \underset{\sim}{H} = -\underset{\sim}{\gamma} \times \delta \underset{\sim}{H} - j\,\delta(\omega\epsilon_0 \underset{\approx}{K}) \cdot \underset{\sim}{E} - j\omega\epsilon_0 \underset{\approx}{K} \cdot \delta \underset{\sim}{E} \tag{7.92}$$

Using Equations 7.91 and 7.92, we can form the expressions $\delta\underset{\sim}{\gamma} \cdot \underset{\sim}{E} \times \underset{\sim}{H}^*$ and $\delta\underset{\sim}{\gamma}^* \cdot \underset{\sim}{E} \times \underset{\sim}{H}^*$, and from these and Equations 7.78 and 7.79 we obtain

$$\delta\underset{\sim}{\alpha} \cdot (\underset{\sim}{p} + j\underset{\sim}{q}) = j\left(\tfrac{1}{4}\mu_0\,\delta\omega\,|\underset{\sim}{H}|^2 - \tfrac{1}{4}\epsilon_0 \underset{\sim}{E}^* \cdot \delta(\omega\underset{\approx}{K}) \cdot \underset{\sim}{E}\right)$$
$$- \underset{\sim}{\gamma} \cdot \mathrm{Re}\,\tfrac{1}{2}\delta\underset{\sim}{E} \times \underset{\sim}{H}^* - \underset{\sim}{\gamma}^* \cdot \mathrm{Re}\,\tfrac{1}{2}\underset{\sim}{E} \times \delta\underset{\sim}{H}^* \tag{7.93}$$

$$\delta\underset{\sim}{\beta} \cdot (\underset{\sim}{p} + j\underset{\sim}{q}) = \tfrac{1}{4}\mu_0\,\delta\omega\,|\underset{\sim}{H}|^2 + \tfrac{1}{4}\epsilon_0 \underset{\sim}{E}^* \cdot \delta(\omega\underset{\approx}{K}) \cdot \underset{\sim}{E}$$
$$- \underset{\sim}{\gamma} \cdot \mathrm{Im}\,\tfrac{1}{2}\delta\underset{\sim}{E} \times \underset{\sim}{H}^* - \underset{\sim}{\gamma}^* \cdot \mathrm{Im}\,\tfrac{1}{2}\underset{\sim}{E}^* \times \delta\underset{\sim}{H} \tag{7.94}$$

From the real and imaginary parts of Equations 7.93 and 7.94 we find

$$\delta(\underset{\sim}{\alpha} \cdot \underset{\sim}{p}) = 0 \tag{7.95}$$

$$\delta\underset{\sim}{\alpha} \cdot \underset{\sim}{q} = \tfrac{1}{4}\mu_0\,\delta\omega\,|\underset{\sim}{H}|^2 - \tfrac{1}{4}\epsilon_0 \underset{\sim}{E}^* \cdot \delta(\omega\underset{\approx}{K}) \cdot \underset{\sim}{E}$$
$$- \underset{\sim}{\beta} \cdot \mathrm{Re}\,\tfrac{1}{2}(\delta\underset{\sim}{E} \times \underset{\sim}{H}^* - \underset{\sim}{E} \times \delta\underset{\sim}{H}^*) \tag{7.96}$$

$$\delta\underset{\sim}{\beta} \cdot \underset{\sim}{p} = \tfrac{1}{4}\mu_0\,\delta\omega\,|\underset{\sim}{H}|^2 + \tfrac{1}{4}\epsilon_0 \underset{\sim}{E}^* \cdot \delta(\omega\underset{\approx}{K}) \cdot \underset{\sim}{E}$$
$$- \underset{\sim}{\alpha} \cdot \mathrm{Im}\,\tfrac{1}{2}(\delta\underset{\sim}{E} \times \underset{\sim}{H}^* - \underset{\sim}{E} \times \delta\underset{\sim}{H}^*) \tag{7.97}$$

$$\delta(\underset{\sim}{\beta} \cdot \underset{\sim}{q}) = 0 \tag{7.98}$$

For a pure propagating wave, $\underset{\sim}{\alpha} = 0$, $\underset{\sim}{\gamma} = j\underset{\sim}{\beta} = j\underset{\sim}{k}$, and for variation in frequency only, Equation 7.97 reduces to Equation 7.73, from which the conclusions of Equation 7.75 and 7.76 follow. Combining Equations 7.76 and 7.89, we obtain

$$\frac{\partial\omega}{\partial\underset{\sim}{\beta}} = \frac{\omega}{\underset{\sim}{\beta}}\,\frac{2\overline{\overline{w}}_m}{(\overline{\overline{w}}_m + \overline{\overline{w}}_e + \overline{\overline{w}}_k)} = \frac{\omega}{\underset{\sim}{\beta}}\,\frac{2\overline{\overline{w}}_m}{(\overline{\overline{w}}_m + \overline{\overline{w}}_{ek})} \tag{7.99}$$

Equation 7.99 shows that the ratio of group velocity to the phase velocity in the direction of the real power density is given by the ratio of twice the time-averaged magnetic energy density to the time-averaged total energy density, and is always positive.

For a pure cutoff wave, $\underset{\sim}{\beta} = 0$, $\underset{\sim}{\gamma} = \underset{\sim}{\alpha}$, and Equation 7.96 gives

$$\frac{\partial\omega}{\partial\underset{\sim}{\alpha}} = \frac{\underset{\sim}{q}}{\overline{\overline{w}}_m - \overline{\overline{w}}_e - \overline{\overline{w}}_k} = \frac{\underset{\sim}{q}}{\overline{\overline{w}}_m - \overline{\overline{w}}_{ek}} \tag{7.100}$$

Combining Equations 7.100 and 7.88, we obtain

$$\frac{\partial \omega}{\partial \underset{\sim}{\beta}} = \frac{\omega}{\underset{\sim}{\beta}} \frac{2\overline{w}_m}{(\overline{w}_m - \overline{w}_e - \overline{w}_k)} = \frac{\omega}{\underset{\sim}{\beta}} \frac{2\overline{w}_m}{(\overline{w}_m - \overline{w}_{ek})} \tag{7.101}$$

7.6 Guided Waves

For plasma systems of finite extent that are uniform in one spatial direction, the fields are most easily analyzed in terms of waves that propagate in that direction. Consider that the plasma waveguide (Figure 7.5) is uniform in the z direction but otherwise has unperturbed properties that may vary with position in a plane transverse to the z direction. The walls of the waveguide will be assumed perfectly conducting (that is, with either the tangential E or

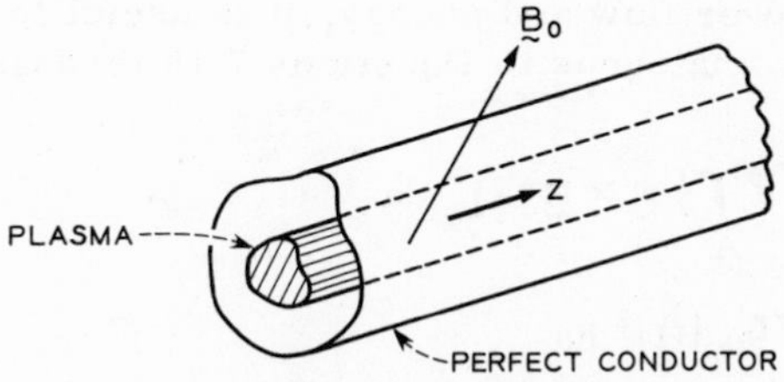

Figure 7.5. Plasma waveguide uniform in the z direction. Both the cross section and the direction of $\underset{\sim}{B}_0$ are arbitrary.

H fields vanishing at the walls). By taking advantage of the uniformity in the z direction, it is convenient to separate the equations of motion into longitudinal and transverse parts. Thus Equations 7.10 through 7.12 become

$$0 = -j\omega m_i \underset{\sim}{v}_{Ti} + e_i \underset{\sim}{E}_T + e_i(\underset{\sim}{v}_i \times \underset{\sim}{B}_0)_T \tag{7.102}$$

$$0 = -j\omega m_i v_{zi} + e_i E_z + e_i(\underset{\sim}{v}_i \times \underset{\sim}{B}_0) \cdot \underset{\sim}{i}_z \tag{7.103}$$

$$\frac{\partial \underset{\sim}{E}_T}{\partial z} = j\omega \underset{\sim}{i}_z \times \mu_0 \underset{\sim}{H}_T + \nabla_T E_z \tag{7.104}$$

$$0 = \underset{\sim}{i}_z \cdot \nabla_T \times \underset{\sim}{E}_T + j\omega\mu_0 H_z \tag{7.105}$$

$$\frac{\partial \underset{\sim}{H}_T}{\partial z} = -j\omega \underset{\sim}{i}_z \times \epsilon_0 \underset{\sim}{E}_T + \nabla_T H_z + \sum_i e_i \underset{\sim}{\Gamma}_{Ti} \times \underset{\sim}{i}_z \tag{7.106}$$

$$0 = \underset{\sim}{i}_z \cdot \nabla_T \times \underset{\sim}{H}_T - j\omega\epsilon_0 E_z - \sum_i e_i \Gamma_{zi} \tag{7.107}$$

where a subscript T on a vector denotes that it is in a plane transverse to the z direction. Likewise the equations with the dielectric tensor description of the plasma, Equations 7.25 to 7.26, become

$$\frac{\partial \underset{\sim}{E}_T}{\partial z} = j\omega \underset{\sim}{i}_z \times \mu_0 \underset{\sim}{H}_T + \nabla_T E_z \tag{7.108}$$

$$0 = \underset{\sim}{i}_z \cdot \nabla_T \times \underset{\sim}{E}_T + j\omega\mu_0 H_z \tag{7.109}$$

$$\frac{\partial \underset{\sim}{H}_T}{\partial z} = -j\omega \underset{\sim}{i}_z \times \underset{\sim}{D}_T + \nabla_T H_z \tag{7.110}$$

$$0 = \underset{\sim}{i}_z \cdot \nabla_T \times \underset{\sim}{H}_T - j\omega D_z \tag{7.111}$$

where we have written $\underset{\sim}{D} = \epsilon_0 \underset{\approx}{K} \cdot \underset{\sim}{E}$.

In the study of power flow and energy, it is useful to introduce the following quantities analogous to Equations 7.18 through 7.21, 7.32, and 7.33:

$$P + jQ = \int_A \tfrac{1}{2} \underset{\sim}{E} \times \underset{\sim}{H}^* \cdot \underset{\sim}{i}_z \, da \tag{7.112}$$

$$U_m = \int_A \tfrac{1}{4} \mu_0 |\underset{\sim}{H}|^2 \, da \tag{7.113}$$

$$U_e = \int_A \tfrac{1}{4} \epsilon_0 |\underset{\sim}{E}|^2 \, da \tag{7.114}$$

$$U_k = \int_A \sum_i \tfrac{1}{4} N_{0i} m_i |\underset{\sim}{v}_i|^2 \, da \tag{7.115}$$

$$U_d = \int_A \sum_i \tfrac{1}{2} N_{0i} e_i \frac{(\underset{\sim}{v}_R \times \underset{\sim}{v}_I)_i}{\omega} \cdot \underset{\sim}{B}_0 \, da \tag{7.116}$$

$$U_\epsilon = \int_A \tfrac{1}{4} \epsilon_0 \underset{\sim}{E}^* \cdot \underset{\approx}{K} \cdot \underset{\sim}{E} \, da$$

$$= U_e - U_k + U_d \tag{7.117}$$

$$U_{ek} = \int_A \tfrac{1}{4} \epsilon_0 \underset{\sim}{E}^* \cdot \frac{\partial(\omega \underset{\approx}{K})}{\partial \omega} \cdot \underset{\sim}{E} \, da$$

$$= U_e + U_k \tag{7.118}$$

$$U = U_m + U_e + U_k \tag{7.119}$$

All the integrals are over the cross section A of the waveguide. Equation 7.112 is the complex power flowing across A in the waveguide; P is the time-averaged power flow, and Q is the reactive power flow. The quantities in Equations 7.113 through 7.119 are all real; U_d and U_ϵ may be positive or negative, but all the other quantities are positive. Equation 7.119 is the time-averaged energy per unit length of waveguide.

When we dot-multiply Equation 7.104 by $(\underset{\sim}{H}_T^* \times \underset{\sim}{i}_z)$, the complex conjugate of Equation 7.106 by $(\underset{\sim}{i}_z \times \underset{\sim}{E}_T)$, add the resulting equations, and integrate the sum over the waveguide cross section bounded by the perfectly conducting walls (identical operations can also be performed on Equations 7.108 and 7.110), we obtain

$$\frac{\partial}{\partial z}(P + jQ) = -j2\omega(U_m - U_e + U_k - U_d)$$

$$= -j2\omega(U_m - U_\epsilon) \qquad (7.120)$$

The real part of Equation 7.120 gives

$$\frac{\partial}{\partial z}\mathrm{Re}\int_A \tfrac{1}{2}\underset{\sim}{E}_T \times \underset{\sim}{H}_T^* \cdot \underset{\sim}{i}_z \, da = 0 \qquad (7.121)$$

which expresses conservation of the time-averaged electromagnetic power flow in this lossless system. Equation 7.121 forms the basis of the orthogonality properties of the waves in such plasma waveguides.

Consider solutions of the fields whose z dependence is of the form exp $(-\gamma z)$, where $\gamma = \alpha + j\beta$ is in general complex and a function of frequency ω; such solutions we shall call complex waves. Detailed field analyses are considered in Chapters 9 and 10; here we are interested only in the properties of such solutions.

If we introduce the exp $(-\gamma z)$ dependence into Equations 7.104, 7.106, 7.108, and 7.110, the derivatives with respect to z become replaced by $-\gamma$. Then, dot-multiplying Equation 7.104 by $-(\underset{\sim}{H}_T^* \times \underset{\sim}{i}_z)$, the complex conjugate of Equation 7.107 by $-E_z$, Equation 7.103 by $-\Gamma_{iz}^*$, adding the resulting three equations, and integrating the result over the waveguide cross section bounded by the perfectly conducting walls, we obtain

$$\gamma(P + jQ) = j2\omega(U_{mT} - U_{ez} + U_{kz}) - 2\omega U_c \qquad (7.122)$$

Similarly, dot-multiplying the complex conjugate of Equation 7.106 by $(\underset{\sim}{E}_T \times \underset{\sim}{i}_z)$, Equation 7.105 by H_z^*, Equation 7.102 by $(\underset{\sim}{i}_z \times \underset{\sim}{\Gamma}_{iT}^*)$, adding the resulting three equations, and integrating the result over the waveguide cross section bounded by the perfectly conducting walls, we obtain

$$\gamma^*(P + jQ) = j2\omega(U_{mz} - U_{eT} + U_{kT} - U_{dz}) + 2\omega U_c^* \qquad (7.123)$$

In Equations 7.122 and 7.123 we have introduced

$$U_c = \int_A \sum_i \tfrac{1}{4} N_{0i} e_i \frac{(v_z \underset{\sim}{i}_z \times \underset{\sim}{v}_T)_i}{\omega} \cdot \underset{\sim}{B}_{0T} \, da \tag{7.124}$$

$$U_{dz} = \int_A \sum_i \tfrac{1}{2} N_{0i} e_i \frac{(\underset{\sim}{v}_{TR} \times \underset{\sim}{v}_{TI})_i}{\omega} \cdot \underset{\sim}{i}_z B_{0z} \, da \tag{7.125}$$

where it is to be noted that

$$U_{dz} + 2 \,\mathrm{Im}\, U_c = U_d \tag{7.126}$$

The subscripts z and T on U_m, U_e, and U_k in Equations 7.122 and 7.123 indicate that only those field components (that is, z components or T components) should enter in their evaluation by Equations 7.113 through 7.115. Note that U_c is in general complex and U_{dz} is real; U_c vanishes when $\underset{\sim}{B}_{0T} = 0$, that is, for the case when the magnetic field is entirely longitudinal. Addition and subtraction of Equations 7.122 and 7.123 give the following relations that must hold for any complex wave solution:

$$\alpha P = 0 \tag{7.127}$$

$$\alpha Q = \omega(U_m - U_e + U_k - U_d) \tag{7.128}$$

$$\beta P = \omega(U_{mT} - U_{mz} + U_{eT} - U_{ez} + U_{kz} - U_{kT} + U_{dz}) \tag{7.129}$$

$$\beta Q = \omega 2 \mathrm{Re}\, U_c \tag{7.130}$$

Analogous expressions can be derived for the tensor medium description of the plasma. From Equations 7.108 through 7.111 and operations analogous to those that lead to Equations 7.122 and 7.123, we obtain

$$\gamma(P + jQ) = j2\omega(U_{mT} - U_{\epsilon z}) - j2\omega U_{\epsilon zT} \tag{7.131}$$

$$\gamma^*(P + jQ) = j2\omega(U_{mz} - U_{\epsilon T}) - j2\omega U_{\epsilon zT}^* \tag{7.132}$$

where we have introduced

$$U_{\epsilon zT} = \int_A \tfrac{1}{4} \epsilon_0 \underset{\sim}{E}_T^* \cdot \underset{\approx}{K}_{Tz} E_z \, da \tag{7.133}$$

$$U_{\epsilon z} = \int_A \tfrac{1}{4} \epsilon_0 K_z |E_z|^2 \, da \tag{7.134}$$

$$U_{\epsilon T} = \int_A \tfrac{1}{4} \epsilon_0 E_T^* \cdot \underset{\approx}{K}_T \cdot \underset{\sim}{E}_T \, da \tag{7.135}$$

and note that

$$U_{\epsilon z} + U_{\epsilon T} + 2 \,\mathrm{Re}\, U_{\epsilon zT} = U_\epsilon \tag{7.136}$$

Here, $U_{\epsilon zT}$ is in general complex, and $U_{\epsilon z}$ and $U_{\epsilon T}$ are real (see Equation 7.28); $U_{\epsilon zT}$ vanishes when $\underset{\approx}{K}_{Tz} = 0$, which holds for the temperate plasma when $\underset{\sim}{B}_{0T} = 0$, that is, when the magnetic field $\underset{\sim}{B}_0$ is entirely longitudinal. The properties of guided waves in temperate plasmas with $\underset{\sim}{B}_{0T} = 0$ were first investigated by Chorney.[25] Equations 7.131 and 7.132 are the generalizations to temperate plasmas in magnetic fields of arbitrary direction.

Addition and subtraction of Equations 7.131 and 7.132 give

$$\alpha P = 0 \tag{7.137}$$

$$\alpha Q = \omega(U_m - U_\epsilon) \tag{7.138}$$

$$\beta P = \omega(U_{mT} - U_{mz} + U_{\epsilon T} - U_{\epsilon z}) \tag{7.139}$$

$$\beta Q = -\omega 2 \text{ Im } U_{\epsilon zT} \tag{7.140}$$

From Equations 7.127 through 7.130 and 7.137 through 7.140, we can derive the following properties for guided waves:

(a) <u>Complex wave:</u> $\gamma = \alpha + j\beta$.

When both α and β are finite,

$$P = 0 \tag{7.141}$$

$$U_{mT} - U_{mz} + U_{eT} - U_{ez} + U_{kz} - U_{kT} + U_{dz} = 0 \tag{7.142a}$$

$$U_{mT} - U_{mz} + U_{\epsilon T} - U_{\epsilon z} = 0 \tag{7.142b}$$

$$\frac{\omega}{\alpha} = \frac{Q}{U_m - U_e + U_k - U_d} = \frac{Q}{U_m - U_\epsilon} \tag{7.143}$$

$$\frac{\omega}{\beta} = \frac{Q}{2 \text{ Re } U_c} = \frac{Q}{-2 \text{ Im } U_{\epsilon zT}} \tag{7.144}$$

Equation 7.141 shows that a complex wave cannot carry any time-averaged power. This is as we should expect on physical grounds for the lossless passive system considered. The reactive power Q is in general finite. Equations 7.142a and 7.142b give the balance condition for a complex wave. Equations 7.143 and 7.144 relate ω/α and ω/β to the reactive power. When the magnetic field $\underset{\sim}{B}_0$ is entirely longitudinal, $U_{\epsilon zT} = 0$, the reactive power must also vanish, Equations 7.143 and 7.144 become indeterminate, and $U_m = U_\epsilon$.

(b) <u>Cutoff wave:</u> $\beta = 0$, $\gamma = \alpha$.

Using Equations 7.127 through 7.130 and Equations 7.137 through 7.140, we find

$$P = 0 \tag{7.145}$$

$$U_{mT} - U_{mz} + U_{eT} - U_{ez} + U_{kz} - U_{kT} + U_{dz} = 0 \quad (7.146a)$$

$$U_{mT} - U_{mz} + U_{\epsilon T} - U_{\epsilon z} = 0 \quad (7.146b)$$

$$\mathrm{Re}\, U_c = 0 \quad (7.147a)$$

$$\mathrm{Im}\, U_{\epsilon zT} = 0 \quad (7.147b)$$

$$\frac{\omega}{\alpha} = \frac{Q}{U_m - U_e + U_k - U_d} = \frac{Q}{U_m - U_\epsilon} \quad (7.148)$$

Similar to the complex wave, a cutoff wave carries no real power and has the same balance condition. On the other hand, the reactive power is finite even when $\underset{\sim}{B}_0$ is entirely longitudinal.

(c) Propagating wave: $\alpha = 0$, $\gamma = j\beta$.
For these waves we find

$$U_m - U_e + U_k - U_d = 0 \quad (7.149a)$$

$$U_m - U_\epsilon = 0 \quad (7.149b)$$

$$u = \frac{\omega}{\beta} = \frac{P}{U_{mT} - U_{mz} + U_{eT} - U_{ez} + U_{kz} - U_{kT} + U_{dz}}$$

$$= \frac{Q}{2\,\mathrm{Re}\, U_c} \quad (7.150a)$$

$$u = \frac{\omega}{\beta} = \frac{P}{U_{mT} - U_{mz} + U_{\epsilon T} - U_{\epsilon z}} = \frac{Q}{-2\,\mathrm{Im}\, U_{\epsilon zT}} \quad (7.150b)$$

Equations 7.149a and 7.149b give the balance condition for a propagating wave; this also follows directly from Equation 7.120. In general, the propagating wave will carry both real and reactive power. When the magnetic field $\underset{\sim}{B}_0$ is entirely longitudinal, the reactive power vanishes. Equations 7.150a and b relate the phase velocity of the wave to the power flow (real or reactive).

Next, we consider the variation theorem for propagating waves, $\gamma = j\beta$. Integrating Equation 7.38 over the waveguide cross section bounded by the perfectly conducting contour of the walls, we obtain

$$\delta\beta\, P = U\, \delta\omega - \omega \sum_i \int_A \left[(\overline{w}_k - \overline{w}_d) \frac{\delta N_{0i}}{N_{0i}} + \tfrac{1}{2} N_{0i} e_i \frac{(\underset{\sim}{v}_R \times \underset{\sim}{v}_I)_i}{\omega} \cdot \delta \underset{\sim}{B}_0 \right] da$$

(7.151)

Equation 7.151 can be used to evaluate the perturbation in propagation constant caused by variations in frequency, density, and magnetic field. When only the frequency is varied, Equation 7.151 gives

$$\frac{\partial\omega}{\partial\beta} = \frac{P}{U} = \frac{P_e}{U_m + U_e + U_k} \tag{7.152}$$

The left-hand side of Equation 7.152 is the group velocity of the guided propagating wave. The right-hand side of Equation 7.152 is the ratio of the time-averaged power flow through the guide cross section to the time-averaged energy per unit length of guide, that is, the energy velocity. Eliminating P between Equations 7.150 and 7.152, we obtain a relation between the group and phase velocity of a guided propagating wave,

$$\frac{\partial\omega}{\partial\beta} = \frac{\omega}{\beta}\,\frac{(U_{mT} - U_{mz} + U_{\epsilon T} - U_{\epsilon z})}{U} \tag{7.153a}$$

$$\frac{\partial\omega}{\partial\beta} = \frac{\omega}{\beta}\,\frac{(U_{mT} - U_{mz} + U_{eT} - U_{ez} + U_{kz} - U_{kT} + U_{dz})}{U} \tag{7.153b}$$

Equations 7.153a and b clearly show the possibility of waves having phase and group velocities of opposite sign; such waves are known as backward waves.

Finally, consider the generalization of the variation theorem to complex waves: Equations 7.108 through 7.111 and their varied forms are

$$0 = \gamma \underset{\approx}{E}_T + j\omega \underset{\sim}{i}_z \times \mu_0 \underset{\approx}{H}_T + \nabla_T E_z \tag{7.154}$$

$$0 = \underset{\sim}{i}_z \cdot \nabla_T \times \underset{\approx}{E}_T + j\omega\mu_0 H_z \tag{7.155}$$

$$0 = \gamma \underset{\approx}{H}_T - j\omega \underset{\sim}{i}_z \times \underset{\approx}{D}_T + \nabla_T H_z \tag{7.156}$$

$$0 = \underset{\sim}{i}_z \cdot \nabla_T \times \underset{\approx}{H}_T - j\omega D_z \tag{7.157}$$

$$\delta\gamma\, \underset{\approx}{E}_T = -\gamma\, \delta \underset{\approx}{E}_T - j \underset{\sim}{i}_z \times \delta(\omega\mu_0 \underset{\approx}{H}_T) - \nabla_T\, \delta E_z \tag{7.158}$$

$$0 = \underset{\sim}{i}_z \cdot \nabla_T \times \delta \underset{\approx}{E}_T + j\delta(\omega\mu_0 H_z) \tag{7.159}$$

$$\delta\gamma\, \underset{\approx}{H}_T = -\gamma\, \delta \underset{\approx}{H}_T + j \underset{\sim}{i}_z \times \delta(\omega \underset{\approx}{D}_T) - \nabla_T\, \delta H_z \tag{7.160}$$

$$0 = \underset{\sim}{i}_z \cdot \nabla_T \times \delta \underset{\approx}{H}_T - j\,\delta(\omega D_z) \tag{7.161}$$

Combining Equations 7.154 through 7.161, we can construct $\delta\gamma\, \underset{\approx}{E}_T \times \underset{\approx}{H}_T^* \cdot \underset{\sim}{i}_z$ and $\delta\gamma^*\, \underset{\approx}{E}_T \times \underset{\approx}{H}_T^* \cdot \underset{\sim}{i}_z$ and integrate these over the waveguide cross section. The real and imaginary parts of these equations then give

$$\delta(\alpha P) = 0 \tag{7.162}$$

$$\begin{aligned}\delta\alpha\, Q = {} & (U_{mT} - U_{mz} - U_{\epsilon T} + U_{\epsilon z})\,\delta\omega \\ & - \tfrac{1}{2}\epsilon_0 \operatorname{Re}\int (\delta E_z^* \,\omega \underset{\approx}{K}_{zT} \cdot \underset{\sim}{E}_T - E_z^* \omega \underset{\approx}{K}_{zT} \cdot \delta \underset{\sim}{E}_T)\, da \\ & - \beta \tfrac{1}{2} \operatorname{Re}\int (\delta \underset{\sim}{E}_T \times \underset{\sim}{H}_T^* - \underset{\sim}{E}_T \times \delta \underset{\sim}{H}_T^*) \cdot \underset{\sim}{i}_z \, da\end{aligned} \tag{7.163}$$

$$\delta\beta\, P = \delta\omega\, U - \alpha \tfrac{1}{2} \operatorname{Im}\int (\delta \underset{\sim}{E}_T \times \underset{\sim}{H}_T^* - \underset{\sim}{E}_T \times \delta \underset{\sim}{H}_T^*) \cdot \underset{\sim}{i}_z \, da \tag{7.164}$$

$$\delta(\beta Q) = -\delta(\omega 2 \operatorname{Im} U_{\epsilon z T}) \tag{7.165}$$

For a propagating wave, $\gamma = j\beta$, $\alpha = 0$, therefore Equation 7.164 reduces to Equation 7.152. If the external magnetic field is entirely in the z direction, $\underset{\approx}{B}_0 = \underset{\sim}{i}_z B_0$, then $\underset{\approx}{K}_{zT} = 0$, and the right-hand side of Equation 7.165 and the second term on the right-hand side of Equation 7.163 vanish. In this case, Equation 7.163 for a cutoff wave $\gamma = \alpha$, $\beta = 0$, gives

$$\frac{\partial\omega}{\partial\alpha} = \frac{Q}{(U_{mT} - U_{mz}) - (U_{\epsilon T} - U_{\epsilon z})} \tag{7.166}$$

Combining Equations 7.148 and 7.166, we obtain

$$\frac{\partial\omega}{\partial\alpha} = \frac{\omega}{\alpha}\, \frac{(U_m - U_\epsilon)}{(U_{mT} - U_{mz}) - (U_{\epsilon T} - U_{\epsilon z})} \tag{7.167}$$

analogous to Equation 7.153 for propagating waves. However, we should note that Equations 7.166 and 7.167 are restricted to the case where $\underset{\approx}{B}_0$ is longitudinal, while Equations 7.152 and 7.153 apply for arbitrary $\underset{\approx}{B}_0$.

Chapter 8

CONSERVATION PRINCIPLES FOR WARM PLASMAS

We now consider the model of a plasma as derived from the Boltzmann transport equations in Chapter 5. In such a model, the equations of motion for the first-order field quantities are

$$\frac{m_i v_{Ti}^2}{N_{0i}} \nabla N_i = -m_i \frac{\partial \underset{\sim}{v}_i}{\partial t} + e_i \underset{\sim}{E} + e_i (\underset{\sim}{v}_i \times \underset{\sim}{B}_0) \tag{8.1}$$

$$\nabla \cdot \underset{\sim}{\Gamma}_i = -\frac{\partial N_i}{\partial t} \tag{8.2}$$

$$\nabla \times \underset{\sim}{E} = -\mu_0 \frac{\partial \underset{\sim}{H}}{\partial t} \tag{8.3}$$

$$\nabla \times \underset{\sim}{H} = \epsilon_0 \frac{\partial \underset{\sim}{E}}{\partial t} + \sum_i e_i \underset{\sim}{\Gamma}_i \tag{8.4}$$

where we have written $v_{Ti}^2 = \gamma_i e_i T_i / m_i$ with T_i in electron volts and γ_i, the gas constant, as in Equation 5.6. In these equations, all first-order field variables N_i, $\underset{\sim}{v}_i$, $\underset{\sim}{\Gamma}_i$, $\underset{\sim}{E}$, and $\underset{\sim}{H}$ are functions of Eulerian spatial coordinates and time. It should be noted from the derivation of Equation 8.1 that it applies to a plasma which in the unperturbed state is neutral and stationary. Hence the unperturbed density N_0 and temperature T are uniform, that is, independent of position.

8.1 Energy and Power

A conservation principle follows from dot-multiplying Equation 8.1 by $\underset{\sim}{\Gamma}$, Equation 8.2 by $(m_i v_{Ti}^2 / N_{0i}) N_i$, Equation 8.3 by $\underset{\sim}{H}$, Equation 8.4 by $-\underset{\sim}{E}$, and subsequently adding the thus modified equations:

$$\nabla \cdot \left(\underset{\sim}{E} \times \underset{\sim}{H} + \sum_i \frac{m_i v_{Ti}^2}{N_{0i}} N_i \underset{\sim}{\Gamma}_i \right)$$

$$= -\frac{\partial}{\partial t} \left(\tfrac{1}{2} \mu_0 H^2 + \tfrac{1}{2} \epsilon_0 E^2 + \sum_i \tfrac{1}{2} N_{0i} m_i v_i^2 + \sum_i \tfrac{1}{2} \frac{m_i v_{Ti}^2}{N_{0i}} N_i^2 \right) \tag{8.5}$$

Comparing Equation 8.5[26] with Equation 7.6, we note that one additional energy and one additional power term have entered in: These are the second term on the left-hand side and the last term on the right-hand side, both of which vanish for $v_{Ti} \to 0$, that is for the temperate plasma. These terms can be identified with the acoustic energy and power density of the gas. In fact, in the absence of the electromagnetic fields, Equation 8.5 is the conservation theorem for acoustic disturbances of the gas. We introduce the following abbreviations for the acoustic power flow out of a volume and the energy stored associated with the pressure:

$$S_p = \oint_S \sum_i \frac{m_i v_{Ti}^2}{N_{0i}} N_i \underset{\sim}{\Gamma}_i \cdot d\underset{\sim}{S} \tag{8.6}$$

$$W_p = \int_V \sum_i \frac{1}{2} \frac{m_i v_{Ti}^2}{N_{0i}} N_i^2 \, dV \tag{8.7}$$

The total power flow and energy stored are then written as

$$S = S_e + S_p \tag{8.8}$$

and

$$W = W_m + W_e + W_k + W_p \tag{8.9}$$

where S_e is the electromagnetic power flow as given by Equation 7.7, and the energy terms are as given by Equations 7.8 and 8.7. Equation 8.5 when integrated over the volume V gives

$$S = -\frac{d}{dt} W \tag{8.10}$$

Consider the system in its unperturbed state at time $t = 0$ when W and S are both zero; integrating Equation 8.10 over some time $t > 0$, we obtain

$$\int_0^t S_e \, dt = -\int_0^t S_p \, dt - W(t) \tag{8.11}$$

The energy W is a positive definite quantity. However, the first term on the right-hand side of Equation 8.11 involving the power flow in the presence of finite temperatures may be negative, indicating a net flow out of electromagnetic power.

If we Fourier analyze the time dependence of the perturbed field quantities, we have

$$\frac{m_i v_{Ti}^2}{N_{0i}} \nabla N_i = -j\omega m \underset{\sim}{v}_i + e_i \underset{\sim}{E}_i + e_i \underset{\sim}{v}_i \times \underset{\sim}{B}_0 \tag{8.12}$$

$$\nabla \cdot \underset{\sim}{\Gamma}_i = -j\omega N_i \tag{8.13}$$

$$\nabla \times \underset{\sim}{E} = -j\omega\mu_0 \underset{\sim}{H} \tag{8.14}$$

$$\nabla \times \underset{\sim}{H} = j\omega\epsilon_0 \underset{\sim}{E} + \sum_i e_i \underset{\sim}{\Gamma}_i \tag{8.15}$$

where all field variables are now the complex Fourier amplitudes, functions of spatial coordinates. The complex power conservation theorem is obtained by dot-multiplying Equation 8.12 by $\underset{\sim}{\Gamma}_i^*$, the complex conjugate of Equation 8.13 by $(m_i v_{Ti}^2/N_{0i})N_i$, Equation 8.14 by $\underset{\sim}{H}^*$, the complex conjugate of Equation 8.15 by $-\underset{\sim}{E}$, and adding the resultant equations:

$$\tfrac{1}{2}\nabla\cdot\left(\underset{\sim}{E}\times\underset{\sim}{H}^* + \sum_i \frac{m_i v_{Ti}^2}{N_{0i}} N_i \underset{\sim}{\Gamma}_i^*\right)$$

$$= -j2\omega\left[\tfrac{1}{4}\mu_0|\underset{\sim}{H}|^2 - \tfrac{1}{4}\epsilon_0|\underset{\sim}{E}|^2 + \sum_i \tfrac{1}{4}N_{0i}m_i|\underset{\sim}{v}_i|^2\right.$$

$$\left. -\sum_i \tfrac{1}{2}N_{0i}e_i\frac{(\underset{\sim}{v}_R \times \underset{\sim}{v}_I)_i}{\omega}\cdot\underset{\sim}{B}_0 - \sum_i \tfrac{1}{4}\frac{m_i v_{Ti}^2}{N_{0i}}|N_i|^2\right] \tag{8.16}$$

For a particular frequency component, the time-averaged power flow through a closed surface is from Equation 8.8:

$$\overline{S} = \overline{S}_e + \overline{S}_p$$

$$= \mathrm{Re}\oint_S \tfrac{1}{2}\underset{\sim}{E}\times\underset{\sim}{H}^*\cdot d\underset{\sim}{S} + \mathrm{Re}\oint_S \sum_i \tfrac{1}{2}\frac{m_i v_{Ti}^2}{N_{0i}} N_i \underset{\sim}{\Gamma}_i^* \cdot d\underset{\sim}{S} \tag{8.17}$$

Since the right-hand side of Equation 8.16 is pure imaginary, we also have

$$\overline{S}_e = -\overline{S}_p \tag{8.18}$$

The time-averaged energy in a volume follows from the time average of Equation 8.9:

$$\overline{W} = \overline{W}_m + \overline{W}_e + \overline{W}_k + \overline{W}_p \tag{8.19}$$

where

$$\overline{W}_p = \int_V \sum_i \tfrac{1}{4}\frac{m_i v_{Ti}^2}{N_{0i}}|N_i|^2\, dV \tag{8.20}$$

and $\overline{W}_m$, $\overline{W}_e$, and $\overline{W}_k$ are as given by Equations 7.18 through 7.20. We note again that the right-hand side of Equation 8.16 contains a term (the fourth one in the bracket) that does not appear in the time-averaged energy density. This is the same term that was discussed in connection with Equation 7.13, $\overline{w}_d$.

The resonance condition for a volume containing a warm plasma is obtained when the complex power flow through the surface enclosing the volume is pure real; from Equation 8.16, this resonance condition is

$$\overline{W}_m - \overline{W}_e + \overline{W}_k - \overline{W}_d - \overline{W}_p = 0 \tag{8.21}$$

where $\overline{W}_d$ is given by Equation 7.21.

8.2 Variation Theorem

Similar to the treatment in Section 7.3, we consider the variations of the field variables in Equations 8.12 through 8.15 with respect to frequency and applied magnetic field. We then find

$$\tfrac{1}{4}\nabla\cdot\left[\delta \underset{\sim}{E}\times \underset{\sim}{H}^* + \underset{\sim}{E}^*\times\delta\underset{\sim}{H} + \sum_i \delta\left(\frac{m_i v_{Ti}^2}{N_{0i}} N_i\right)\underset{\sim}{\Gamma}_i^* + \sum_i \frac{m_i v_{Ti}^2}{N_{0i}} N_i^*\,\delta\underset{\sim}{\Gamma}_i\right]$$

$$= -j\omega\left[\overline{w}\,\frac{\delta\omega}{\omega} + \sum_i \tfrac{1}{2} N_{0i} q_i \,\frac{(\underset{\sim}{v}_R\times\underset{\sim}{v}_I)_i}{\omega}\cdot\delta\underset{\sim}{B}_0\right] \tag{8.22}$$

where the time-averaged energy density $\overline{w}$ includes the acoustic term as given by the integrand of Equation 8.20.

The frequency variation of a volume enclosing the plasma with perfectly conducting walls (Figure 7.2), to which the plasma does not lose particles, is as given by Equation 7.39, where only the total time-averaged energy term in the denominator must be modified to include $\overline{W}_p$ of Equation 8.20, and where δN_{0i} may be set equal to zero.

Similarly, the reactance or susceptance variations at some reference plane in a waveguide coupled to the perfectly conducting enclosure (Figure 7.3) and propagating the dominant mode (Equations 7.40 through 7.44) is again as given by Equation 7.45, with δN_0 equal to zero. The reactance and susceptance functions at the reference plane follow from Equation 8.16 integrated over the perfectly conducting walls and the waveguide reference plane:

$$\left.\begin{array}{l} X|I|^2 \\ \\ B|V|^2 \end{array}\right\} = \pm 4\omega(\overline{W}_m - \overline{W}_e + \overline{W}_k - \overline{W}_d - \overline{W}_p) \tag{8.23}$$

Their frequency variations are similarly found from Equation 8.22:

$$\left.\begin{array}{l} \dfrac{\partial X}{\partial\omega}|I|^2 \\ \\ \dfrac{\partial B}{\partial\omega}|V|^2 \end{array}\right\} = 4\overline{W} = 4(\overline{W}_m + \overline{W}_e + \overline{W}_k + \overline{W}_p) \tag{8.24}$$

Combining Equations 8.23 and 8.24, we find

$$\frac{\dfrac{\partial X}{\partial \omega}}{\dfrac{|X|}{\omega}} = \frac{\dfrac{\partial B}{\partial \omega}}{\dfrac{|B|}{\omega}} = \frac{(\overline{W}_m + \overline{W}_k + \overline{W}_p + \overline{W}_e)}{|\overline{W}_m + \overline{W}_k - \overline{W}_p - \overline{W}_e - \overline{W}_d|} \tag{8.25}$$

Hence, if $\overline{W}_d < 0$ and $|\overline{W}_d| > (\overline{W}_e + \overline{W}_p)$, we have $\partial X/\partial\omega < |X|/\omega$ and $(\partial B/\partial\omega) < |B|/\omega$. In the absence of the applied magnetic field, $\overline{W}_d = 0$, the system is isotropic and Equation 8.25 is always greater than unity.

8.3 Free Waves

For free waves having a spatial dependence $\exp(-j\underset{\sim}{k}\cdot\underset{\sim}{r})$, Equations 8.12 through 8.15 become

$$\underset{\sim}{k}\,\frac{m_i v_{Ti}^2}{N_{0i}} N_i = \omega m_i \underset{\sim}{v}_i + j e_i \underset{\sim}{E} + j e_i \underset{\sim}{v}_i \times \underset{\sim}{B}_0 \tag{8.26}$$

$$\underset{\sim}{k}\cdot\underset{\sim}{\Gamma}_i = \omega N_i \tag{8.27}$$

$$\underset{\sim}{k}\times\underset{\sim}{E} = \omega\mu_0 \underset{\sim}{H} \tag{8.28}$$

$$\underset{\sim}{k}\times\underset{\sim}{H} = -\omega\epsilon_0\underset{\sim}{E} - j\sum_i e_i \underset{\sim}{\Gamma}_i \tag{8.29}$$

We shall first assume that $\underset{\sim}{k}$ is real. The complex power density in the direction of $\underset{\sim}{k}$ is

$$\underset{\sim}{k}\cdot\tfrac{1}{2}\left(\underset{\sim}{E}\times\underset{\sim}{H}^* + \sum_i \frac{m_i v_{Ti}^2}{N_{0i}} N_i \underset{\sim}{\Gamma}_i^*\right) = 2\omega\left[\overline{w}_m + \overline{w}_p\right] \tag{8.30}$$

and in a direction perpendicular to $\underset{\sim}{k}$

$$\underset{\sim}{k}\times\tfrac{1}{2}\left(\underset{\sim}{E}\times\underset{\sim}{H}^* + \sum_i \frac{m_i v_{Ti}^2}{N_{0i}} N_i \underset{\sim}{\Gamma}_i^*\right) = -\tfrac{1}{2}(\underset{\sim}{k}\cdot\underset{\sim}{E})\underset{\sim}{H}^* + \tfrac{1}{2}\sum_i \frac{m_i v_{Ti}^2}{N_{0i}} N_i \underset{\sim}{k}\times\underset{\sim}{\Gamma}_i^* \tag{8.31}$$

With Equations 8.30 and 8.31, the complex power density can be written as

$$\left(\underset{\sim}{E}\times\underset{\sim}{H}^* + \sum_i \frac{m_i v_{Ti}^2}{N_{0i}} N_i \underset{\sim}{\Gamma}_i^*\right) = \frac{2\omega(\overline{w}_m + \overline{w}_p)}{k^2}\underset{\sim}{k} + \frac{(\underset{\sim}{k}\cdot\underset{\sim}{E})}{2k^2}(\underset{\sim}{k}\times\underset{\sim}{H}^*) + \tfrac{1}{2}\sum_i \frac{m_i v_{Ti}^2}{N_{0i}} \underset{\sim}{\Gamma}_{\perp i} \tag{8.32}$$

where $\underset{\sim}{\Gamma}_\perp$ is the particle current density in a direction perpendicular to $\underset{\sim}{k}$. The complex power density in the direction of $\underset{\sim}{k}$ is real as in the temperate plasma. However, the complex power density in a direction perpendicular to $\underset{\sim}{k}$ is modified by the current flow, as shown in Equation 8.31.

The phase velocity is related by Equation 8.30 to the time-averaged power density in the direction of $\underset{\sim}{k}$:

$$\frac{\underset{\sim}{k}\cdot\frac{1}{2}\,\mathrm{Re}\left(\underset{\sim}{E}\times\underset{\sim}{H}^{*}+\sum_{i}\frac{m_i v_{Ti}^{2}}{N_{0i}}N_i\underset{\sim}{\Gamma}_i^{*}\right)}{k2(\overline{w}_m+\overline{w}_p)}=\frac{\omega}{k}=u \qquad (8.33)$$

The group velocity can be found as in Equations 7.66 through 7.69; however, the dispersion relation, Equation 7.68, is now much more complicated because $\underset{\approx}{K}$ is now a function of $\underset{\sim}{k}$ (Equations 5.13 through 5.15); that is, the plasma is now space-dispersive.

We can extend our considerations to complex waves as in Section 7.5. If we represent the warm plasma by the tensor $\underset{\approx}{K}$ of Chapter 5, Maxwell's equations have again the form of Equations 7.78 through 7.81. The properties of the waves, Equations 7.87 through 7.90 and 7.95 through 7.98, follow as before. However, the interpretation of these results is now different since $\underset{\approx}{K}$ is a function of both ω and γ.

Thus for pure propagating waves, $\underset{\sim}{\alpha}=0$ and $\underset{\sim}{\gamma}=j\underset{\sim}{\beta}=j\underset{\sim}{k}$, Equation 7.97 becomes

$$\delta\underset{\sim}{k}\cdot\left[\underset{\sim}{p}_e-\tfrac{1}{4}\epsilon_0\underset{\sim}{E}^{*}\cdot\frac{\partial(\omega\underset{\approx}{K})}{\partial\underset{\sim}{k}}\cdot\underset{\sim}{E}\right]=\delta\omega\left[\tfrac{1}{4}\mu_0|\underset{\sim}{H}|^{2}+\tfrac{1}{4}\epsilon_0\underset{\sim}{E}^{*}\cdot\frac{\partial(\omega\underset{\approx}{K})}{\partial\omega}\cdot\underset{\sim}{E}\right] \qquad (8.34)$$

In the brackets on the left-hand side, the first term, $\underset{\sim}{p}_e$, is the time-averaged electromagnetic power density vector (Equation 7.74); the second term is the power density vector associated with the medium, in our case what we have called the acoustic power density vector. The term in brackets on the right-hand side is the time-averaged energy density. For a fixed frequency,

$$\delta\underset{\sim}{k}\cdot(\underset{\sim}{p}_e+\underset{\sim}{p}_M)=0 \qquad (8.35)$$

where we have written

$$\underset{\sim}{p}_M=-\tfrac{1}{4}\epsilon_0\underset{\sim}{E}^{*}\cdot\frac{\partial(\omega\underset{\approx}{K})}{\partial\underset{\sim}{k}}\cdot\underset{\sim}{E} \qquad (8.36)$$

Hence, the total power density vector, electromagnetic and acoustic in this case, is perpendicular to the index surface (Figure 8.1). The group velocity vector follows from Equation 8.34:

$$\frac{\partial\omega}{\partial\underset{\sim}{k}}=\frac{\underset{\sim}{p}_e+\underset{\sim}{p}_M}{\overline{w}} \qquad (8.37)$$

(see also Section 8.5) where we have written

$$\overline{w} = \tfrac{1}{4}\mu_0|\underset{\sim}{H}|^2 + \tfrac{1}{4}\epsilon_0\underset{\sim}{E}^* \cdot \frac{\partial(\omega \underset{\approx}{K})}{\partial\omega} \cdot \underset{\sim}{E} \tag{8.38}$$

Equation 8.37 is clearly also the energy velocity.

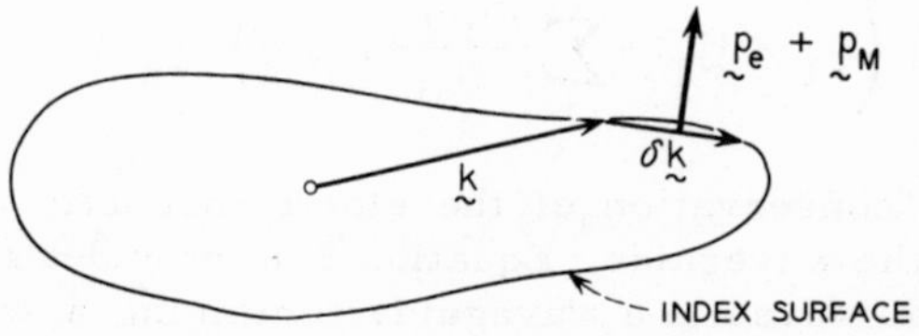

Figure 8.1. Total time-averaged power density vector (electromagnetic and medium) perpendicular to the index surface (Equation 8.35).

8.4 Guided Waves

For plasma systems of finite extent and uniform in one spatial direction, for example, the z direction, it is convenient to separate the equations as we have done in Section 7.6 for the temperate plasma, Equations 7.102 through 7.107. For the warm plasma, the only equation that is changed is the force equation, and its longitudinal and transverse components are

$$\frac{\partial}{\partial z}\left(\frac{m_i v_{Ti}^2}{N_{0i}} N_i\right) = -j\omega m_i v_{zi} + e_i E_z + e_i(\underset{\sim}{v}_i \times \underset{\approx}{B}_0) \cdot \underset{\sim}{i}_z \tag{8.39}$$

$$0 = -\underset{\sim}{i}_z \times \nabla_T\left(\frac{m_i v_{Ti}^2}{N_{0i}} N_i\right) - j\omega m_i \underset{\sim}{i}_z \times \underset{\sim}{v}_{Ti} + e_i \underset{\sim}{i}_z \times \underset{\sim}{E}_T + \underset{\sim}{i}_z \times e_i(\underset{\sim}{v}_i \times \underset{\approx}{B}_0)_T \tag{8.40}$$

We now introduce the complex power flow through the waveguide cross section

$$P + jQ = \int_A \tfrac{1}{2}\left(\underset{\sim}{E} \times \underset{\sim}{H}^* + \sum_i \frac{m_i v_{Ti}^2}{N_{0i}} N_i \underset{\sim}{\Gamma}_i^*\right) \cdot \underset{\sim}{i}_z \, da \tag{8.41}$$

and the real quantity

$$U_p = \int_A \sum_i \tfrac{1}{4} \frac{m_i v_{Ti}^2}{N_{0i}} |N_i|^2 \, da \tag{8.42}$$

Combining (see operations preceding Equation 7.120) Equations 7.104 through 7.107, 8.39, and 8.40, and integrating over the waveguide cross section, we obtain

$$\frac{\partial}{\partial z}(P + jQ) = -j2\omega(U_m - U_e + U_k - U_p - U_d) \tag{8.43}$$

where all quantities are as defined by Equations 7.113 through 7.116, 8.41, and 8.42. The real part of Equation 8.43,

$$\frac{\partial}{\partial z} \operatorname{Re} \int_A \frac{1}{2} \left(\underset{\sim}{E}_T \times \underset{\sim}{H}_T^* + \sum_i \frac{m_i v_{Ti}^2}{N_{0i}} N_i \underset{\sim}{\Gamma}_i^* \right) \cdot \underset{\sim}{i}_z \; da = 0 \tag{8.44}$$

expresses the conservation of the electromagnetic and acoustic power flow in the waveguide. Equation 8.44 provides a basis for the orthogonality of waves in a waveguide containing a warm plasma.

For complex waves with z dependence $\exp(-\gamma z)$, where $\gamma = \alpha + j\beta$ Equations 7.104 through 7.107, 8.39, and 8.40 can be used to derive the relationships between the propagation constant and the complex power flow. By techniques analogous to those used in Section 7.6, we find

$$\gamma(P + jQ) = j2\omega\,(U_{mT} - U_{ez} + U_{kz} - 2\omega U_c) \tag{8.45}$$

$$\gamma^*(P + jQ) = j2\omega(U_{mz} - U_{eT} + U_{kT} - U_p) + 2\omega\, U_c^* - j2\omega U_{dz} \tag{8.46}$$

From Equations 8.45 and 8.46, we find

$$\alpha P = 0 \tag{8.47}$$

$$\alpha Q = \omega(U_m - U_e + U_k - U_p - U_d) \tag{8.48}$$

$$\beta P = \omega(U_{mT} - U_{mz} + U_{eT} - U_{ez} + U_{kz} - U_{kT} + U_p + U_{dz}) \tag{8.49}$$

$$\beta Q = 2\omega \operatorname{Re} U_c \tag{8.50}$$

where all the terms are as defined in Section 7.6 and by Equations 8.41 and 8.42. We can now discuss the properties of the various possible waves.

(a) <u>Complex wave</u>: $\gamma = \alpha + j\beta$

For such a wave, both α and β are finite, and hence Equations 8.47 through 8.50 give

$$P = 0 \tag{8.51}$$

$$U_{mT} - U_{mz} + U_{eT} - U_{ez} + U_{kz} - U_{kT} + U_p + U_{dz} = 0 \tag{8.52}$$

$$\frac{\omega}{\alpha} = \frac{Q}{U_m - U_e + U_k - U_p - U_d} \tag{8.53}$$

$$\frac{\omega}{\beta} = \frac{Q}{2 \operatorname{Re} U_c} \tag{8.54}$$

Equation 8.51 implies that

$$\mathrm{Re} \int_A \frac{1}{2} \left(\underset{\sim}{E} \times \underset{\sim}{H}^* + \sum_i \frac{m_i v_{Ti}^2}{N_{0i}} N_i \underset{\sim}{\Gamma}_i^* \right) \cdot \underset{\sim}{i}_z \, da = 0 \qquad (8.55a)$$

$$P_e + P_p = 0 \qquad (8.55b)$$

Hence, the exponential behavior of P_e and P_p must be identical and of opposite sign. If $\underset{\approx}{B}_0$ is entirely longitudinal, the reactive power vanishes and Equations 8.53 and 8.54 become indeterminate.

(b) Cutoff wave: $\beta = 0; \quad \gamma = \alpha$

For this case we obtain from Equations 8.47 through 8.50

$$P = 0; \quad P_e + P_p = 0 \qquad (8.56)$$

$$\mathrm{Re}\, U_c = 0 \qquad (8.57)$$

$$U_{mT} - U_{mz} + U_{eT} - U_{ez} + U_{kz} - U_{kT} + U_p + U_{dz} = 0 \qquad (8.58)$$

$$\frac{\omega}{\alpha} = \frac{Q}{U_m - U_e + U_k - U_p - U_d} \qquad (8.59)$$

Here the reactive power is finite. The total real power is balanced as in Equation 8.55a.

(c) Propagating waves: $\gamma = j\beta; \quad \alpha = 0$

Equations 8.47 through 8.50 give

$$U_m - U_e + U_k - U_p - U_d = 0 \qquad (8.60)$$

$$u = \frac{\omega}{\beta} = \frac{P_e + P_p}{U_{mT} - U_{mz} + U_{eT} - U_{ez} + U_{kz} - U_{kT} + U_p + U_{dz}}$$

$$= \frac{Q}{2\, \mathrm{Re}\, U_c} \qquad (8.61)$$

Equation 8.60 is the balance condition for the propagating wave and also follows from Equation 8.43. Equation 8.61 relates the phase velocity of the wave to the time-averaged power flow. In general, the propagating wave will carry both real and reactive power. As in the case of a temperate plasma, the reactive power vanishes when the magnetic field $\underset{\approx}{B}_0$ is entirely longitudinal.

The variation theorem, Equation 8.22, can be applied to a propagating wave in the waveguide. Integrating Equation 8.22 over the waveguide cross section bounded by the perfectly conducting contour, and assuming that the plasma particle current to the walls is zero, we obtain

$$\delta\beta\ P = U\ \delta\omega + \omega \sum_i \int_A \tfrac{1}{2} N_{0i} e_i \frac{(\underset{\sim}{v}_R \times \underset{\sim}{v}_I)_i}{\omega} \cdot \delta \underset{\approx}{B}_0\ da \tag{8.62}$$

where P is the time-averaged power flow, electromagnetic and acoustic, as given by the real part of Equation 8.41, and U is the time-averaged energy per unit length:

$$U = U_m + U_e + U_k + U_p \tag{8.63}$$

When variations in only frequency are considered, Equation 8.62 gives the group velocity

$$\frac{\partial\omega}{\partial\beta} = \frac{P}{U} = \frac{P_e + P_p}{U_m + U_e + U_k + U_p} \tag{8.64}$$

which can also be identified as the energy velocity in the waveguide. Finally, combining Equations 8.61 and 8.64, we obtain

$$\frac{\partial\omega}{\partial\beta} = \frac{\omega}{\beta} \frac{(U_{mT} - U_{mz} + U_{eT} - U_{ez} + U_{kz} - U_{kT} + U_p + U_{dz})}{U_m + U_e + U_k + U_p} \tag{8.65}$$

which relates the group and phase velocity of the wave. The possibility of backward waves can also arise in this case as in the temperate case; the conditions are, however, different. (Compare Equations 8.65 and 7.153.)

8.5 General Properties of Dispersive Media

We now generalize our considerations of energy and power so that they apply to a general linear description of the medium, as for example the description in Chapter 6 based on the linearized Boltzmann equation. Maxwell's equations for the macroscopic electric and magnetic fields in the presence of material media may be written as

$$\nabla \times \underset{\approx}{E} + \mu_0 \frac{\partial \underset{\approx}{H}}{\partial t} = -\underset{\sim}{J}_m \tag{8.66}$$

$$\nabla \times \underset{\approx}{H} - \epsilon_0 \frac{\partial \underset{\approx}{E}}{\partial t} = \underset{\sim}{J}_e \tag{8.67}$$

Poynting's theorem then takes on the form

$$-\nabla \cdot (\underset{\approx}{E} \times \underset{\approx}{H}) - \frac{\partial}{\partial t}(\tfrac{1}{2}\mu_0 H^2 + \tfrac{1}{2}\epsilon_0 E^2) = \underset{\approx}{H} \cdot \underset{\sim}{J}_m + \underset{\approx}{E} \cdot \underset{\sim}{J}_e \tag{8.68}$$

In these equations the presence of the medium is represented by the vectors $\underset{\sim}{J}_m$ and $\underset{\sim}{J}_e$. In general, $\underset{\sim}{J}_m$ and $\underset{\sim}{J}_e$ are nonlinear functions of both $\underset{\sim}{E}$ and $\underset{\sim}{H}$. In a linearized description of the medium, only the first-order fields enter into Equations 8.66 and 8.67, and $\underset{\sim}{J}_m$ and $\underset{\sim}{J}_e$ are linear functions of the first-order fields. We shall introduce some restrictions on the linear description that will focus attention on the particular gaseous plasma medium of interest here. These restrictions can be easily removed so that the following development can be generalized to an arbitrary linear description of a medium.[27]

Assume that for the linearized description of the plasma medium of interest (described by the linearized Boltzmann equation) $\underset{\sim}{J}_m = 0$, and $\underset{\sim}{J}_e$ is a linear function of $\underset{\sim}{E}$ only. Further assume that the unperturbed parameters of the medium are time-independent and homogeneous. Then the most general linear anisotropic relation between $\underset{\sim}{J}_e$ and $\underset{\sim}{E}$ is of the form

$$\underset{\sim}{J}_e(\underset{\sim}{r}, t) = \iint \underset{\approx}{\sigma}(\underset{\sim}{r} - \underset{\sim}{r}'; t - t') \cdot \underset{\sim}{E}(\underset{\sim}{r}', t')\, d^3r'\, dt' \qquad (8.69)$$

where the convolution kernel $\underset{\approx}{\sigma}$ is the Green's function characteristic of the integrodifferential equation relating $\underset{\sim}{J}_e$ to $\underset{\sim}{E}$. Both the time dependence and space dependence of the fields can be Fourier analyzed. Consider components with time dependence $e^{j\omega t}$ and space dependence $\exp(-\underset{\sim}{\gamma} \cdot \underset{\sim}{r})$, where $\underset{\sim}{\gamma} = \underset{\sim}{\alpha} + j\underset{\sim}{\beta}$ is in general complex to allow for fields that decay or grow. In terms of Fourier amplitudes, Equation 8.69 becomes

$$\underset{\sim}{J}_e(\underset{\sim}{\gamma}, \omega) = \underset{\approx}{\sigma}(\underset{\sim}{\gamma}, \omega) \cdot \underset{\sim}{E}(\underset{\sim}{\gamma}, \omega) \qquad (8.70)$$

The time average of Equation 8.68 (with $\underset{\sim}{J}_m = 0$) can be expressed in terms of the Fourier amplitudes and Equation 8.70:

$$2\underset{\sim}{\alpha} \cdot (\mathrm{Re}\ \tfrac{1}{2} \underset{\sim}{E} \times \underset{\sim}{H}^*) = \mathrm{Re}\ \tfrac{1}{2} \underset{\sim}{E} \cdot \underset{\sim}{J}_e^*$$

$$= \tfrac{1}{2} \underset{\sim}{E}^* \cdot \underset{\approx}{\sigma}_h \cdot \underset{\sim}{E} \qquad (8.71)$$

where $\underset{\approx}{\sigma}_h$ is the Hermitian part of $\underset{\approx}{\sigma}(\underset{\sim}{\gamma}, \omega)$. If $\underset{\approx}{\sigma}_h$ is positive definite, the right-hand side of Equation 8.71 represents the time-averaged power-density loss, and the medium is passive. If $\underset{\approx}{\sigma}_h$ is negative definite, the medium is active in the sense that time-averaged electromagnetic power is being generated. If $\underset{\approx}{\sigma}_h = 0$, the medium is "lossless." However, it may still be potentially either active or passive (or even unstable), depending upon its energy state.

The energy and power in a lossless medium may be determined as follows. In the absence of loss, the Fourier components of the fields may be chosen with time dependence $e^{j\omega t}$ and space dependence $\exp(-j\underset{\sim}{\beta} \cdot \underset{\sim}{r})$, and the medium tensor is $\underset{\approx}{\sigma}(\underset{\sim}{\beta}, \omega)$. Consider now a very slightly "lossy" (either active or passive) medium. In general, the time and space dependence of the Fourier components of the fields will now be $\exp st$, with $s = \nu + j\omega$, and $\exp(-\underset{\sim}{\gamma} \cdot \underset{\sim}{r})$, with $\underset{\sim}{\gamma} = \underset{\sim}{\alpha} + j\underset{\sim}{\beta}$, where

$$|\nu| << |\omega| \tag{8.72}$$

and

$$|\alpha| << |\beta| \tag{8.73}$$

Causality, when applied to the relation of Equation 8.69, allows us to continue analytically $\underset{\approx}{\sigma}(\underset{\sim}{\gamma}, \omega)$ into the complex s plane, giving $\underset{\approx}{\sigma}(\underset{\sim}{\gamma}, s)$. Maxwell's equations for the Fourier amplitudes become

$$\underset{\sim}{\gamma} \times \underset{\sim}{E} - s\mu_0 \underset{\sim}{H} = 0 \tag{8.74}$$

$$\underset{\sim}{\gamma} \times \underset{\sim}{H} + s\epsilon_0 \underset{\sim}{E} = -\underset{\approx}{\sigma} \cdot \underset{\sim}{E} \tag{8.75}$$

(where again only for simplicity we have chosen to continue with the assumption $\underset{\sim}{J}_m = 0$). In the limit of Equation 8.72, $\nu \to 0$, the time-averaged electromagnetic power flow can be identified with $2\underset{\sim}{\alpha} \cdot (\text{Re}\,\frac{1}{2}\underset{\sim}{E} \times \underset{\sim}{H}^*) \equiv 2\underset{\sim}{\alpha} \cdot \underset{\sim}{p}_e$ as in Equation 8.71. Dot-multiplying Equation 8.74 by $\underset{\sim}{H}^*$, the complex conjugate of Equation 8.75 by $-\underset{\sim}{E}$, adding the resultant equations, and taking the real part, we obtain

$$2\underset{\sim}{\alpha} \cdot \underset{\sim}{p}_e - 2\nu(\tfrac{1}{4}\mu_0 |\underset{\sim}{H}|^2 + \tfrac{1}{4}\epsilon_0 |\underset{\sim}{E}|^2) = \tfrac{1}{2}\underset{\sim}{E}^* \cdot \underset{\approx}{\sigma}_h(\underset{\sim}{\gamma}, s) \cdot \underset{\sim}{E} \tag{8.76}$$

Under our assertion of very small "loss" and Equations 8.72 and 8.73, we now evaluate the right-hand side of Equation 8.76 to first order in both ν and α. Thus

$$\underset{\approx}{\sigma}(\underset{\sim}{\gamma}, s) \approx \underset{\approx}{\sigma}(\underset{\sim}{\beta}, \omega) + \left(\frac{\partial \underset{\approx}{\sigma}}{\partial \omega}\right)(-j\nu) + \left(\frac{\partial \underset{\approx}{\sigma}}{\partial \underset{\sim}{\beta}}\right) \cdot (-j\underset{\sim}{\alpha}) \tag{8.77}$$

and $\underset{\approx}{\sigma}_h = (\underset{\approx}{\sigma} + \underset{\approx}{\sigma}^+)/2$. Substituting $\underset{\approx}{\sigma}_h$ from Equation 8.77 into Equation 8.76 and rearranging, we find

$$2\underset{\sim}{\alpha} \cdot (\underset{\sim}{p}_e + \underset{\sim}{p}_M) = 2\nu\left[\tfrac{1}{4}\mu_0 |\underset{\sim}{H}|^2 + \tfrac{1}{4}\epsilon_0 |\underset{\sim}{E}|^2 + \overline{w}_M\right] + p_d \tag{8.78}$$

where

$$\underset{\sim}{p}_M = -\tfrac{1}{4}\underset{\sim}{E}^* \cdot \frac{\partial(-j\underset{\approx}{\sigma}_a)}{\partial \underset{\sim}{\beta}} \cdot \underset{\sim}{E} \tag{8.79}$$

$$\overline{w}_M = \tfrac{1}{4}\underset{\sim}{E}^* \cdot \frac{\partial(-j\underset{\approx}{\sigma}_a)}{\partial \omega} \cdot \underset{\sim}{E} \tag{8.80}$$

$$p_d = \tfrac{1}{2}\underset{\sim}{E}^* \cdot \underset{\approx}{\sigma}_h \cdot \underset{\sim}{E} \tag{8.81}$$

and $\underset{\approx}{\sigma}_a$ is the anti-Hermitian part of $\underset{\approx}{\sigma}$. Equation 8.78 expresses the conservation of energy to first order in the quantities $\underset{\approx}{\sigma}_h$, ν, and α. The last term in Equation 8.78 is readily identified with the time-averaged power density in the small "loss" $\underset{\approx}{\sigma}_h$. It also follows that $\overline{w}_M$ and $\underset{\sim}{p}_M$ must be associated with the time-averaged energy density and power density, respectively, of the medium.† Using Equation 1.9, we can rewrite Equations 8.79 through 8.81 in terms of the dielectric coefficient tensor $\underset{\approx}{K}$:

$$\underset{\sim}{p}_M = -\tfrac{1}{4}\epsilon_0 \underset{\sim}{E}^* \cdot \frac{\partial(\omega \underset{\approx}{K}_h)}{\partial \underset{\sim}{\beta}} \cdot \underset{\sim}{E} \tag{8.82}$$

$$\tfrac{1}{4}\epsilon_0 |\underset{\sim}{E}|^2 + \overline{w}_M = \tfrac{1}{4}\epsilon_0 \underset{\sim}{E}^* \cdot \frac{\partial(\omega \underset{\approx}{K}_h)}{\partial \omega} \cdot \underset{\sim}{E} \tag{8.83}$$

$$p_d = \tfrac{1}{2}\omega\epsilon_0 \underset{\sim}{E}^* \cdot (j\underset{\approx}{K}_a) \cdot \underset{\sim}{E} \tag{8.84}$$

where, as before, p_d and $\underset{\approx}{K}_a$ are small.

We now consider the equations for a dispersive plasma with loss. We split $\underset{\approx}{K}$ into a Hermitian and an anti-Hermitian part:

$$\underset{\approx}{K} = \underset{\approx}{K}_h + \underset{\approx}{K}_a \tag{8.85}$$

Maxwell's equations are as before

$$\underset{\sim}{\gamma} \times \underset{\sim}{E} = j\omega\mu_0 \underset{\sim}{H} \tag{8.86}$$

$$\underset{\sim}{\gamma} \times \underset{\sim}{H} = -j\omega\epsilon_0 \underset{\approx}{K} \cdot \underset{\sim}{E} \tag{8.87}$$

where we have taken Fourier components $\exp(j\omega t)$ and $\exp(-\underset{\sim}{\gamma} \cdot \underset{\sim}{r})$ with $\underset{\sim}{\gamma}$ complex. Using Equations 8.86 and 8.87, we form $\underset{\sim}{\gamma} \cdot \underset{\sim}{E} \times \underset{\sim}{H}^*$ and $\underset{\sim}{\gamma}^* \cdot \underset{\sim}{E} \times \underset{\sim}{H}^*$ and take the real and imaginary parts of the sum and difference of these expressions

$$2\underset{\sim}{\alpha} \cdot \underset{\sim}{p}_e = \tfrac{1}{2}\underset{\sim}{E}^* \cdot (j\omega\epsilon_0 \underset{\approx}{K}_a) \cdot \underset{\sim}{E} \tag{8.88}$$

$$\underset{\sim}{\alpha} \cdot \underset{\sim}{q}_e = \omega(\tfrac{1}{4}\mu_0 |\underset{\sim}{H}|^2 - \tfrac{1}{4}\epsilon_0 \underset{\sim}{E}^* \cdot \underset{\approx}{K}_h \cdot \underset{\sim}{E}) \tag{8.89}$$

$$\underset{\sim}{\beta} \cdot \underset{\sim}{p}_e = \omega(\tfrac{1}{4}\mu_0 |\underset{\sim}{H}|^2 + \tfrac{1}{4}\epsilon_0 \underset{\sim}{E}^* \cdot \underset{\approx}{K}_h \cdot \underset{\sim}{E}) \tag{8.90}$$

$$2\underset{\sim}{\beta} \cdot \underset{\sim}{q}_e = \tfrac{1}{2}\underset{\sim}{E}^* \cdot (j\omega\epsilon_0 \underset{\approx}{K}_a) \cdot \underset{\sim}{E} \tag{8.91}$$

The presence of finite loss $\underset{\approx}{K}_a$ gives rise to components of real electromagnetic power in the direction of $\underset{\sim}{\alpha}$, Equation 8.88, and of reactive electromagnetic power in the direction $\underset{\sim}{\beta}$, Equation 8.91. The right-hand side of Equations 8.88 and 8.91 is the time-averaged power density dissipated as in Equations 8.71 and 8.84. Using the

† Note added in proof: The author recently learned of an independent identification of $\underset{\sim}{p}_M$, by a different technique, by M. E. Gertsenstein, J. Exptl. Theoret. Phys. (U.S.S.R.), 26, 680 (1954).

variation of Equations 8.86 and 8.87 to form $\delta \underset{\sim}{\gamma} \cdot \underset{\sim}{E} \times \underset{\sim}{H}^*$ and $\delta \underset{\sim}{\gamma}^* \cdot \underset{\sim}{E} \times \underset{\sim}{H}^*$, we obtain the following:

$$\delta(\underset{\sim}{\alpha} \cdot \underset{\sim}{p}_e) = \tfrac{1}{4}\underset{\sim}{E}^* \cdot \delta(j\omega\epsilon_0 \underset{\approx}{K}_a) \cdot \underset{\sim}{E} - \tfrac{1}{2}\,\mathrm{Im}\,(\delta\underset{\sim}{E}^* \cdot \omega\epsilon_0 \underset{\approx}{K}_a \cdot \underset{\sim}{E}) \tag{8.92}$$

$$\delta\underset{\sim}{\alpha} \cdot \underset{\sim}{q}_e = \tfrac{1}{4}\delta(\omega\mu_0)|\underset{\sim}{H}|^2 - \tfrac{1}{4}\epsilon_0\underset{\sim}{E}^* \cdot \delta(\omega\underset{\approx}{K}_h) \cdot \underset{\sim}{E}$$
$$- \underset{\sim}{\beta} \cdot \tfrac{1}{2}\,\mathrm{Re}\,(\delta\underset{\sim}{E} \times \underset{\sim}{H}^* - \underset{\sim}{E} \times \delta\underset{\sim}{H}^*) + \tfrac{1}{2}\,\mathrm{Re}\,(\delta\underset{\sim}{E}^* \cdot \omega\epsilon_0\underset{\approx}{K}_a \cdot \underset{\sim}{E}) \tag{8.93}$$

$$\delta\underset{\sim}{\beta} \cdot \underset{\sim}{p}_e = \tfrac{1}{4}\delta(\omega\mu_0)|\underset{\sim}{H}|^2 + \tfrac{1}{4}\epsilon_0\underset{\sim}{E}^* \cdot \delta(\omega\underset{\approx}{K}_h) \cdot \underset{\sim}{E}$$
$$- \underset{\sim}{\alpha} \cdot \tfrac{1}{2}\,\mathrm{Im}\,(\delta\underset{\sim}{E} \times \underset{\sim}{H}^* - \underset{\sim}{E} \times \delta\underset{\sim}{H}^*) - \tfrac{1}{2}\,\mathrm{Re}\,(\delta\underset{\sim}{E}^* \cdot \omega\epsilon_0\underset{\approx}{K}_a \cdot \underset{\sim}{E}) \tag{8.94}$$

$$\delta(\underset{\sim}{\beta} \cdot \underset{\sim}{q}_e) = \tfrac{1}{4}\underset{\sim}{E}^* \cdot \delta(j\omega\epsilon_0\underset{\approx}{K}_a) \cdot \underset{\sim}{E} - \tfrac{1}{2}\,\mathrm{Im}\,(\delta\underset{\sim}{E}^* \cdot \underset{\approx}{K}_a \cdot \underset{\sim}{E}) \tag{8.95}$$

Consider now that the variation is caused by a small loss in the medium so that $\underset{\approx}{K}_a$ and $\underset{\sim}{\alpha}$ are of first order. All terms of second order in Equations 8.92 through 8.95 can be neglected. In particular, the third and fourth terms on the right-hand side of Equation 8.94 can be disregarded, and we obtain

$$\delta\underset{\sim}{\beta} \cdot \underset{\sim}{p}_e = \tfrac{1}{4}\delta(\omega\mu_0)|\underset{\sim}{H}|^2 + \tfrac{1}{4}\epsilon_0\underset{\sim}{E}^* \cdot \delta(\omega\underset{\approx}{K}_h) \cdot \underset{\sim}{E} \tag{8.96}$$

where

$$\delta(\omega\underset{\approx}{K}_h) = \frac{\partial(\omega\underset{\approx}{K}_h)}{\partial\omega}\,\delta\omega + \frac{\partial(\omega\underset{\approx}{K}_h)}{\partial\underset{\sim}{\beta}} \cdot \delta\underset{\sim}{\beta} \tag{8.97}$$

Hence

$$\delta\underset{\sim}{\beta} \cdot \left(\underset{\sim}{p}_e - \tfrac{1}{4}\epsilon_0\underset{\sim}{E}^* \cdot \frac{\partial(\omega\underset{\approx}{K}_h)}{\partial\underset{\sim}{\beta}} \cdot \underset{\sim}{E}\right)$$
$$= \delta\omega\left(\tfrac{1}{4}\mu_0|\underset{\sim}{H}|^2 + \tfrac{1}{4}\epsilon_0\underset{\sim}{E}^* \cdot \frac{\partial(\omega\underset{\approx}{K}_h)}{\partial\omega} \cdot \underset{\sim}{E}\right) \tag{8.98}$$

or

$$\delta\underset{\sim}{\beta} \cdot (\underset{\sim}{p}_e + \underset{\sim}{p}_M) = \delta\omega\,(\overline{w}_m + \overline{w}_e + \overline{w}_M) \tag{8.99}$$

where $\underset{\sim}{p}_M$ and $\overline{w}_M$ are as given previously by Equations 8.82 and 8.83. From Equation 8.99, which is exact for a lossless medium, we obtain at a fixed frequency, $\delta\omega = 0$,

$$\delta\underset{\sim}{\beta} \cdot (\underset{\sim}{p}_e + \underset{\sim}{p}_M) = 0 \tag{8.100}$$

and for $\delta\omega$ finite

$$\frac{\partial\omega}{\partial\underset{\sim}{\beta}} = \frac{\underset{\sim}{p}_e + \underset{\sim}{p}_M}{\overline{w}_m + \overline{w}_e + \overline{w}_M} \tag{8.101}$$

Equation 8.100 shows that the total time-averaged power density, electromagnetic and medium, is perpendicular to the index surface. Equation 8.101 shows the group velocity equal to the energy velocity, where the latter must be taken as the ratio of total time-averaged power density (electromagnetic and medium) to the total time-averaged energy density (electromagnetic and medium).

Finally, when the loss is not small, it is convenient to introduce the fields of an adjoint system. The adjoint system is characterized by fields $\underset{\sim}{E}^+$ and $\underset{\sim}{H}^+$ that satisfy Maxwell's equations with an adjoint dielectric tensor $\underset{\approx}{K}^+(\omega, \underset{\sim}{\gamma})$, which is the Hermitian-transpose of $\underset{\approx}{K}(\omega, \underset{\sim}{\gamma})$, and with propagation constant equal to $-\underset{\sim}{\gamma}^*$. Using the variation of Equations 8.86 and 8.87 and the corresponding variation of Maxwell's equations for the adjoint system, we find

$$\delta\underset{\sim}{\gamma} \cdot (\underset{\sim}{E} \times \underset{\sim}{H}^+ + \underset{\sim}{E}^+ \times \underset{\sim}{H}) = j\delta(\omega\mu_0)\, \underset{\sim}{H} \cdot \underset{\sim}{H}^+ + j\, \underset{\sim}{E}^+ \cdot \delta(\omega\epsilon_0 \underset{\approx}{K}) \cdot \underset{\sim}{E} \tag{8.102}$$

and other relations analogous to Equations 7.95 through 7.98 that do not contain the additional terms with field variations as in Equations 8.92 through 8.95. Equation 8.102 can be rearranged in the form

$$\delta\underset{\sim}{\gamma} \cdot \left[\tfrac{1}{4}(\underset{\sim}{E} \times \underset{\sim}{H}^+ + \underset{\sim}{E}^+ \times \underset{\sim}{H}) - \tfrac{1}{4}\epsilon_0 j \underset{\sim}{E}^+ \cdot \frac{\partial(\omega\underset{\approx}{K})}{\partial\underset{\sim}{\gamma}} \cdot \underset{\sim}{E}\right]$$

$$= j\delta\omega\left(\tfrac{1}{4}\mu_0\, \underset{\sim}{H} \cdot \underset{\sim}{H}^+ + \tfrac{1}{4}\epsilon_0 \underset{\sim}{E}^+ \cdot \frac{\partial(\omega\underset{\approx}{K})}{\partial\underset{\sim}{\gamma}} \cdot \underset{\sim}{E}\right) \tag{8.103}$$

where the terms inside the bracket on the left-hand side may be regarded as complex cross-power densities (electromagnetic and medium) and the terms inside the parentheses on the right-hand side as complex cross-energy densities. The solutions of the adjoint system and its properties, Equations 8.102 and 8.103, can be used to establish variational principles for the complex propagation constant.

Chapter 9

FIELD ANALYSIS OF PLASMA WAVEGUIDES

The waveguide systems that we shall consider are illustrated in Figure 9.1a and b. They consist of a plasma inside an enclosure whose walls we shall assume to be perfectly conducting. These systems are assumed to be uniform in the z direction, and situated in a constant magnetic field whose direction is either coincident with the z direction or at right angles to the z direction. In the first case (a longitudinally magnetized plasma waveguide, from here on abbreviated LMG), the cross-sectional geometry of the system will be left arbitrary (Figure 9.1a). In the second case (a transversely magnetized plasma waveguide, from here on abbreviated TMG), the cross section will be of a simple rectangular or circular form, and the applied magnetic field is in the direction of one of the transverse orthogonal coordinates (Figure 9.1b).

Throughout this chapter we shall assume that the plasma is a temperate one, so that it is describable by a dielectric tensor as, for example, given by Equation 2.13 for the LMG system. In general, we shall allow the properties of the medium to vary in a plane transverse to the z direction. For example, the unperturbed parameters of the plasma (such as the plasma frequency, or the collision frequency, or the cyclotron frequency) might vary with transverse position; or, for example, the plasma might not fill the entire space of the waveguide cross section.

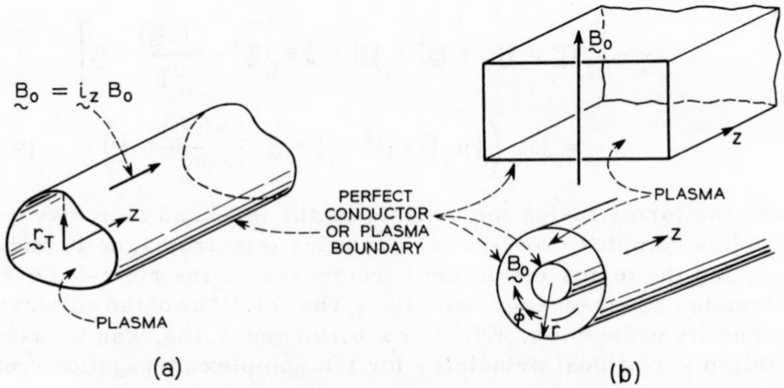

Figure 9.1. Plasma waveguides: (a) longitudinally magnetized (LMG); (b) transversely magnetized (TMG).

We begin again with Maxwell's equations for the complex vector electric and magnetic fields:

$$\nabla \times \underset{\sim}{E} = -j\omega\mu_0 \underset{\sim}{H} \tag{9.1}$$

$$\nabla \times \underset{\sim}{H} = j\omega\epsilon_0 \underset{\approx}{K} \cdot \underset{\sim}{E} \tag{9.2}$$

The divergence equations, although not independent of Equations 9.1 and 9.2, provide useful relationships. They are

$$\nabla \cdot (\epsilon_0 \underset{\approx}{K} \cdot \underset{\sim}{E}) = 0 \tag{9.3}$$

$$\nabla \cdot (\mu_0 \underset{\sim}{H}) = 0 \tag{9.4}$$

From Equations 9.1 and 9.2, we derive the wave equation for the electric field:

$$\nabla^2 \underset{\sim}{E} + k_0^2 \underset{\approx}{K} \cdot \underset{\sim}{E} - \nabla(\nabla \cdot \underset{\sim}{E}) = 0 \tag{9.5}$$

where $k_0 = \omega/c$ and $c = (\mu_0\epsilon_0)^{-\frac{1}{2}}$.

The source-free solutions for the electric and magnetic fields are obtainable from Equations 9.1 and 9.5 subject to boundary conditions on the fields at the finite transverse limits of the system. The appropriate boundary conditions at surfaces of transition between a plasma and a metal, dielectric, or free space involve considerations (such as the plasma sheath) that are beyond the scope of this monograph. We shall consider only two types of boundary conditions:

(a) Perfectly conducting surfaces at which either the tangential electric field or the tangential magnetic field vanishes

$$\underset{\sim}{n} \times \underset{\sim}{E} = 0 \qquad \text{electric wall} \tag{9.6}$$

or

$$\underset{\sim}{n} \times \underset{\sim}{H} = 0 \qquad \text{magnetic wall} \tag{9.7}$$

where $\underset{\sim}{n}$ is a unit vector normal to the surface. At such a boundary the motion of the plasma is somewhat restricted.

(b) Free boundaries at which the plasma motion is replaced by an equivalent surface charge across which the tangential components of the electric and magnetic fields are continuous, and the component of electric field normal to the surface is discontinuous.

In this monograph we shall use only these two simple boundary conditions. This chapter will be devoted to a study of the field solutions and some of the dispersion characteristics. The details of the determinantal equation from the boundary conditions will be illustrated for some examples in Chapter 10.

The basic difficulty in the solution of Equation 9.5 arises from the second and third terms (due to the anisotropy and inhomogeneity of the medium), which couple all the components of the electric field vector. There are many ways in which Equation 9.5 can be solved. One approach is to write the electric field as a superposition of free waves. In the general anisotropic case this becomes rather cumbersome, particularly since not all the solutions are known (for example, the complex waves). We shall choose an analysis that is similar to the normal mode approach in isotropic waveguides. This will allow us to compare the solutions with known simple waveguide solutions, and at the resonances and cutoffs we shall be able to relate the solutions to the resonances and cutoffs of the free waves. We now turn our attention to the LMG plasma waveguide.

9.1 Longitudinally Magnetized, Inhomogeneous Plasma Waveguides

It can be shown readily that for a system with uniformity in the z direction, and otherwise arbitrary tensor medium, the solutions of the electromagnetic fields, satisfying Maxwell's equations, have a z dependence of the form $\exp(-\gamma z)$, where γ is an arbitrary complex function of frequency only and independent of position. Hence, it is convenient to make the following separation in the vector fields and vector and tensor operations:

$$\underset{\sim}{E} = \underset{\sim}{E}_T + \underset{\sim}{i}_z E_z \tag{9.8}$$

$$\underset{\sim}{H} = \underset{\sim}{H}_T + \underset{\sim}{i}_z H_z \tag{9.9}$$

$$\nabla = \nabla_T + \underset{\sim}{i}_z \frac{\partial}{\partial z} \tag{9.10}$$

$$= \nabla_T + \underset{\sim}{i}_z(-\gamma) \tag{9.11}$$

$$\underset{\approx}{K} \cdot \underset{\sim}{E} = \underset{\approx}{K}_T \cdot \underset{\sim}{E}_T + \underset{\sim}{i}_z K_{\parallel} E_z \tag{9.12}$$

where

$$\underset{\approx}{K}_T \cdot \underset{\sim}{E}_T = K_{\perp} \underset{\sim}{E}_T + K_{\times} \underset{\sim}{i}_z \times \underset{\sim}{E}_T \tag{9.13}$$

In these equations, a subscript T on a vector quantity indicates that the vector is in a plane perpendicular to the z direction. Equations 9.12 and 9.13 are a direct consequence of the special form assumed for the dielectric tensor as given in Equation 2.13. This form is adequate for representing a temperate plasma. It is now convenient to replace Equations 9.1 and 9.2 by their longitudinal and transverse components; that is, we dot- and cross-multiply $\underset{\sim}{i}_z$, respectively, into Equations 9.1 and 9.2, and obtain

$$\underset{\sim}{i}_z \cdot \nabla_T \times \underset{\sim}{E}_T = -j\omega\mu_0 H_z \tag{9.14}$$

$$\underset{\sim}{i}_z \cdot \nabla_T \times \underset{\sim}{H}_T = j\omega\epsilon_0 K_{\parallel} E_z \tag{9.15}$$

$$\nabla_T E_z + \gamma \underset{\sim}{E}_T = -j\omega\mu_0 \underset{\sim}{i}_z \times \underset{\sim}{H}_T \tag{9.16}$$

$$\nabla_T H_z + \gamma \underset{\sim}{H}_T = j\omega\epsilon_0 K_\perp \underset{\sim}{i}_z \times \underset{\sim}{E}_T - j\omega\epsilon_0 K_\times \underset{\sim}{E}_T \tag{9.17}$$

Similarly using Equations 9.8 through 9.11 in Equations 9.3 and 9.4, we obtain

$$K_\perp \nabla_T \cdot \underset{\sim}{E}_T + \underset{\sim}{E}_T \cdot \nabla_T K_\perp + \nabla_T K_\times \cdot \underset{\sim}{i}_z \times \underset{\sim}{E}_T - K_\times \underset{\sim}{i}_z \cdot \nabla_T \times \underset{\sim}{E}_T - \gamma K_\parallel E_z = 0 \tag{9.18}$$

$$\nabla_T \cdot \underset{\sim}{H}_T - \gamma H_z = 0 \tag{9.19}$$

We note in particular that Equations 9.14 through 9.19 apply to a system containing an inhomogeneous plasma. The tensor elements $K_\times$, $K_\perp$, and $K_\parallel$ may be functions of the transverse coordinates.

Equations 9.16 and 9.17 can be conveniently summarized in matrix form:

$$\begin{bmatrix} -\gamma & 0 & 0 & -j\omega\mu_0 \\ -j\omega\epsilon_0 K_\times & -\gamma & j\omega\epsilon_0 K_\perp & 0 \\ 0 & j\omega\mu_0 & -\gamma & 0 \\ -j\omega\epsilon_0 K_\perp & 0 & -j\omega\epsilon_0 K_\times & -\gamma \end{bmatrix} \begin{bmatrix} \underset{\sim}{E}_T \\ \underset{\sim}{H}_T \\ \underset{\sim}{i}_z \times \underset{\sim}{E}_T \\ \underset{\sim}{i}_z \times \underset{\sim}{H}_T \end{bmatrix} = \begin{bmatrix} \nabla_T E_z \\ \nabla_T H_z \\ \underset{\sim}{i}_z \times \nabla_T E_z \\ \underset{\sim}{i}_z \times \nabla_T H_z \end{bmatrix} \tag{9.20}$$

By inverting the matrix in Equation 9.20, we obtain a solution for the transverse fields in terms of the longitudinal fields:

$$\begin{bmatrix} P & R & Q & S \\ T & P & U & Q \\ -Q & -S & P & R \\ -U & -Q & T & P \end{bmatrix} \begin{bmatrix} \nabla_T E_z \\ \nabla_T H_z \\ \underset{\sim}{i}_z \times \nabla_T E_z \\ \underset{\sim}{i}_z \times \nabla_T H_z \end{bmatrix} = \begin{bmatrix} \underset{\sim}{E}_T \\ \underset{\sim}{H}_T \\ \underset{\sim}{i}_z \times \underset{\sim}{E}_T \\ \underset{\sim}{i}_z \times \underset{\sim}{H}_T \end{bmatrix} \tag{9.21}$$

where

$$P = \frac{-\gamma(\gamma^2 + k_0^2 K_\perp)}{D} \tag{9.22}$$

$$R = \frac{j\omega\mu_0 k_0^2 K_\times}{D} \tag{9.23}$$

$$Q = \frac{\gamma k_0^2 K_\times}{D} \tag{9.24}$$

$$S = \frac{j\omega\mu_0(\gamma^2 + k_0^2K_\perp)}{D} \tag{9.25}$$

$$T = \frac{\gamma^2 j\omega\epsilon_0 K_\times}{D} \tag{9.26}$$

$$U = \frac{-j\omega\epsilon_0(\gamma^2 K_\perp + k_0^2 K_r K_\ell)}{D} \tag{9.27}$$

and

$$\begin{aligned} D &= (\gamma^2 + k_0^2K_r)(\gamma^2 + k_0^2K_\ell) \\ &= (\gamma^2 + k_0^2 n_r^2)(\gamma^2 + k_0^2 n_\ell^2) \\ &= (\gamma^2 + k_0^2K_\perp)^2 + (k_0^2K_\times)^2 \end{aligned} \tag{9.28}$$

$$k_0^2 = \omega^2\mu_0\epsilon_0 = \frac{\omega^2}{c^2} \tag{9.29}$$

The equations describing E_z and H_z are obtained from the transverse divergence of Equations 9.16 and 9.17 together with Equations 9.14, 9.15, 9.18, 9.19, and 9.21. The results are

$$\begin{aligned} \nabla_T^2 E_z + aE_z = {} & bH_z + \underset{\sim}{b}_1 \cdot \nabla_T E_z + \underset{\sim}{b}_2 \cdot \nabla_T H_z + \underset{\sim}{b}_3 \cdot \underset{\sim}{i}_z \times \nabla_T E_z \\ & + \underset{\sim}{b}_4 \cdot \underset{\sim}{i}_z \times \nabla_T H_z \end{aligned} \tag{9.30}$$

$$\begin{aligned} \nabla_T^2 H_z + cH_z = {} & dE_z + \underset{\sim}{d}_1 \cdot \nabla_T E_z + \underset{\sim}{d}_2 \cdot \nabla_T H_z + \underset{\sim}{d}_3 \cdot \underset{\sim}{i}_z \times \nabla_T E_z \\ & + \underset{\sim}{d}_4 \cdot \underset{\sim}{i}_z \times \nabla_T H_z \end{aligned} \tag{9.31}$$

where

$$a = (\gamma^2 + k_0^2K_\perp)\frac{K_\parallel}{K_\perp} \tag{9.32}$$

$$b = j\omega\mu_0\gamma\frac{K_\times}{K_\perp} \tag{9.33}$$

$$c = \gamma^2 + k_0^2\frac{K_r K_\ell}{K_\perp} \tag{9.34}$$

$$d = -j\omega\epsilon_0\gamma\frac{K_\times K_\parallel}{K_\perp} \tag{9.35}$$

$$\begin{bmatrix} \underset{\sim}{b}_1 \\ \hline \underset{\sim}{b}_2 \\ \hline \underset{\sim}{b}_3 \\ \hline \underset{\sim}{b}_4 \end{bmatrix} = \gamma \left[\begin{array}{c|c} P & -Q \\ \hline R & -S \\ \hline Q & P \\ \hline S & R \end{array}\right] \begin{bmatrix} \dfrac{\nabla_T K_\perp}{K_\perp} \\ \hline \dfrac{K_\times}{K_\perp}\dfrac{\nabla_T K_\times}{K_\times} \end{bmatrix} \qquad (9.36)$$

$$\begin{bmatrix} \underset{\sim}{d}_1 \\ \hline \underset{\sim}{d}_2 \\ \hline \underset{\sim}{d}_3 \\ \hline \underset{\sim}{d}_4 \end{bmatrix} = j\omega\epsilon_0 \left\{ \frac{\nabla_T K_\perp}{K_\perp} \left[\begin{array}{c|c} -Q & P \\ \hline -S & R \\ \hline P & Q \\ \hline R & S \end{array}\right] \begin{bmatrix} K_\perp \\ \hline K_\times \end{bmatrix} + \frac{\nabla_T K_\times}{K_\times} \left[\begin{array}{c|c} -Q & -P \\ \hline -S & -R \\ \hline P & -Q \\ \hline R & -S \end{array}\right] \begin{bmatrix} \dfrac{K_\times^2}{K_\perp} \\ \hline K_\times \end{bmatrix} \right\} \qquad (9.37)$$

Equations 9.21, 9.30, and 9.31 are explicit forms of Maxwell's equations (9.1 through 9.4) for the inhomogeneous and anisotropic plasma described by the dielectric tensor $\underset{\approx}{K}$ of Equation 2.13. In any particular problem, Equations 9.30 and 9.31 are solved for E_z and H_z as functions of the transverse coordinates and γ. These solutions are then used in Equation 9.21 to determine the transverse fields. For the case of a source-free field problem, the solutions must be regular everywhere inside the system. In addition, these solutions must be chosen so as to satisfy the boundary conditions (as, for example, at discontinuities in the medium, or at the perfectly conducting walls, Equations 9.6 and 9.7). This then leads to a determinantal equation for γ as a function of frequency and characteristic transverse dimensions of the system. We have thus outlined, in principle, the solution of the source-free field problem.

From Equations 9.30 and 9.31, we note that in general the equations for E_z and H_z are coupled. This coupling results from the anisotropy ($K_\times \neq 0$), and the inhomogeneity ($\nabla_T K_\perp \neq 0$ and $\nabla_T K_\times \neq 0$) of the medium. At cutoff, however, $\gamma = 0$, and $\bar{b}$, d, $\underset{\sim}{b}_1$ through $\underset{\sim}{b}_4$, P, Q, T, $\underset{\sim}{d}_1$, and $\underset{\sim}{d}_3$ vanish, and the solutions for E_z and H_z are independent of each other. This is also clear directly from Equations 9.14 through 9.19 with $\gamma = 0$. Thus at cutoff we can classify the waves as either E waves or H waves.

Rigorous solutions of the given equations for the anisotropic, inhomogeneous waveguide have not yet been carried out. An understanding of such solutions can be gained by considering some simpler cases. We first summarize the well-known field solutions and properties of the free-space waveguide bounded by perfectly conducting walls. Then we consider the homogeneous and inhomogeneous, isotropic plasma waveguide and finally the homogeneous, anisotropic plasma waveguides.

9.2 Free-Space Waveguide

In the absence of the plasma, $K_\times = 0$ and $K_\perp = K_\parallel = 1$. The right-hand sides of Equations 9.30 and 9.31 vanish, and we obtain two uncoupled wave equations for E_z and H_z. These wave equations and associated transverse fields are now summarized.

<u>H waves:</u>

$$E_z = 0 \tag{9.38}$$

$$\nabla_T^2 H_z + p_h^2 H_z = 0; \qquad p_h^2 = \gamma^2 + k_0^2 \tag{9.39}$$

$$\underset{\sim}{H}_T = \frac{-\gamma}{p_h^2} \nabla_T H_z \tag{9.40}$$

$$\underset{\sim}{E}_T = \frac{j\omega\mu_0}{p_h^2} \underset{\sim}{i}_z \times \nabla_T H_z \tag{9.41}$$

<u>E waves:</u>

$$H_z = 0 \tag{9.42}$$

$$\nabla_T^2 E_z + p_e^2 E_z = 0; \qquad p_e^2 = \gamma^2 + k_0^2 \tag{9.43}$$

$$\underset{\sim}{E}_T = \frac{-\gamma}{p_e^2} \nabla_T E_z \tag{9.44}$$

$$\underset{\sim}{H}_T = \frac{-j\omega\epsilon_0}{p_e^2} \underset{\sim}{i}_z \times \nabla_T E_z \tag{9.45}$$

The solutions of Equations 9.39 and 9.43 subject to the perfectly conducting boundary conditions, Equations 9.6 and 9.7, are well known.[28] For perfectly conducting waveguide walls, p^2 is an eigenvalue of Equations 9.39 and 9.43, determined by geometry and independent of frequency. It therefore follows that for each eigenvalue p^2 the field patterns in the waveguide are also independent of frequency. By applying Green's theorem to the longitudinal fields of Equations 9.39 and 9.43, subject to the perfectly conducting boundary conditions, it can be shown that the eignevalues p^2 are positive real numbers.† Hence, the propagation constant γ is either pure real $(\gamma = \alpha)$, signifying a cutoff wave, or pure imaginary $(\gamma = j\beta)$, representing a propagating wave. The cutoff frequencies, for which $\gamma = 0$, are functions of the geometry only:

$$\omega_{co} = pc \tag{9.46}$$

For perfectly conducting electric walls, the boundary conditions are $E_z = 0$ and $\underset{\sim}{n} \cdot \nabla_T H_z = 0$. The boundary conditions on H_z are then of the Neumann type, and hence the lowest cutoff frequency is that of an H wave.

The dispersion characteristics of the waves are the following: For $\omega < \omega_{co}$,

$$\alpha^2 = p^2 - k_0^2 \tag{9.47}$$

For $\omega > \omega_{co}$,

$$\beta^2 = k_0^2 - p^2 \tag{9.48}$$

These are illustrated in Figure 9.2.

The energy and power flow relations follow from the considerations in Chapter 7. In Equations 7.112 through 7.119, $U_k = U_d = 0$ and $U_\epsilon = U_{ek} = U_e$. The properties of a propagating wave follow from Equations 7.149a through 7.150b and 7.152 through 7.153b:

$$Q = 0 \tag{9.49}$$

$$U_m = U_e \tag{9.50}$$

---

† Let F be either E_z or H_z, satisfying the boundary condition $F = 0$ or $\underset{\sim}{n} \cdot \nabla_T F = 0$ at the wall, where $\underset{\sim}{n}$ is a unit vector normal to the boundary of the wall. Green's theorem in two dimensions then gives

$$p^2 = \frac{\int_A |\nabla_T F|^2 \, da}{\int_A |F|^2 \, da}$$

where A is the cross-sectional area of the waveguide.

$$u = \frac{\omega}{\beta} = \frac{P}{2(U_{eT} - U_{mz})} = \frac{P}{2(U_{mT} - U_{ez})} \tag{9.51}$$

$$u_g = \frac{\partial\omega}{\partial\beta} = \frac{P}{U_m + U_e} \tag{9.52}$$

$$\frac{\partial\omega}{\partial\beta} = \frac{\omega}{\beta}\,\frac{2(U_{mT} - U_{ez})}{(U_m + U_e)} = \frac{\omega}{\beta}\,\frac{2(U_{eT} - U_{mz})}{(U_m + U_e)} \tag{9.53}$$

Equation 9.50 is the "resonance" condition for a propagating wave. Since the group velocity and phase velocity have the same signs (see Equation 9.48 and Figure 9.2), it follows from Equations 9.50 through 9.53 that

$$U_{mT} > U_{ez} \tag{9.54}$$

$$U_{eT} > U_{mz} \tag{9.55}$$

These equations also can be verified directly from Equations 9.38 through 9.45.

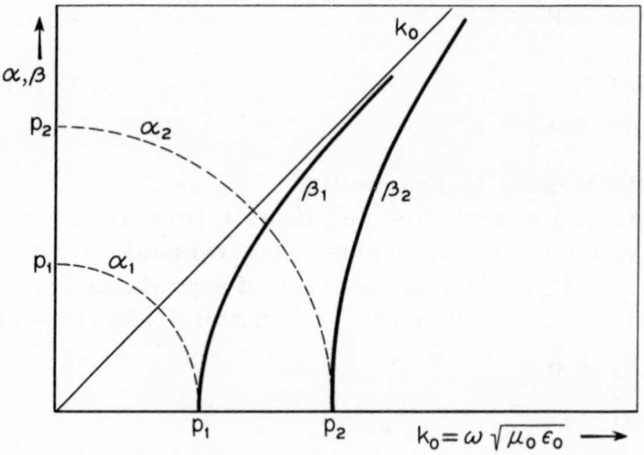

Figure 9.2 Dispersion characteristic for waves in free-space waveguide for two distinct eigenvalues p_1 and p_2. The figure is symmetric about the $\alpha = \beta = 0$ line.

For a cutoff wave, we have from Equations 7.145 through 7.148 and 7.166

$$P = 0 \tag{9.56}$$

$$U_{mT} + U_{eT} = U_{mz} + U_{ez} \tag{9.57}$$

$$\frac{\omega}{\alpha} = \frac{Q}{U_m - U_e} \tag{9.58}$$

$$\frac{\partial\omega}{\partial\alpha} = \frac{Q}{(U_{mT} - U_{mz}) - (U_{eT} - U_{ez})} \tag{9.59}$$

$$\frac{\partial\omega}{\partial\alpha} = \frac{\omega}{\alpha}\,\frac{2(U_{mz} - U_{eT})}{(U_{mT} - U_{mz}) - (U_{eT} - U_{ez})} = \frac{\omega}{\alpha}\,\frac{2(U_{mT} - U_{ez})}{(U_{mT} - U_{mz}) - (U_{eT} - U_{ez})} \tag{9.60}$$

From Equation 9.47 (see also Figure 9.2), we note that $\partial\omega/\partial\alpha$ and ω/α are always of opposite sign. Using Equations 9.38 through 9.45, we deduce the following properties for cutoff waves.

For H waves below cutoff:

$$U_m > U_e; \qquad Q > 0 \tag{9.61}$$

and hence by Equation 9.57

$$U_{mz} > U_{eT} \tag{9.62}$$

and therefore

$$U_{mz} > U_{mT} \tag{9.63}$$

For E waves below cutoff:

$$U_e > U_m; \qquad Q < 0 \tag{9.64}$$

and hence by Equation 9.57

$$U_{ez} > U_{mT} \tag{9.65}$$

and therefore

$$U_{ez} > U_{eT} \tag{9.66}$$

At cutoff, Equations 9.49, 9.50, 9.56, and 9.57 apply, and

$$\alpha = \beta = 0 \tag{9.67}$$

$$\frac{\partial\omega}{\partial\alpha} = \frac{\partial\omega}{\partial\beta} = 0 \tag{9.68}$$

Finally, we note that Equations 9.49 through 9.68 apply to single waves either propagating or cutoff. In general, however, at a single frequency we have at least two waves $(\pm\gamma)$ for each value of p (Equations 9.47 and 9.48); each pair is usually called a waveguide mode. When the complex power flow in such a mode is evaluated, we find that both propagating and cutoff modes will in general carry both real and reactive power.

9.3 Homogeneous, Isotropic Plasma Waveguides

When the waveguide is filled with a homogeneous plasma and $\underset{\sim}{B}_0 = 0$, we have $K_\times = 0$ and $K_\perp = K_\parallel$. The equations for E_z and H_z, Equations 9.30 and 9.31, are again independent, and the solutions for the E and H waves are of the same form as Equations 9.38 through 9.45 for the empty guide with ϵ_0 replaced by $\epsilon_0 K_\parallel$. The dispersion relation is now

$$p^2 = \gamma^2 + k_0^2 K_\parallel \tag{9.69}$$

and depends upon the characteristics of the plasma medium. For the perfectly conducting boundary conditions on the waveguide walls, Equations 9.6 and 9.7, p^2 is again a positive real number determined by the transverse geometry (see footnote, page 139).

9.3.1 Lossless Plasma. In the absence of collisions, $K_\parallel$ is real for all frequencies, and Equation 9.69 shows that the cutoffs $(\gamma = 0)$ occur at frequencies above the plasma frequency

$$\omega_{co}^2 = \omega_p^2 + (pc)^2 \tag{9.70}$$

For a wave associated with a particular value of p, the propagation constant γ is pure imaginary $(\gamma = j\beta)$ for frequencies above ω_{co}, and pure real $(\gamma = \alpha)$ below ω_{co}. For $K_\parallel = 1 - (\omega_p^2/\omega^2)$, the dispersion characteristics for the waves, $\beta(\omega)$ and $\alpha(\omega)$, have a frequency dependence similar to that in an empty waveguide:

$$\omega < \omega_{co}, \qquad \alpha^2 = (p^2 + k_p^2) - k_0^2 \tag{9.71}$$

$$\omega > \omega_{co}, \qquad \beta^2 = k_0^2 - (p^2 + k_p^2) \tag{9.72}$$

where $k_p = (\omega_p/c)$. The difference is in the value of the cutoff frequency. Hence, the presence of the plasma simply shifts all the free-space-waveguide dispersion curves to higher frequencies and propagation constants, as illustrated in Figure 9.3.

Energy and power flow characteristics of this plasma waveguide differ markedly from the free-space waveguide. This arises from the dispersive character of the medium, $K_\parallel(\omega)$. In Equations 7.112 through 7.119 we now have $U_d = 0$, and

$$U_\epsilon = \tfrac{1}{4}\epsilon_0 K_\parallel \int_A |\underset{\sim}{E}|^2 \, da$$

$$= \tfrac{1}{4}\epsilon_0 \left(1 - \frac{\omega_p^2}{\omega^2}\right) \int_A |\underset{\sim}{E}|^2 \, da \tag{9.73}$$

$$U_{ek} = \tfrac{1}{4}\epsilon_0 \frac{\partial(\omega K_{\|})}{\partial\omega} \int_A |\underset{\sim}{E}|^2 \, da$$

$$= \tfrac{1}{4}\epsilon_0 \left(1 + \frac{\omega_p^2}{\omega^2}\right) \int_A |\underset{\sim}{E}|^2 \, da \qquad (9.74)$$

Equations 9.73 and 9.74 clearly show that U_ϵ is the difference between the free-space electric and kinetic energy per unit length, while U_{ek} is the sum of these energies. We also note that U_ϵ is negative for $\omega < \omega_p$, while U_{ek} is positive for all ω.

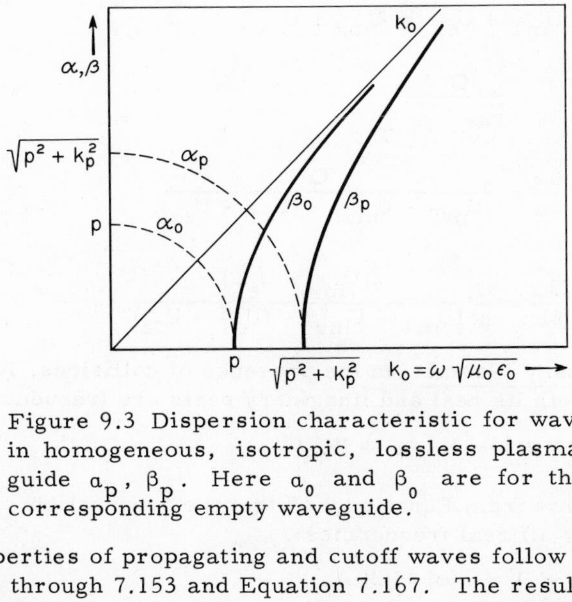

Figure 9.3 Dispersion characteristic for waves in homogeneous, isotropic, lossless plasma guide α_p, β_p. Here α_0 and β_0 are for the corresponding empty waveguide.

The properties of propagating and cutoff waves follow from Equations 7.145 through 7.153 and Equation 7.167. The results are similar to Equations 9.49 through 9.68 with U_e replaced by U_ϵ everywhere except in terms associated with the total energy (denominators of Equations 9.52 and 9.53), where U_e must be replaced by U_{ek}. We summarize some of these equations.

<u>Propagating wave:</u> $\gamma = j\beta$

$$Q = 0 \qquad (9.75)$$

$$U_m = U_\epsilon \qquad (9.76)$$

$$u = \frac{\omega}{\beta} = \frac{P}{2(U_{\epsilon T} - U_{mz})} = \frac{P}{2(U_{mT} - U_{\epsilon z})} \qquad (9.77)$$

$$\frac{\partial\omega}{\partial\beta} = \frac{P}{U_m + U_{ek}} \qquad (9.78)$$

$$\frac{\partial\omega}{\partial\beta} = \frac{\omega}{\beta}\,\frac{2(U_{mT} - U_{\epsilon z})}{(U_m + U_{ek})} = \frac{\omega}{\beta}\,\frac{2(U_{\epsilon T} - U_{mz})}{(U_m + U_{ek})} \tag{9.79}$$

and for $\omega > \omega_p$,

$$U_{mT} > U_{\epsilon z} \tag{9.80}$$

$$U_{\epsilon T} > U_{mz} \tag{9.81}$$

Cutoff wave: $\gamma = \alpha$

$$P = 0 \tag{9.82}$$

$$U_{mT} + U_{\epsilon T} = U_{mz} + U_{\epsilon z} \tag{9.83}$$

$$\frac{\omega}{\alpha} = \frac{Q}{U_m - U_\epsilon} \tag{9.84}$$

$$\frac{\partial\omega}{\partial\alpha} = \frac{Q}{(U_{mT} - U_{mz}) - (U_{\epsilon T} - U_{\epsilon z})} \tag{9.85}$$

$$\frac{\partial\omega}{\partial\alpha} = \frac{\omega}{\alpha}\,\frac{2(U_{mz} - U_{\epsilon T})}{(U_{mT} - U_{mz}) - (U_{\epsilon T} - U_{\epsilon z})} \tag{9.86}$$

9.3.2 Lossy Plasma. In the presence of collisions, $K_{\parallel}$ is complex, and both its real and imaginary parts are frequency-dependent:

$$K_{\parallel}(\omega) = K_{\parallel r}(\omega) + jK_{\parallel i}(\omega) \tag{9.87}$$

It then follows from Equation 9.69 that the propagation constant is complex for all real frequencies:

$$\gamma(\omega) = \alpha(\omega) + j\beta(\omega) \tag{9.88}$$

Using Equation 9.87, we find

$$\alpha^2 - \beta^2 = p^2 - k_0^2 K_{\parallel r} \tag{9.89}$$

$$2\alpha\beta = -k_0^2 K_{\parallel i} \tag{9.90}$$

Equations 9.89 and 9.90 can be solved simultaneously for the attenuation constant α and the phase constant β. It is convenient to introduce normalized variables:

$$\frac{\beta}{p} = y \tag{9.91}$$

$$\frac{\alpha}{p} = x \tag{9.92}$$

$$\frac{k_0^2 K_{\parallel r}}{p^2} = R \tag{9.93}$$

$$\frac{K_{\parallel i}}{2K_{\parallel r}} = \delta \tag{9.94}$$

From Equations 9.91 through 9.94 we find

$$y = \sqrt{\tfrac{1}{2}(R-1) + \tfrac{1}{2}\sqrt{(R-1)^2 + 4(R\delta)^2}} \tag{9.95}$$

$$x = \sqrt{\tfrac{1}{2}(1-R) + \tfrac{1}{2}\sqrt{(1-R)^2 + 4(R\delta)^2}} \tag{9.96}$$

As an example, where the macroscopic damped motion may be described by an effective collision frequency ν, then

$$K_{\parallel} = 1 - \frac{\omega_p^2}{(\omega^2 + \nu^2)} - j\frac{\nu}{\omega}\frac{\omega_p^2}{(\omega^2 + \nu^2)} \tag{9.97}$$

The parameters R and $R\delta$ can be expressed in terms of normalized plasma parameters. Let

$$\frac{\omega}{\omega_p} = \Omega \tag{9.98}$$

$$\frac{pc}{\omega_p} = \Pi \tag{9.99}$$

$$\frac{\nu}{\omega_p} = \Upsilon \tag{9.100}$$

Then

$$R = \frac{\Omega^2}{\Pi^2}\left(\frac{\Omega^2 + \Upsilon^2 - 1}{\Omega^2 + \Upsilon^2}\right) \tag{9.101}$$

$$R\delta = \frac{1}{2}\frac{-\Omega\Upsilon}{\Pi^2(\Omega^2 + \Upsilon^2)} \tag{9.102}$$

Dispersion curves of the propagation constant for some normalized parameters of waveguide cutoff Π and collision frequency Υ are shown in Figures 9.4 and 9.5a through c. We note the following general behavior:

(a) At high frequencies, $\Omega \gg 1$, the propagation constant approaches asymptotically the free-space values: $\beta \to k_0$, $\alpha \to 0$.
(b) At low frequencies, $\Omega \ll 1$, the propagation characteristics approach asymptotically $\beta \to 0$, $\alpha \to p$.

(c) The attenuation constant and phase constant become equal at a frequency given by the positive real Ω^2 root of

$$\Omega^4 - (\Pi^2 - \Upsilon^2 + 1)\Omega^2 - \Pi^2\Upsilon^2 = 0 \tag{9.103}$$

and are given by

$$\alpha^2 = \beta^2 = p^2|\delta| \tag{9.104}$$

where δ is evaluated from Equation 9.94 at the positive real Ω^2 root of Equation 9.103.

(d) The case $\Pi = 0$ corresponds to a one-dimensional medium (waveguide walls at infinity). Extensive calculations of the propagation constants for this case are available.[29] These may be used conveniently to calculate the propagation constant for finite values of p as follows. Let the solutions for the one-dimensional case be denoted by α_1 and β_1; then for $p^2 = 0$, Equation 9.69 gives

$$(\alpha_1 + j\beta_1)^2 = -k_0^2K_{\parallel} \tag{9.105}$$

Using Equations 9.87, 9.93, and 9.94, we find

$$R = -\frac{\alpha_1^2 - \beta_1^2}{p^2} \tag{9.106}$$

$$R\delta = -\frac{\alpha_1\beta_1}{2p^2} \tag{9.107}$$

The propagation constant for any particular waveguide geometry is then obtained from Equations 9.95 and 9.96.

(e) Finally, we consider the case of small loss, $|\delta| \ll 1$. We have approximately a cutoff $x \approx 0$ and $y \approx 0$, at a frequency cor-

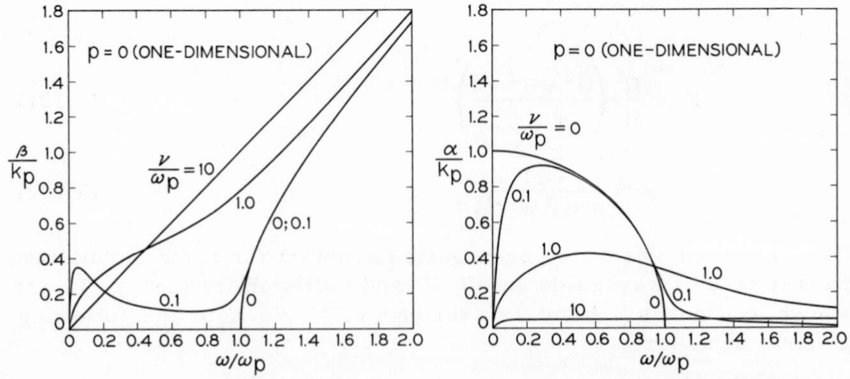

Figure 9.4. Dispersion characteristic for waves in one-dimensional, isotropic plasma with collisions. The fields have a space dependence of $\exp(-\alpha - j\beta)z$.

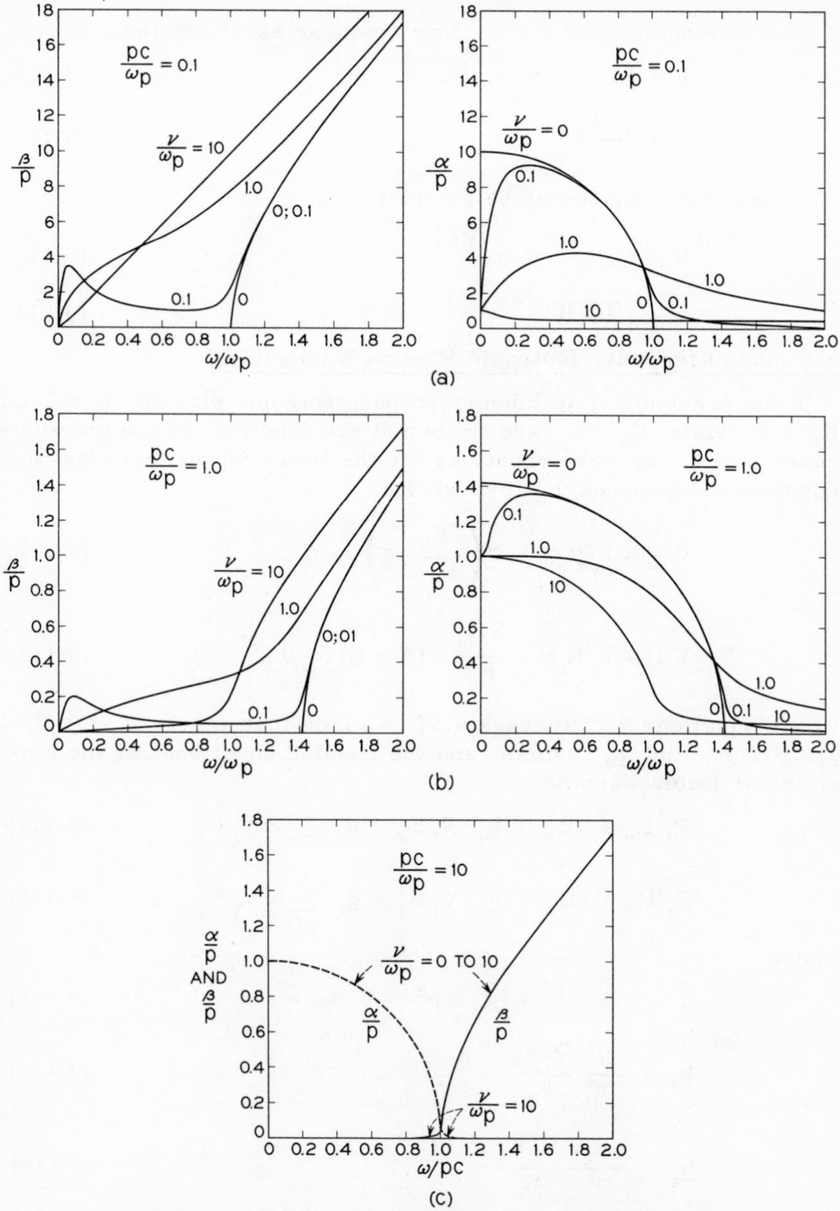

Figure 9.5. Dispersion characteristic for waves in a homogeneous, isotropic plasma waveguide with collisions. The z dependence of the fields is $\exp(-\alpha - j\beta)z$. The free-space waveguide with the same perfectly conducting boundaries has a cutoff frequency pc.

responding to $R \approx 1$. For frequencies above the cutoff,

$$R > 1, \qquad y \approx \sqrt{R-1} \tag{9.108}$$

$$x \approx \frac{|R\delta|}{y} \tag{9.109}$$

and for frequencies below the cutoff

$$R < 1, \qquad y \approx \frac{|R\delta|}{x} \tag{9.110}$$

$$x \approx \sqrt{1-R} \tag{9.111}$$

9.4 Inhomogeneous, Isotropic Plasma Waveguides

In the presence of an inhomogeneous, isotropic plasma, $\underset{\sim}{B}_0 = 0$ and $K_\times = 0$, while $K_\parallel = K_\perp$ are finite and are functions of the transverse coordinates. The wave equations for the fields follow from Maxwell's equations, Equations 9.1 through 9.4,

$$\nabla^2 \underset{\sim}{E} + k_0^2 K_\parallel \underset{\sim}{E} + \nabla\left(\frac{\nabla K_\parallel}{K_\parallel} \cdot \underset{\sim}{E}\right) = 0 \tag{9.112}$$

$$\nabla^2 \underset{\sim}{H} + k_0^2 K_\parallel \underset{\sim}{H} + \frac{\nabla K_\parallel}{K_\parallel} \times (\nabla \times \underset{\sim}{H}) = 0 \tag{9.113}$$

From Equations 9.22 through 9.37, we note that b, d, R, Q, T, $\underset{\sim}{b}_2$, $\underset{\sim}{b}_3$, $\underset{\sim}{d}_1$, and $\underset{\sim}{d}_4$ vanish, and the coupled equations for the longitudinal fields become

$$\nabla_T^2 E_z + aE_z = \underset{\sim}{b}_1 \cdot \nabla_T E_z + \underset{\sim}{b}_4 \cdot \underset{\sim}{i}_z \times \nabla_T H_z \tag{9.114}$$

$$\nabla_T^2 H_z + cH_z = \underset{\sim}{d}_2 \cdot \nabla_T H_z + \underset{\sim}{d}_3 \cdot \underset{\sim}{i}_z \times \nabla_T E_z \tag{9.115}$$

where

$$a = c = \gamma^2 + k_0^2 K_\parallel \equiv p^2 \tag{9.116}$$

$$\underset{\sim}{b}_1 = -\frac{\gamma^2}{p^2}\,\frac{\nabla_T K_\parallel}{K_\parallel} \tag{9.117}$$

$$\underset{\sim}{b}_4 = \frac{j\omega\mu_0\gamma}{p^2}\,\frac{\nabla_T K_\parallel}{K_\parallel} \tag{9.118}$$

$$\underset{\sim}{d}_2 = \frac{k_0^2 K_\parallel}{p^2}\,\frac{\nabla_T K_\parallel}{K_\parallel} \tag{9.119}$$

$$\underset{\sim}{d}_3 = -\frac{j\omega\epsilon_0 K_\parallel \gamma}{p^2}\,\frac{\nabla_T K_\parallel}{K_\parallel} \tag{9.120}$$

Equations 9.21 for the transverse fields in terms of the longitudinal fields also take on the simpler form

$$\underset{\sim}{E}_T = \frac{-\gamma}{p^2}\nabla_T E_z + \frac{j\omega\mu_0}{p^2}\underset{\sim}{i}_z \times \nabla_T H_z \tag{9.121}$$

$$\underset{\sim}{H}_T = \frac{-\gamma}{p^2}\nabla_T H_z + \frac{-j\omega\epsilon_0 K_{\parallel}}{p^2}\underset{\sim}{i}_z \times \nabla_T E_z \tag{9.122}$$

Equations 9.114 through 9.122 have been discussed for nondispersive media.[30] In general, Equations 9.114 and 9.115 are coupled, and E and H waves cannot exist separately. However, at cutoff, when $\gamma = 0$, these equations for E_z and H_z become uncoupled, and the waves may be classified there as either E or H. We also note that, unlike the homogeneous waveguide of Section 9.3, p^2 can no longer be regarded as an eigenvalue. Because of the uniformity in the z direction, γ is a function of only frequency, but $K_{\parallel}$ is a function of both frequency and transverse coordinates. It follows from Equation 9.116 that p^2 must likewise be a function of transverse coordinates and frequency. It therefore also follows that all field patterns will be, in general, functions of frequency. Very few problems of this kind have been solved explicitly.

9.4.1 Uniformity in One Transverse Dimension. The solution of the field problem is markedly simplified when (a) the plasma medium is inhomogeneous in only one of the transverse orthogonal directions, and (b) the field solutions of interest are independent of the other transverse orthogonal dimension. Specifically, let u_1 and u_2 be the coordinates of an orthogonal system in a plane transverse to the z axis, and let h_1 and h_2 be their metric coefficients. Let $K_{\parallel}$ be a function of u_1 and independent of u_2, and consider field solutions that are independent of u_2. From Equations 9.118 and 9.120, it follows that

$$\underset{\sim}{b}_4 \cdot \underset{\sim}{i}_z \times \nabla_T H_z = 0 \tag{9.123}$$

$$\underset{\sim}{d}_3 \cdot \underset{\sim}{i}_z \times \nabla_T E_z = 0 \tag{9.124}$$

and hence the equations for E_z and H_z, Equations 9.114 and 9.115 become uncoupled. We can separate the solutions into E waves and H waves.

E waves: $E_z(u_1)$, independent of u_2

$$H_z = 0 \tag{9.125}$$

$$\nabla_T^2 E_z - \underset{\sim}{b}_1 \cdot \nabla_T E_z + p^2 E_z = 0 \tag{9.126}$$

$$\underset{\sim}{E}_T = \frac{-\gamma}{p^2}\nabla_T E_z \equiv \underset{\sim}{i}_1 E_1 \tag{9.127}$$

$$\underset{\sim}{H}_T = -\frac{j\omega\epsilon_0 K_{\parallel}}{p^2}\, \underset{\sim}{i}_z \times \nabla_T E_z \equiv \underset{\sim}{i}_2 H_2 \qquad (9.128)$$

and H_2 satisfies the wave equation, Equation 9.113:

$$\nabla_T^2 H_2 + p^2 H_2 + \underset{\sim}{i}_2 \cdot \left[\frac{\nabla_T K_{\parallel}}{K_{\parallel}} \times (\nabla \times \underset{\sim}{i}_2 H_2)\right] = 0 \qquad (9.129)$$

It should be noted that p^2 and $\underset{\sim}{b}_1$ are functions of u_1 (Equations 9.116 and 9.117), and therefore the differential equations for E_z or H_2, Equations 9.126 or 9.129, have nonconstant coefficients. For some special forms of $K_{\parallel}(u_1)$, series solutions to these differential equations can be found.

<u>H waves:</u> $H_z(u_1)$, independent of u_2

$$E_z = 0 \qquad (9.130)$$

$$\nabla_T^2 H_z - \underset{\sim}{d}_2 \cdot \nabla_T H_z + p^2 H_z = 0 \qquad (9.131)$$

$$\underset{\sim}{H}_T = \frac{-\gamma}{p^2} \nabla_T H_z \equiv \underset{\sim}{i}_1 H_1 \qquad (9.132)$$

$$\underset{\sim}{E}_T = \frac{j\omega\mu_0}{p^2} \underset{\sim}{i}_z \times \nabla_T H_z \equiv \underset{\sim}{i}_2 E_2 \qquad (9.133)$$

Since $K_{\parallel}$ is a function of only u_1, we have

$$\nabla K_{\parallel} \cdot \underset{\sim}{E} = 0 \qquad (9.134)$$

and hence the wave equation for the electric field, Equation 9.112, takes on the simple form

$$\nabla_T^2 E_2 + p^2 E_2 = 0 \qquad (9.135)$$

where p^2 is a function of u_1 and is given by Equation 9.116. Using Equation 9.134 in Equation 9.3, we find that for these waves

$$\nabla \cdot \underset{\sim}{E} = 0 \qquad (9.136)$$

The lack of space-charge perturbations results from the fact that the first-order electric field is everywhere perpendicular to the unperturbed charge-density gradients. Thus the E field is transverse despite the inhomogeneity, and the wave equation for E, Equation 9.135, also follows from Equation 9.5. The solutions to Equation 9.135 for some simple variations of $K_{\parallel}(u_1)$ can be given in terms of hypergeometric functions or Mathieu functions. We shall illustrate this in the chapter on boundary-value problems.

<u>9.4.2 Uniform Plasma.</u> Other examples of inhomogeneous plasma waveguides consist of homogeneous plasmas that partially fill the waveguide cross section. Two such examples are shown in Figure

9.6. Both inside the plasma and outside in the homogeneous dielectric the field solutions can be written in terms of E waves and H waves. Inside the plasma the field solutions are of the form given by Equations 9.38 through 9.45 with p^2 given by Equation 9.69, and ϵ_0 in Equation 9.45 replaced by $\epsilon_0 K_{\parallel}$. Outside the plasma, in the homogeneous dielectric, the fields are again of the same form (Equations 9.38 through 9.45) with ϵ_0 and μ_0 in general replaced by ϵ and μ of the dielectric medium.

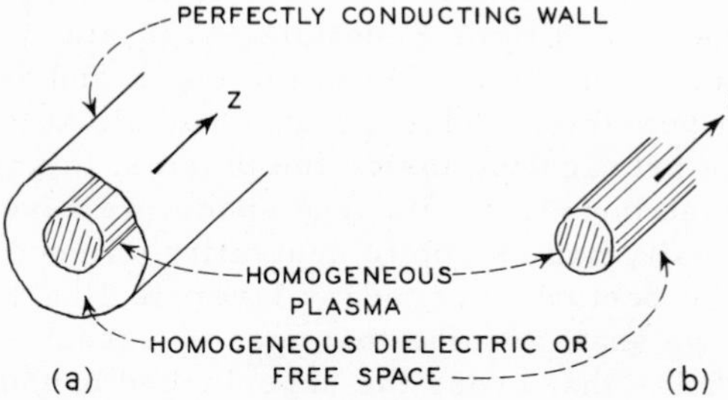

Figure 9.6. Inhomogeneous, isotropic plasma waveguides consisting of a homogeneous plasma that either (a) partially fills a waveguide, or (b) is surrounded by an unbounded dielectric.

The specific field solutions of the systems shown in Figure 9.6 depend upon the geometry and the boundary conditions. At the perfectly conducting wall the boundary condition is as given by Equations 9.6 and 9.7. At the plasma boundary the charges are free to move. Hence, in the presence of first-order fields the boundary is perturbed. Following a technique first suggested by Hahn,[31] we shall replace the first-order perturbation in the boundary by an equivalent first-order surface-charge density on the unperturbed boundary. We shall briefly outline the conditions under which this is valid.

Let $\underset{\sim}{r}_{bi}$ be the first-order perturbation in the position of the particle (of the ith species) at the unperturbed boundary between the plasma and free space. If $\underset{\sim}{n}$ is a unit vector normal to the unperturbed boundary, then the charge per unit area of boundary that has moved across the unperturbed boundary is, to first order,

$$\rho_s = \sum_i \rho_{0i} \underset{\sim}{n} \cdot \underset{\sim}{r}_{bi} \tag{9.137}$$

where ρ_{0i} is the unperturbed charge density of the ith species of particles. For the temperate plasma, the first-order perturbation in position $\underset{\sim}{r}_{bi}$ is simply related to the first-order Eulerian velocity of the particle by

$$j\omega \underset{\sim}{r}_{bi} = \underset{\sim}{v}_{bi} \tag{9.138}$$

which in turn is related to the total electric field at the perturbed position by the force equation

$$j\omega m_i \underset{\sim}{v}_{bi} = e\underset{\sim}{E}_b \tag{9.139}$$

If the plasma is neutral in the unperturbed state and its positive and negative particles have different mobilities, then in the perturbed boundary layer the variation of the electric field $\underset{\sim}{E}_b$ is appreciable. Under these conditions Equations 9.138 and 9.139 must be solved for the individual particles in the layer, and an Eulerian description is not possible. This greatly complicates the over-all solution of wave propagation inside the plasma. In any real plasma at the unperturbed boundary with free space, we have neither a discontinuity in density nor complete neutrality. Hence, the variation of $\underset{\sim}{E}_b$ within the perturbed boundary layer is likely to be small. For simplicity we shall assume that $\underset{\sim}{E}_b$ is essentially the same for all the particles that cross the unperturbed boundary. Then from Equations 9.137 through 9.139 we obtain

$$\rho_s = \sum_i - \frac{\omega_{pi}^2}{\omega^2} \epsilon_0 \underset{\sim}{E}^p \cdot \underset{\sim}{n} \tag{9.140}$$

where $\underset{\sim}{E}^p$ is the first-order electric field in the plasma at the unperturbed boundary. In effect, we may now replace the perturbed boundary of the plasma by the first-order surface charge density of Equation 9.137 (Equation 9.140) placed at the unperturbed boundary of the plasma. Under these conditions the normal component of the electric field at the unperturbed boundary is discontinuous by ρ_s/ϵ_0, where ρ_s is as given by Equation 9.140. Letting $\underset{\sim}{E}^a$ be the electric field in the free space at the boundary of the plasma, we obtain

$$\underset{\sim}{n} \cdot (\epsilon_0 \underset{\sim}{E}^a - \epsilon_0 K_{\parallel} \underset{\sim}{E}^p) = 0 \tag{9.141}$$

Equation 9.141 is the boundary condition on the normal electric field, provided that the assumptions leading to Equation 9.140 are valid. The tangential components of the electric and magnetic fields are continuous across this boundary.

The field solutions in the inhomogeneous waveguides containing a homogeneous isotropic plasma (Figure 9.6) have features that are not encountered in the homogeneous isotropic plasma waveguide treated in Section 9.3. Let the field solutions inside and outside the plasma be denoted by superscripts i and o, respectively. We then have

$$\nabla_T^2 E_z^i + p^2 E_z^i = 0 \tag{9.142}$$

$$\nabla_T^2 H_z^i + p^2 H_z^i = 0 \tag{9.143}$$

$$p^2 = \gamma^2 + k_0^2 K_{\parallel} \tag{9.144}$$

$$\underset{\sim}{E}_T^i = \frac{-\gamma}{p^2}\nabla_T E_z^i + \frac{j\omega\mu_0}{p^2}\underset{\sim}{i}_z \times \nabla_T H_z^i \tag{9.145}$$

$$\underset{\sim}{H}_T^i = \frac{-\gamma}{p^2}\nabla_T H_z^i + \frac{-j\omega\epsilon_0 K_{\parallel}}{p^2}\underset{\sim}{i}_z \times \nabla_T E_z^i \tag{9.146}$$

and

$$\nabla_T^2 E_z^o + q^2 E_z^o = 0 \tag{9.147}$$

$$\nabla_T^2 H_z^o + q^2 H_z^o = 0 \tag{9.148}$$

$$q^2 = \gamma^2 + k^2; \qquad k^2 = \omega^2\mu\epsilon \tag{9.149}$$

$$\underset{\sim}{E}_T^o = \frac{-\gamma}{q^2}\nabla_T E_z^o + \frac{j\omega\mu}{q^2}\underset{\sim}{i}_z \times \nabla_T H_z^o \tag{9.150}$$

$$\underset{\sim}{H}_T^o = \frac{-\gamma}{q^2}\nabla_T H_z^o + \frac{-j\omega\epsilon}{q^2}\underset{\sim}{i}_z \times \nabla_T E_z^o \tag{9.151}$$

As an example, consider the homogeneous, isotropic plasma in free space. Field solutions that are guided along the plasma $(\gamma = j\beta)$ will have decaying fields outside the plasma so that

$$q^2 < 0 \tag{9.152}$$

and hence, by Equation 9.149,

$$\beta^2 > k^2 \tag{9.153}$$

These are slow waves (phase velocity less than velocity of light) that do not exist in the homogeneously filled plasma waveguide. From Equations 9.153 and 9.144, it also follows that for $\mu\epsilon \geq \mu_0\epsilon_0$

$$p^2 < 0 \tag{9.154}$$

and hence, E_z, Equations 9.142 and 9.147, is largest at the plasma boundary where its slope is discontinuous so as to satisfy Equation 9.141. These solutions are surface waves propagating along the plasma.

The energy and power flow relations in the inhomogeneous plasma waveguide may also have new features. Thus

$$U_\epsilon = \int_A \tfrac{1}{4}\epsilon_0 K_{\parallel} |E|^2 \, da \tag{9.155}$$

$$U_{ek} = \int_A \tfrac{1}{4}\epsilon_0 |\underset{\sim}{E}|^2 \frac{\partial(\omega K_{\parallel})}{\partial\omega}\, da \tag{9.156}$$

and the properties of the waves are as given by Equations 9.75 through 9.79 and 9.82 through 9.86. However, Equations 9.80 and 9.81 can no longer be guaranteed, and backward waves (group velocity opposite to phase velocity) may be present (see Equation 9.79).

9.5 Longitudinally Magnetized, Homogeneous Plasma Waveguides

In the presence of an externally applied magnetic field, $\underset{\sim}{B}_0 \neq 0$, the plasma medium is anisotropic, $K_\times \neq 0$ and $K_\perp \neq K_\parallel$. When $K_\times$, $K_\perp$, and $K_\parallel$ are independent of transverse coordinates, $\underset{\sim}{b}_1$ through $\underset{\sim}{b}_4$ and $\underset{\sim}{d}_1$ through $\underset{\sim}{d}_4$ vanish, and the coupled wave equations for E_z and H_z, Equations 9.30 and 9.31, become

$$\nabla_T^2 E_z + aE_z = bH_z \tag{9.157}$$

$$\nabla_T^2 H_z + cH_z = dE_z \tag{9.158}$$

where a, b, c, and d are as given by Equations 9.32 through 9.35. This coupled set of second-order equations can be transformed to an uncoupled set of fourth-order equations:

$$[\nabla_T^4 + (a + c)\nabla_T^2 + (ac - bd)]E_z = 0 \tag{9.159}$$

$$[\nabla_T^4 + (a + c)\nabla_T^2 + (ac - bd)]H_z = 0 \tag{9.160}$$

The solutions for E_z and H_z are of the same form. Equations 9.159 and 9.160 admit solutions of the form $\exp(-j\underset{\sim}{p}\cdot\underset{\sim}{r}_T)$ for

$$p^4 - (a + c)p^2 + (ac - bd) = 0 \tag{9.161}$$

For arbitrary geometries and boundary conditions, an infinite set of solutions of the form $\exp(-j\underset{\sim}{p}\cdot\underset{\sim}{r}_T)$ is required. It is often more convenient to seek solutions for each value of p^2, of the form

$$(\nabla_T^2 + p_i^2)E_{zi} = 0 \tag{9.162}$$

which for most common geometries has solutions that are in terms of well-known functions. Since Equation 9.161 is of fourth order, two independent solutions of the form of Equation 9.162 will in general be required. Thus

$$E_z = E_{z1} + E_{z2} \tag{9.163}$$

$$H_z = h_1 E_{z1} + h_2 E_{z2} \tag{9.164}$$

where

$$h = \frac{a - p^2}{b} = \frac{d}{c - p^2} \tag{9.165}$$

In Equations 9.163 and 9.164, E_{z1} and E_{z2} are two independent solutions of Equation 9.162 corresponding to two distinct values of p^2 that must satisfy Equation 9.161. To each value of p^2 there corresponds a value for h, given by Equation 9.165. Equation 9.161 is the dispersion relation that relates the propagation constant to the transverse wave numbers p_1^2 and p_2^2, the frequency ω, and the plasma parameters ω_p and ω_b's.

The transverse fields can now be determined in terms of E_{z1} and E_{z2}. Using Equations 9.163 and 9.164 in Equation 9.21, and after some algebraic manipulations, we obtain

$$\begin{bmatrix} \underset{\sim}{E}_T \\ \underset{\sim}{H}_T \\ \underset{\sim}{i}_z \times \underset{\sim}{E}_T \\ \underset{\sim}{i}_z \times \underset{\sim}{H}_T \end{bmatrix} = \begin{bmatrix} -\frac{1}{\gamma} & \frac{R}{b} & -\frac{1}{\gamma}\frac{K_\perp}{K_\times} & \frac{S}{b} \\ \frac{\gamma}{b} & \frac{P}{b} & 0 & \frac{Q}{b} \\ \frac{1}{\gamma}\frac{K_\perp}{K_\times} & -\frac{S}{b} & -\frac{1}{\gamma} & \frac{R}{b} \\ 0 & -\frac{Q}{b} & \frac{\gamma}{b} & \frac{P}{b} \end{bmatrix} \left[\begin{pmatrix} \nabla_T \\ p_2^2 \nabla_T \\ \underset{\sim}{i}_z \times \nabla_T \\ p_2^2 \underset{\sim}{i}_z \times \nabla_T \end{pmatrix} E_{z1} + \begin{pmatrix} \nabla_T \\ p_1^2 \nabla_T \\ \underset{\sim}{i}_z \times \nabla_T \\ p_1^2 \underset{\sim}{i}_z \times \nabla_T \end{pmatrix} E_{z2} \right] \tag{9.166}$$

The solution of any particular source-free problem then proceeds as follows: For a particular plasma geometry, appropriate solutions to Equation 9.162 are chosen, and Equations 9.163, 9.164, and 9.166 determine the complete fields. These fields are required to satisfy the boundary conditions imposed by the perfectly conducting walls (Equation 9.6 or 9.7). This leads to a determinantal equation that relates the propagation constant γ, the transverse wave numbers p^2, the frequency ω, the plasma parameters ω_p and ω_b's, and the transverse dimensions of the waveguide. The determinantal equation, together with the dispersion relation, Equation 9.161, can in principle be solved for the propagation constant as a function of frequency when the other parameters are known.

<u>9.5.1 Characteristics of the Dispersion Relation.</u> The explicit form of the dispersion relation, Equation 9.161, can be found with the aid of Equations 9.32 through 9.35 (see also Section 3.8):

$$p^4 - \left[\gamma^2\left(\frac{K_\parallel}{K_\perp} + 1\right) + k_0^2\left(K_\parallel + \frac{K_r K_\ell}{K_\perp}\right)\right]p^2 + \frac{K_\parallel}{K_\perp}(\gamma^2 + k_0^2 K_r)(\gamma^2 + k_0^2 K_\ell) = 0 \tag{9.167}$$

We note that for $p = 0$, corresponding to field solutions with no transverse variations, Equation 9.167 gives the propagation constants for the right and left circularly polarized waves. For each value of γ^2, Equation 9.167 gives two values of p^2:

$$p_{1,2}^{\,2} = \tfrac{1}{2}\left[\gamma^2\left(\frac{K_\parallel}{K_\perp}+1\right) + k_0^{\,2}\left(K_\parallel + \frac{K_r K_\ell}{K_\perp}\right)\right]$$

$$\pm \tfrac{1}{2}\sqrt{\left[\gamma^2\left(\frac{K_\parallel}{K_\perp}-1\right) + k_0^{\,2}\left(K_\parallel - \frac{K_r K_\ell}{K_\perp}\right)\right]^2 + 4\gamma^2 k_0^{\,2} K_\parallel \frac{K_\times^{\,2}}{K_\perp^{\,2}}} \qquad (9.168)$$

Conversely, we can regard Equation 9.161 as giving γ in terms of p:

$$K_\parallel\gamma^4 - (K_\parallel + K_\perp)\left(p^2 - 2k_0^{\,2}\frac{K_\parallel K_\perp}{K_\parallel + K_\perp}\right)\gamma^2 + K_\perp(p^2 - k_0^{\,2}K_\parallel)\left(p^2 - k_0^{\,2}\frac{K_r K_\ell}{K_\perp}\right) = 0 \qquad (9.169)$$

Here we note that for $\gamma = 0$, corresponding to field solutions with no z variation, the values of p from Equation 9.169 are the propagation constants of the ordinary and extraordinary waves. In Equation 9.169, for each value of p^2 we obtain two values of γ^2:

$$\gamma_{1,2}^{\,2} = \tfrac{1}{2}\left(\frac{K_\perp}{K_\parallel}+1\right)\left(p^2 - 2k_0^{\,2}\frac{K_\parallel K_\perp}{K_\parallel + K_\perp}\right)$$

$$\pm \tfrac{1}{2}\sqrt{\left(\frac{K_\perp}{K_\parallel}+1\right)^2\left(p^2 - 2k_0^{\,2}\frac{K_\parallel K_\perp}{K_\parallel + K_\perp}\right)^2 - 4\frac{K_\perp}{K_\parallel}\left(p^2 - k_0^{\,2}K_\parallel\right)\left(p^2 - k_0^{\,2}\frac{K_r \ldots}{\ldots}\right.} \qquad (9.170)$$

Equation 9.167 shows that there are three possibilities for which one of the p^2 roots is zero: (a) $\gamma^2 = -k_0^{\,2}K_\ell$, (b) $\gamma^2 = -k_0^{\,2}K_r$, and (c) $K_\parallel = 0$, giving

$$p_1^{\,2} = 0 \qquad (9.171)$$

$$p_2^{\,2} = \gamma^2\left(\frac{K_\parallel}{K_\perp}+1\right) + k_0^{\,2}\left(K_\parallel + \frac{K_r K_\ell}{K_\perp}\right) \qquad (9.172)$$

In cases (a) and (b) Equation 9.28 vanishes, and hence for finite transverse fields both E_z and H_z must vanish. However, for the perfectly conducting boundary condition Equations 9.16 and 9.17 then also require that the transverse fields vanish. Thus the purely transverse fields of the right and left circularly polarized waves cannot exist in the homogeneous waveguide with perfectly conducting walls. In case (c) Equation 9.172 becomes

$$p_2^2 = \gamma^2 + k_0^2 \frac{K_r K_\ell}{K_\perp} \tag{9.173}$$

and Equation 9.169 shows that γ may either approach infinity or remain finite and arbitrary. The case $\gamma \to \infty$ (resonance) we shall treat in Section 9.5.3. For finite values of γ we have $a = 0$, $d = 0$, $c = p_2^2$, and Equations 9.157, 9.158, and 9.166 show that for perfectly conducting walls all fields must vanish. We conclude that for finite values of the propagation constant γ and frequency ω the field solutions in the waveguide require two nonzero transverse wave numbers p_1^2 and p_2^2.

A situation similar to $K_\parallel = 0$ arises at a frequency when $K_\perp = 0$. Equations 9.157 and 9.158 must then be rederived from Equations 9.16 through 9.19, and all fields are found to vanish for perfectly conducting walls.

We also note from Equations 9.167 and 9.168 that the transverse wave numbers p^2 can become complex even when both ω and γ^2 are real. The conditions for which this can occur are

$$4\gamma^2 k_0^2 K_\parallel \frac{K_\times^2}{K_\perp} < 0 \tag{9.174}$$

and

$$\left| 4\gamma^2 k_0^2 K_\parallel \frac{K_\times^2}{K_\perp^2} \right| > \left[\gamma^2 \left(\frac{K_\parallel}{K_\perp} - 1 \right) + k_0^2 \left(K_\parallel - \frac{K_r K_\ell}{K_\perp} \right) \right]^2 \tag{9.175}$$

or equivalently

$$4 \frac{K_\parallel}{K_\perp} (\gamma^2 + k_0^2 K_r)(\gamma^2 + k_0^2 K_\ell) > \left[\gamma^2 \left(\frac{K_\parallel}{K_\perp} + 1 \right) + k_0^2 \left(K_\parallel + \frac{K_r K_\ell}{K_\perp} \right) \right]^2 \tag{9.176}$$

From Equations 9.174 and 9.175 we see that the region of the real (γ^2, ω) plane where p^2 can be complex is bounded by $K_\parallel = 0$, $\gamma^2 = -k_0^2 K_\ell$, and $\gamma^2 = -k_0^2 K_r$. On these bounds one of the p^2 roots is zero. Necessary (but not also sufficient) conditions for p^2 to be complex follow from Equation 9.174: (1) $\gamma^2 < 0$ and $\omega < \omega_p$; (2) $\gamma^2 > 0$ and $\omega > \omega_p$.

In a similar manner, we can find regions in the plane of real p^2 and ω for which the propagation constant γ may become complex even though the medium be lossless. Such waves carry no time-averaged power and were discussed in Chapter 7. The condition for such complex γ follows from Equation 9.170:

$$4 \frac{K_\perp}{K_\parallel} (p^2 - k_0^2 K_\parallel) \left(p^2 - k_0^2 \frac{K_r K_\ell}{K_\perp} \right) > \left(\frac{K_\parallel + K_\perp}{K_\parallel} \right)^2 \left(p^2 - 2k_0^2 \frac{K_\parallel K_\perp}{K_\parallel + K_\perp} \right)^2 \tag{9.177}$$

Finally, for very high frequencies, $\omega >> \omega_p$ and $\omega >> \omega_b$, we have $K_\times \to 0$, $K_\perp \to 1$, and $K_\parallel \to 1$, and the dispersion characteristic of the waves approaches that of a free-space waveguide, Equations 9.47 and 9.48. (See also the $n_\parallel^2$, $n_\perp^2$ diagrams in Section 3.8; $\gamma^2 \equiv -k_0^2 n_\parallel^2$, $p^2 \equiv k_0^2 n_\perp^2$.)

9.5.2 Cutoff Conditions and Fields. When the propagation constant vanishes, $\gamma = 0$, Equation 9.169 gives two values for p^2:

$$p_e^2 = k_0^2 K_\parallel \tag{9.178}$$

$$p_h^2 = k_0^2 \frac{K_r K_\ell}{K_\perp} \tag{9.179}$$

Equations 9.178 and 9.179, together with the boundary-condition determinantal equation for $\gamma = 0$, determine the cutoff frequencies and the transverse wave numbers. On the other hand, from Equation 9.170 we see that to each of the values of p^2 in Equations 9.178 and 9.179 there corresponds only one cutoff propagation constant. We find for p_e^2

$$\left.\begin{aligned} \gamma_1^2 &= 0 \\ \gamma_2^2 &= -k_0^2 K_\parallel \left(1 - \frac{K_\parallel}{K_\perp}\right) \end{aligned}\right\} \tag{9.180}$$

and for p_h^2

$$\left.\begin{aligned} \gamma_1^2 &= 0 \\ \gamma_2^2 &= -k_0^2 \left[2K_\perp - \left(1 + \frac{K_\perp}{K_\parallel}\right) \frac{K_r K_\ell}{K_\perp}\right] \end{aligned}\right\} \tag{9.181}$$

It is interesting to note that Equations 9.178 and 9.179 also give the critical values of p^2 for complex γ^2 (see Equation 9.177).

The fields at cutoff split into E waves and H waves and are readil determined. From Equations 9.32 through 9.35 we have $b = 0$, $d = 0$, and

$$a = k_0^2 K_\parallel = p_e^2 \tag{9.182}$$

$$c = k_0^2 \frac{K_r K_\ell}{K_\perp} = p_h^2 \tag{9.183}$$

Hence, p_e^2 is the eigenvalue for cutoff E waves, and p_h^2 is the eigenvalue for cutoff H waves. The equations giving the transverse fields are also considerably simplified. Thus for $K_\perp$ and $K_\times$ finite, the quantities P, Q, and T (Equations 9.22, 9.24, and 9.26) vanish. We summarize the fields at cutoff $\gamma = 0$:

E waves

$$\nabla_T^2 E_z + p_e^2 E_z = 0; \qquad p_e^2 = k_0^2 K_{\parallel} \tag{9.184}$$

$$H_z = 0 \tag{9.185}$$

$$\underset{\sim}{E}_T = 0 \tag{9.186}$$

$$\underset{\sim}{H}_T = \frac{1}{j\omega\mu_0} \underset{\sim}{i}_z \times \nabla_T E_z \tag{9.187}$$

H waves

$$\nabla_T^2 H_z + p_h^2 H_z = 0; \qquad p_h^2 = k_0^2 \frac{K_r K_\ell}{K_\perp} \tag{9.188}$$

$$E_z = 0 \tag{9.189}$$

$$\underset{\sim}{H}_T = 0 \tag{9.190}$$

$$\underset{\sim}{E}_T = \frac{j\omega\mu_0}{p_h^2}\left(\frac{K_\times}{K_\perp} \nabla_T H_z + \underset{\sim}{i}_z \times \nabla_T H_z\right) \tag{9.191}$$

The waveguide fields at cutoff are independent of z and correspond to waves propagating in a direction transverse to the z direction. In the longitudinally magnetized plasma waveguide, this corresponds to propagation across the magnetic field. Hence the transverse wave numbers of Equations 9.184 and 9.188 are the propagation constants of the ordinary and extraordinary waves (Equations 3.6 and 3.7). The field solutions for the E waves at cutoff are the same as the fields of the ordinary wave, and the H-wave fields at cutoff are the same as the fields of the extraordinary wave.

For perfectly conducting walls the boundary conditions on the E waves are (see Equations 9.6, 9.7, 9.184, and 9.187)

$$E_z = 0 \qquad \text{(electric wall)} \tag{9.192}$$

$$\underset{\sim}{n} \cdot \nabla_T E_z = 0 \qquad \text{(magnetic wall)} \tag{9.193}$$

where $\underset{\sim}{n}$ is a unit vector normal to the wall. Hence, by Green's theorem (see footnote, page 139), the values of p_e^2 are real and positive, and the cutoff frequencies of the E waves are

$$\omega_{co}^2 = \omega_p^2 + (p_e c)^2 \tag{9.194}$$

where $p_e c$ is any of the cutoff frequencies of the E waves in the free-space waveguide. The boundary conditions on the H waves are (see Equations 9.6, 9.7, 9.188, and 9.191)

$$\frac{K_\times}{K_\perp}\underset{\sim}{n} \times \nabla_T H_z + (\underset{\sim}{n} \cdot \nabla_T H_z)\underset{\sim}{i}_z = 0 \qquad \text{(electric wall)} \qquad (9.195)$$

In general the values for p_h^2 and the cutoff frequencies must be obtained from a simultaneous solution of Equations 9.188 and 9.195. The peculiar behavior of p_h^2 is here caused by the electric field of the extraordinary wave, which is in general elliptically polarized in the plane transverse to the z axis.

In many simple geometries it is possible to choose solutions to Equation 9.188 that have $\nabla_T H_z$ in the direction of $\underset{\sim}{n}$ (for example, circular waveguide fields with no ϕ variation). For these simple solutions the boundary conditions on the perfectly conducting electric wall simplifies

$$\underset{\sim}{n} \cdot \nabla_T H_z = 0 \qquad (9.196)$$

and the values of p_h^2 are real, positive, and identical to the eigenvalues of H waves in the free-space waveguide. The solutions for the H-wave cutoff frequencies in the LMG plasma waveguide can then be obtained from the values of $\omega n_\times = p_h c$, where $p_h c$ are the cutoff frequencies of H waves in the free-space waveguide. Figure 9.7 illustrates how these cutoff frequencies can be determined. A plot of $\omega n_\times$ vs. ω can be readily obtained from Figure 3.5 for a given ratio of m_+/m_- and $\omega_p^2/\omega_{b+}\omega_{b-}$. A typical example is shown in Figure 3.2 in the curves denoted by $\times$. The poles of $n_\times$ occur at the roots of $K_\perp = 0$:

$$\omega^4 - (\omega_{b-}^2 + \omega_{b+}^2 + \omega_p^2)\omega^2 + \omega_{b-}\omega_{b+}(\omega_p^2 + \omega_{b-}\omega_{b+}) = 0 \qquad (9.197)$$

and the zeros of $n_\times$ occur at the roots of

$$\omega^4 - (\omega_{b-}^2 + \omega_{b+}^2 + 2\omega_p^2)\omega^2 + (\omega_p^2 + \omega_{b+}\omega_{b-})^2 = 0 \qquad (9.198)$$

The intersections marked H_0 give the cutoff frequencies for the H-wave solutions having $\nabla_T H_z$ in the direction of $\underset{\sim}{n}$. The value of $p_h c$ in Figure 9.7 was chosen equal to ω_p. Assuming this to be the lowest cutoff frequency of the free-space waveguide, we note that there are three regions in which an infinite number of H-wave cutoff frequencies will exist. The extent of these regions is indicated below the figure. For reference, the lowest E-wave cutoff is shown, and the extent of the frequency region where E-wave cutoffs occur is also given below the figure.

The energy and power relations at cutoff follow from Equations 7.137 through 7.140. For the LMG temperate plasma U_c and $U_{\epsilon zT}$ vanish (Equations 7.124 and 7.133), and we obtain

$$P = 0 \qquad (9.199)$$

$$Q = 0 \qquad (9.200)$$

$$U_m = U_\epsilon \qquad (9.201)$$

$$U_{mT} + U_{\epsilon T} = U_{mz} + U_{\epsilon z} \tag{9.202}$$

$$\frac{\partial\omega}{\partial\beta} = 0 \tag{9.203}$$

These relations apply at any of the finite, nonzero, cutoff frequencies. Throughout this section we have ignored losses in the plasma and found sharp cutoffs in the propagation. The effect of slight losses is to blur the cutoffs, similar to what was found in Section 9.3.

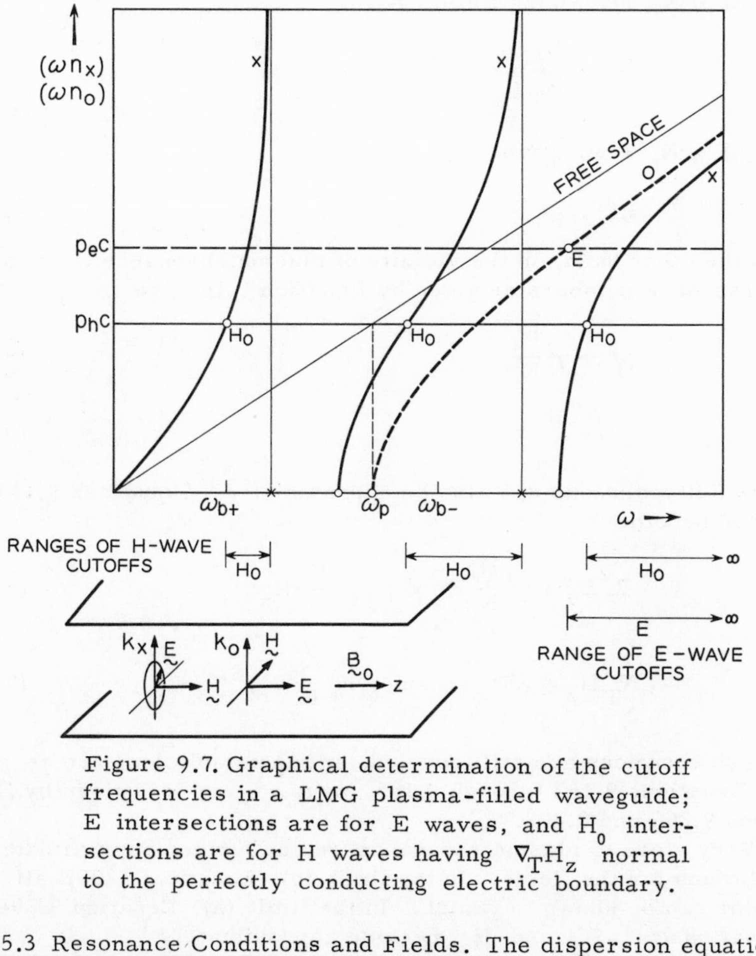

Figure 9.7. Graphical determination of the cutoff frequencies in a LMG plasma-filled waveguide; E intersections are for E waves, and H_0 intersections are for H waves having $\nabla_T H_z$ normal to the perfectly conducting electric boundary.

9.5.3 Resonance Conditions and Fields. The dispersion equation, Equation 9.167, for $\gamma \to \infty$ gives two possible solutions for the two values of p^2. In one, both values of $p^2 \to \infty$ as γ^2, and in the other, one value of $p^2 \to \gamma^2$ and the other value of p^2 remains finite. In this section we shall consider the resonances for which one of the p^2 values remains finite. The corresponding field solutions are appropriate to the case of a homogeneous plasma that fills the waveguide with perfectly conducting walls. The resonances corresponding to both p^2 values infinite are associated with surface waves. These will be considered in Chapter 10, where the waveguide is only partially filled with a homogeneous plasma.

Plasma resonance: $\omega \to \omega_p$, $\gamma \to \infty$

From Equation 9.170 we note that plasma resonance can be approached in two limits:

(a) $\gamma^2 K_{\parallel}$ remaining finite, giving

$$\gamma^2 \to \frac{K_{\perp}}{K_{\parallel}} p^2 \tag{9.204}$$

(b) $\gamma^2 K_{\parallel} \to \infty$, giving

$$\gamma^2 \to p^2 \tag{9.205}$$

On the other hand, in the vicinity of plasma resonance the two transverse wave numbers as given by Equation 9.168 are

$$p_1^2 \to \gamma^2 \frac{K_{\parallel}}{K_{\perp}} \tag{9.206}$$

$$p_2^2 \to \gamma^2 \tag{9.207}$$

The field solutions can also be approximated. Equations 9.157 and 9.158 become

$$\nabla_T^2 E_z + \gamma^2 \frac{K_{\parallel}}{K_{\perp}} E_z = j\omega\mu_0\gamma \frac{K_{\times}}{K_{\perp}} H_z \tag{9.208}$$

$$\nabla_T^2 H_z + \gamma^2 H_z = -j\omega\epsilon_0\gamma K_{\parallel} \frac{K_{\times}}{K_{\perp}} E_z \tag{9.209}$$

These equations are still coupled, and their solutions are as given by Equations 9.163 through 9.165, with p_1^2 and p_2^2 given by Equations 9.206 and 9.207.

Very close to plasma resonance we can further approximate the solutions for the fields. In the limit (b), Equation 9.205, all the fields can be shown to vanish. In the limit (a), Equation 9.204, the equations for E_z and H_z become approximately

$$H_z = 0 \tag{9.210}$$

$$\nabla_T^2 E_z + \gamma^2 \frac{K_{\parallel}}{K_{\perp}} E_z = 0 \tag{9.211}$$

In this limit the transverse fields become

$$\begin{bmatrix} \underset{\sim}{E}_T \\ \underset{\sim}{H}_T \end{bmatrix} \rightarrow \begin{bmatrix} P & Q \\ T & U \end{bmatrix} \begin{bmatrix} \nabla_T E_z \\ \underset{\sim}{i}_z \times \nabla_T E_z \end{bmatrix} \tag{9.212}$$

where

$$P \rightarrow -\frac{1}{\gamma} \tag{9.213}$$

$$Q \rightarrow \frac{k_0^2 K_\times}{\gamma^3} \tag{9.214}$$

$$T \rightarrow \frac{j\omega\epsilon_0 K_\times}{\gamma^2} \tag{9.215}$$

$$U \rightarrow \frac{-j\omega\epsilon_0 K_\perp}{\gamma^2} \tag{9.216}$$

Since $\nabla_T E_z$ is proportional to $\gamma^2 K_\parallel$, the transverse electric and magnetic fields very near plasma resonance become vanishingly small; $\underset{\sim}{E}_T$ vanishes as $\gamma K_\parallel$, and $\underset{\sim}{H}_T$ approaches zero as $K_\parallel$. The only field of importance is then E_z from Equation 9.211. The solutions of Equation 9.211 with boundary conditions pertaining to perfectly conducting walls have eigenvalues

$$p_e^2 = \gamma^2 \frac{K_\parallel}{K_\perp} \tag{9.217}$$

that are real and positive (see footnote, page 139). Since the transverse fields are negligible, the values of p_e^2 are independent of frequency and pertain to E-wave eigenvalues in a free-space waveguide of the same geometry. For a better approximation to the fields near plasma resonance, Equations 9.208 and 9.209 must be used. These show that in general both $\underset{\sim}{E}$ and $\underset{\sim}{H}$ waves must exist.

We note that near plasma resonance the solutions are quasi-static E waves. A quasi-static analysis which assumes from the start that the electric field is derivable from a potential, and that only E waves exist, was first carried out by Smullin and Chorney[32] and by Trivelpiece and Gould.[33] The resulting dispersion relation from their analysis is identical with Equation 9.217 (see also Section 10.5).

The limit of a one-dimensional system corresponds to letting p_e approach zero. Thus, from Equation 9.217

$$\gamma^2 = \frac{K_\perp}{K_\parallel} p_e^2 \rightarrow \text{arbitrary} \tag{9.218}$$

which corresponds to the solution of Equation 2.27 for $\theta \rightarrow 0$ and $K_\parallel \rightarrow 0$.

Cyclotron resonances: $\omega \rightarrow \omega_{b+}$ or $\omega \rightarrow \omega_{b-}$, $\gamma \rightarrow \infty$
The cyclotron resonances can also be approached in two distinct limits of Equation 9.170:

(a) $\dfrac{\gamma^2}{K_\perp}$ remaining finite, and

$$\gamma^2 \rightarrow \frac{K_\perp}{K_\parallel}(p^2 - 2k_0^2K_\parallel) \tag{9.219}$$

(b) $\dfrac{\gamma^2}{K_\perp} \rightarrow \infty$, and

$$\gamma^2 \rightarrow p^2 \tag{9.220}$$

From Equation 9.168, near the cyclotron frequencies, we obtain for the two transverse wave numbers

$$p_1^2 \rightarrow \frac{K_\parallel}{K_\perp}\gamma^2 + 2k_0^2K_\parallel \tag{9.221}$$

$$p_2^2 \rightarrow \gamma^2 \tag{9.222}$$

The coupled equations for E_z and H_z become

$$\nabla_T^2 E_z + \left(\gamma^2\frac{K_\parallel}{K_\perp} + k_0^2K_\parallel\right)E_z = j\omega\mu_0\gamma\frac{K_\times}{K_\perp}H_z \tag{9.223}$$

$$\nabla_T^2 H_z + \gamma^2 H_z = -j\omega\epsilon_0\gamma K_\parallel\frac{K_\times}{K_\perp}E_z \tag{9.224}$$

and their solutions with p_1^2 and p_2^2 of Equations 9.221 and 9.222 are as given by Equations 9.163 through 9.165.

Very close to the cyclotron resonances, Equations 9.221 and 9.222 can be simplified. In the limit (b), Equation 9.220, all the fields can be shown to vanish. In the limit (a), Equation 9.219, we obtain from Equation 9.224

$$H_z \rightarrow -\frac{j\omega\epsilon_0}{\gamma}\frac{K_\times}{K_\perp}K_\parallel E_z \tag{9.225}$$

Hence H_z vanishes, and using Equation 9.225 in Equation 9.223, we obtain

$$H_z = 0 \tag{9.226}$$

$$\nabla_T^2 E_z + \left(\gamma^2\frac{K_\parallel}{K_\perp} + 2k_0^2K_\parallel\right)E_z = 0 \tag{9.227}$$

In this limit the transverse fields are also of the form of Equation 9.212 but with

$$P \to -\frac{1}{\gamma} \frac{(\gamma^2 + k_0^2 K_\perp)}{(\gamma^2 + 2k_0^2 K_\perp)} \to -\frac{1}{p^2} \frac{(p^2 - k_0^2 K_{||})}{\gamma} \tag{9.228}$$

$$Q \to \frac{1}{\gamma} \frac{K_\times}{K_\perp} \frac{k_0^2 K_\perp}{(\gamma^2 + 2k_0^2 K_\perp)} \to \pm j \frac{K_{||}}{p^2} \frac{k_0^2}{\gamma} \tag{9.229}$$

$$T \to \frac{K_\times}{K_\perp} \frac{j\omega\epsilon_0 K_\perp}{(\gamma^2 + 2k_0^2 K_\perp)} \to \mp \omega\epsilon_0 \frac{K_{||}}{p^2} \tag{9.230}$$

$$U \to \frac{-j\omega\epsilon_0 K_\perp}{(\gamma^2 + 2k_0^2 K_\perp)} \to -j\omega\epsilon_0 \frac{K_{||}}{p^2} \tag{9.231}$$

In contrast to the case of plasma resonance, the transverse fields do not vanish, and in particular the transverse magnetic field is finite. The solutions of Equation 9.227 with perfectly conducting electric walls have eigenvalues

$$p_{eb}^2 = \gamma^2 \frac{K_{||}}{K_\perp} + 2k_0^2 K_{||} \tag{9.232}$$

that are real and positive. Equations 9.228 through 9.231 show that the transverse fields are in general elliptically polarized. Hence the boundary conditions in general require that p_{eb} be a function of frequency. A better approximation to the fields near the cyclotron resonances must be obtained from the coupled set of Equations 9.223 and 9.224.

A quasi-static field analysis is in this case not a valid approximation. Such an analysis yields a dispersion relation identical to Equation 9.217 rather than Equation 9.232. This dispersion relation is therefore approximately valid only when $2k_0^2 K_{||}$ at $\omega = \omega_{b\pm}$ can be neglected in comparison with p_{eb}^2.

In the one-dimensional limit, $p_{eb} \to 0$, Equation 9.232 gives

$$\gamma^2 \to -2k_0^2 K_\perp \tag{9.233}$$

which corresponds exactly to the limit of the cyclotron resonances of the left and right circularly polarized principal waves propagating along $\underset{\sim}{B}_0$, Table 3.1.

The characteristics of $\gamma(\omega)$ near the resonances

We shall now consider in detail the frequency dependence of Equations 9.217 and 9.232 in the vicinity of the plasma and cyclotron resonances, respectively. The values of p_e^2 and p_{eb}^2 are real and positive.

Near plasma resonance Equation 9.217 gives

$$\gamma_p^2 \approx -p_e^2 \left[\frac{C\omega_p^2 \omega_{b-}^2(\omega_p^2 - \omega_z^2)}{(\omega_p^2 - \omega_{b+}^2)(\omega_p^2 - \omega_{b-}^2)} \right] \frac{1}{(\omega^2 - \omega_p^2)} \qquad (9.234)$$

where

$$C = 1 - \frac{m_-}{m_+} + \left(\frac{m_-}{m_+}\right)^2 \qquad (9.235)$$

$$\omega_z^2 = \frac{\omega_{b+}^2}{C} \qquad \text{(zero of } K_\perp \text{ at } \omega = \omega_p) \qquad (9.236)$$

For the usual situation of $m_- < m_+$, we have $C < 1$ and $\omega_z > \omega_{b+}$. Equation 9.234 shows that near $\omega = \omega_p$ the propagation constant changes sign at the cyclotron frequencies and at ω_z. In Table 9.1 we give the two frequency ranges that have different propagation characteristics.

Table 9.1. Characteristics of the Propagation Constant in the Vicinity of $\omega = \omega_p$

I. $\omega_{b+} < \omega_p < \omega_z$ or $\omega_p > \omega_{b-}$			
	γ_p	$\left(\frac{\partial\omega}{\partial\beta}\right)_p$	$\left(\frac{\partial\omega}{\partial\alpha}\right)_p$
$\omega < \omega_p$	α_p		> 0
$\omega > \omega_p$	$j\beta_p$	< 0	
II. $\omega_z < \omega_p < \omega_{b-}$ or $\omega_p < \omega_{b+}$			
	γ_p	$\left(\frac{\partial\omega}{\partial\beta}\right)_p$	$\left(\frac{\partial\omega}{\partial\alpha}\right)_p$
$\omega < \omega_p$	$j\beta_p$	> 0	
$\omega > \omega_p$	α_p		< 0

In the ranges specified by I the propagation at frequencies slightly above ω_p is that of a backward wave (phase and group velocities of opposite sign), while in the ranges specified by II the propagation at frequencies slightly below ω_p is that of a traveling wave. It should be noted that for a one-component plasma ($m_+ \to \infty$ and $\omega_z \to 0$)

the change from backward wave to traveling wave occurs as the plasma frequency becomes less than the cyclotron frequency. In the special cases when either $\omega_p = \omega_{b+}$ or $\omega_p = \omega_{b-}$, we find from Equations 9.217 that $\gamma^2 < 0$ (propagating wave) for frequencies both slightly above and slightly below ω_p.

Near the cyclotron resonances Equation 9.232 gives

$$\gamma_\pm^2 \approx \left\{ \frac{\omega_p^2 \omega_{b+}^3 \left[(p_{eb} c)^2 + 2(\omega_p^2 - \omega_{b\pm}^2) \right]}{c^2 (\omega_{b-} + \omega_{b+})(\omega_p^2 - \omega_{b\pm})} \right\} \frac{1}{(\omega^2 - \omega_{b\pm}^2)} \tag{9.237}$$

where the double subscript $\pm$ is to be taken as + at $\omega \to \omega_{b+}$ and - at $\omega \to \omega_{b-}$. The character of the waves is seen to depend on both the magnitudes of the plasma frequency and the characteristic cutoff frequency $p_{eb} c \equiv \omega_{co}$. In Table 9.2 we show the three sets of parameters that produce all possible characteristics of the propagation constants.

Table 9.2. Characteristics of the Propagation Constant in the Vicinity of $\omega = \omega_{b\pm}$

I. $\omega_p > \omega_{b\pm}$ and p_{eb} arbitrary			
	$\gamma_\pm$	$\left(\frac{\partial\omega}{\partial\beta}\right)_\pm$	$\left(\frac{\partial\omega}{\partial\alpha}\right)_\pm$
$\omega < \omega_{b\pm}$	$j\beta_\pm$	> 0	
$\omega > \omega_{b\pm}$	$\alpha_\pm$		< 0
II. $\omega_p < \omega_{b\pm}$ and $(p_{eb}c)^2 > 2\lvert\omega_p^2 - \omega_{b\pm}^2\rvert$			
	$\gamma_\pm$	$\left(\frac{\partial\omega}{\partial\beta}\right)_\pm$	$\left(\frac{\partial\omega}{\partial\alpha}\right)_\pm$
$\omega < \omega_{b\pm}$	$\alpha_\pm$		> 0
$\omega > \omega_{b\pm}$	$j\beta_\pm$	< 0	
III. $\omega_p < \omega_{b\pm}$ and $(p_{eb}c)^2 < 2\lvert\omega_p^2 - \omega_{b\pm}^2\rvert$			
	$\gamma_\pm$	$\left(\frac{\partial\omega}{\partial\beta}\right)_\pm$	$\left(\frac{\partial\omega}{\partial\alpha}\right)_\pm$
$\omega < \omega_{b\pm}$	$j\beta_\pm$	> 0	
$\omega > \omega_{b\pm}$	$\alpha_\pm$		< 0

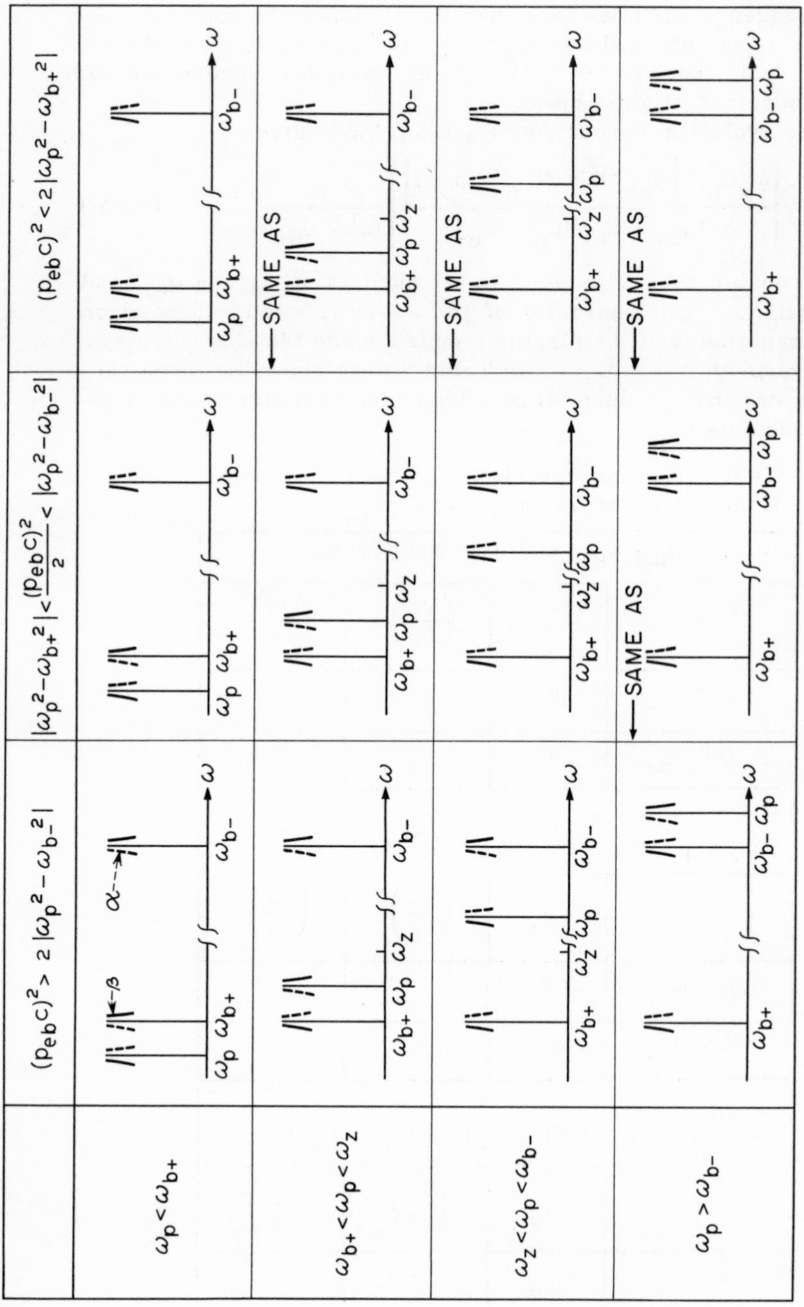

Figure 9.8. Dispersion of propagation constant $\gamma = \alpha + j\beta$ near resonance, $\gamma \to \infty$, for an LMG, homogeneous, lossless plasma waveguide with a perfectly conducting wall; the ordinates for each set are β (solid line) and α (dashed line). For clarity, ω_z/ω_{b+} is shown exaggerated.

The propagating waves in cases I and III of Table 9.2 are traveling waves, while those in case II are backward waves. For the special cases when either $\omega_p = \omega_{b+}$ or $\omega_p = \omega_{b-}$, we find as before that $\gamma^2 < 0$ (propagating wave) both slightly above and below ω_{b+} or ω_{b-}, respectively. In Figure 9.8 we illustrate the possible dispersion characteristics of the propagation constant in the vicinity of the resonances as given in Tables 9.1 and 9.2. The figures are arranged with the plasma frequency increasing from top to bottom and the effective empty-waveguide cutoff frequency decreasing from left to right. We note that the existence of slow backward waves is a new feature that can occur in a temperate plasma only when the plasma is bounded (compare Equations 7.99 and 7.153).

9.5.4 Low-Frequency, Magnetohydrodynamic Regime. When the frequency is low enough, or the density and magnetic field high enough, so that $\omega_p/\omega \gg 1$ and $\omega_{b+}/\omega \gg 1$, the displacement current becomes small compared with the convection current, and the charges are constrained to the magnetic lines of force. This is the magnetohydrodynamic limit. In this limit the elements of the dielectric tensor for the temperate plasma are approximately given by

$$K_{\parallel} \to -a^2 = -\frac{\omega_p^2}{\omega^2} \tag{9.238}$$

$$K_{\perp} \to 1 + \frac{a^2}{\beta_- \beta_+} = 1 + \frac{c^2}{u_a^2} \tag{9.239}$$

$$K_{\times} \to -j\frac{c^2}{u_a^2}\frac{(\beta_- - \beta_+)}{\beta_- \beta_+} \to 0 \tag{9.240}$$

where

$$u_a^2 = \frac{\dfrac{B_0^2}{\mu_0}}{N(m_+ + m_-)} \tag{9.241}$$

is the Alfvén speed. Equation 9.239 is obtained by neglecting terms of order $(c^2/u_a^2)/\beta_+$, that is, of the order of $K_\times$. With these approximations the anisotropic temperate plasma is describable by a diagonal dielectric tensor, and the analysis of wave propagation simplifies.

For the limiting forms given by Equations 9.238 through 9.240, the coefficients of the coupled wave equations, Equations 9.157 and 9.158, become

$$a \to \left[\gamma^2 + k_0^2\left(1 + \frac{c^2}{u_a^2}\right)\right]\frac{-a^2}{\left(1 + \dfrac{c^2}{u_a^2}\right)} \tag{9.242}$$

$$b \rightarrow -\omega\mu_0\gamma \frac{\frac{c^2}{u_a^2}}{\left(1 + \frac{c^2}{u_a^2}\right)} \frac{(\beta_- - \beta_+)}{\beta_-\beta_+} \tag{9.243}$$

$$c \rightarrow \gamma^2 + k_0^2 \left(1 + \frac{c^2}{u_a^2}\right) \tag{9.244}$$

$$d \rightarrow -\omega\epsilon_0\gamma \frac{\left(\frac{c^2}{u_a^2}\right)^2}{\left(1 + \frac{c^2}{u_a^2}\right)} (\beta_- - \beta_+) \tag{9.245}$$

We shall consider only the low-frequency limit for which we have

$$b \rightarrow 0 \tag{9.246}$$

$$d \rightarrow 0 \tag{9.247}$$

according to $K_\times \rightarrow 0$. Hence, the wave equations for E_z and H_z become uncoupled, and we can study separately E waves and H waves. The coefficients that determine the transverse fields, Equation 9.21, become

$$P \rightarrow \frac{-\gamma}{(\gamma^2 + k_0^2 K_\perp)} \tag{9.248}$$

$$S \rightarrow \frac{j\omega\mu_0}{(\gamma^2 + k_0^2 K_\perp)} \tag{9.249}$$

$$U \rightarrow \frac{-j\omega\epsilon_0 K_\perp}{(\gamma^2 + k_0^2 K_\perp)} \tag{9.250}$$

and R, Q, and T vanish.

E waves: $H_z = 0$

The wave equation for the longitudinal E field is given by Equations 9.157 and 9.242 (Equation 9.32):

$$\nabla_T^2 E_z + p_e^2 E_z = 0 \tag{9.251}$$

$$p_e^2 = (\gamma^2 + k_0^2 K_\perp) \frac{K_\parallel}{K_\perp} \tag{9.252}$$

For the perfectly conducting boundary conditions at the waveguide wall, p_e^2 is positive and real. For a particular geometry we have an infinite set of distinct values for p_e^2 corresponding to the E-wave eigenvalues of a free-space waveguide of the same geometry. Equation 9.252 can be solved for the propagation constant, giving

$$\gamma^2 = -\frac{\omega^2}{u_a^2}\left(1+\frac{u_a^2}{c^2}\right)\left(1+\frac{p_e^2c^2}{\omega_p^2}\right) \qquad (9.253)$$

Hence, the magnetohydrodynamic E waves have no low-frequency cutoff and have their phase and group velocity equal. Propagation in such a wave is free of distortions. The phase velocity of these waves is seen to be less than the Alfvén speed, Equation 9.241. In the high-density limit, $\omega_p^2 >> \omega_{b+}\omega_{b-}$, $u_a^2 << c^2$, and the expression in the first parentheses in Equation 9.253 is approximately equal to unity. The expression in the second parentheses depends upon the relative magnitude of the empty-waveguide cutoff frequency $p_e c$ to the plasma frequency ω_p. A sketch of Equation 9.253 is shown in Figure 9.9. For each value of p_e the dispersion characteristic of the propagation constant is a straight line from the origin with slope that increases with increasing values of p_e.

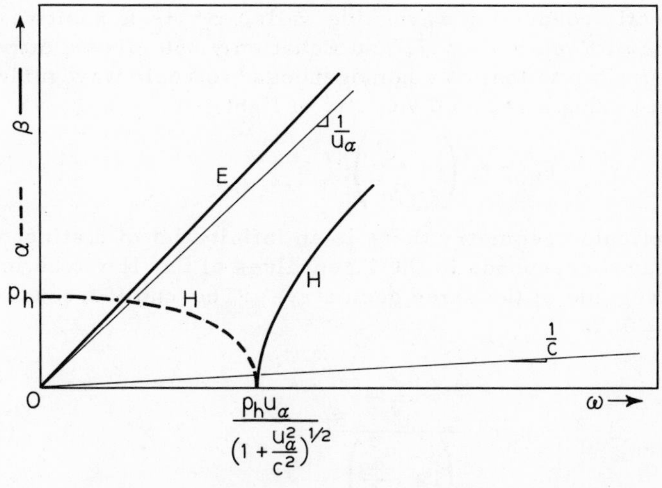

Figure 9.9. Dispersion characteristic of the E and H waves in the low-frequency magnetohydrodynamic regime, $\omega_p/\omega >> 1$ and $\omega_{b+}/\omega >> 1$.

The transverse fields follow from Equations 9.21 and 9.248 through 9.250:

$$\underset{\sim}{E}_T = \frac{-\gamma K_{\parallel}}{p_e^2 K_{\perp}} \nabla_T E_z \qquad (9.254)$$

$$\underset{\sim}{H}_T = \frac{-j\omega\epsilon_0 K_{\parallel}}{p_e^2} \underset{\sim}{i}_z \times \nabla_T E_z \qquad (9.255)$$

where $K_{\parallel}$ and $K_{\perp}$ are as given by Equations 9.238 and 9.239. The frequency dependence of these fields is seen to be related to γ and $K_{\parallel}$. The wave impedance, given by the ratio $|\underset{\sim}{i}_z \times \underset{\sim}{E}_T| / |\underset{\sim}{H}_T|$, is frequency-independent:

$$Z_{we} = \sqrt{\frac{\mu_0}{\epsilon_0}} \frac{u_a}{c} \sqrt{\frac{1 + \dfrac{p_e^2 c^2}{\omega_p^2}}{1 + \dfrac{u_a^2}{c^2}}} \tag{9.256}$$

This is generally less than the free-space value and depends upon p_e.

H waves: $E_z = 0$

The wave equation for the longitudinal H field follows from Equations 9.158 and 9.244:

$$\nabla_T^2 H_z + p_h^2 H_z = 0 \tag{9.257}$$

$$p_h^2 = \gamma^2 + k_0^2 K_\perp \tag{9.258}$$

For perfectly conducting waveguide walls, p_h^2 is a positive real eigenvalue of Equation 9.257, and Equation 9.258 gives a dispersion relation similar to that of a homogeneous isotropic waveguide with a medium having a reduced velocity of light:

$$\gamma^2 = p_h^2 - k_0^2 \left(1 + \frac{c^2}{u_a^2}\right) \tag{9.259}$$

For a particular geometry there is an infinite set of distinct values of p_h^2 that corresponds to the eigenvalues of the H waves in a free-space waveguide of the same geometry. The cutoff frequency for which $\gamma = 0$ is

$$\omega_{co}^2 = p_h^2 c^2 \frac{\dfrac{u_a^2}{c^2}}{\left(1 + \dfrac{u_a^2}{c^2}\right)} \tag{9.260}$$

In the high-density limit $(\omega_p^2 \gg \omega_{b+}\omega_{b-})$ the cutoff frequency is approximately $p_h u_a$. It should be noted that the description of this H wave is valid only if $p_h u_a \ll \omega_{b+}$. In fact, the correct value of the cutoff frequency must correspond to the field solutions of Equations 9.188 through 9.191. For frequencies above ω_{co}, the propagation constant is pure imaginary $\gamma = j\beta$, and

$$\beta^2 = \frac{\omega^2}{u_a^2}\left(1 + \frac{u_a^2}{c^2}\right) - p_h^2 \tag{9.261}$$

The high-frequency asymptote has a phase velocity $u_a/[1+(u_a^2/c^2)]^{\frac{1}{2}}$, which is larger than the phase velocity of the E waves, Equation 9.253. For frequencies below ω_{co}, we have a cutoff wave $\gamma = \alpha$,

$$\alpha^2 = p_h^2 - \frac{\omega^2}{u_a^2}\left(1 + \frac{u_a^2}{c^2}\right) \tag{9.262}$$

The attenuation constant is maximum at $\omega = 0$ and equal to p_h, as in an empty waveguide. The dispersion characteristic of the H wave for one value of p_h is also shown in Figure 9.9.

The transverse fields for the H waves follow from Equations 9.21 and 9.248 through 9.250.

$$\underset{\sim}{H}_T = -\frac{\gamma}{p_h^2}\nabla_T H_z \tag{9.263}$$

$$\underset{\sim}{E}_T = \frac{j\omega\mu_0}{p_h^2}\underset{\sim}{i}_z \times \nabla_T H_z \tag{9.264}$$

In the propagating region the wave impedance, $|\underset{\sim}{E}_T|/|\underset{\sim}{i}_z \times \underset{\sim}{H}_T|$, is

$$Z_{wh} = \frac{\sqrt{\dfrac{\mu_0}{\epsilon_0}}\left(\dfrac{u_a}{c}\right)}{\sqrt{\left(1 + \dfrac{u_a^2}{c^2}\right) - \dfrac{p_h^2 u_a^2}{\omega^2}}} \tag{9.265}$$

The dispersion characteristic of the filled hydromagnetic waveguide (a waveguide filled with a perfectly conducting plasma) in the absence of pressure forces[34] follows in the limit of $\omega_p \to \infty$. (In particular this requires $\omega_p^2 \gg \omega_{b+}\omega_{b-}$ and $\omega_p \gg pc$.) In this limit the E waves become a degenerate set of TEM waves, the Alfvén wave, with phase velocity u_a. The H waves persist and have their cutoff frequency at $p_h u_a$. In the presence of pressure forces in the perfectly conducting plasma, the above TEM Alfvén and H waves become modified and coupled to the acoustic waves.[35]

9.5.5 Summary and Energy and Power Relations (LMG). The results of Section 9.5 are summarized in Table 9.3. For the LMG, homogeneous, temperate plasma, Maxwell's equations lead to a coupled set of wave equations for the longitudinal fields E_z and H_z. These are shown at the top of Table 9.3, and the coefficients are listed in the column marked General. Associated with wave solutions to Maxwell's equations is the dispersion relation. This functional equation relating the propagation constant to the frequency and plasma parameters was already derived in Part I (Equation 2.24). There, in the absence of boundaries, this equation related the propagation constant in an arbitrary direction to the frequency (Equation 2.27).

Table 9.3 The Coupled Wave Equations and the Dispersion Relation for an LMG, Homogeneous, Plasma Waveguide

$$\left.\begin{aligned} \nabla_T^2 E_z + aE_z &= bH_z \\ \nabla_T^2 H_z + cH_z &= dE_z \end{aligned}\right\} \quad p^4 + (a + c)\, p^2 + (ac - bd) = 0$$

	General, ω Arbitrary	$\omega \to 0$, MHD $\omega << \omega_{b\pm}$; $\omega << \omega_p$	$\gamma = 0$ Cutoff	$\gamma \to \infty$, Resonance $\omega \to \omega_p$	$\gamma \to \infty$, Resonance $\omega \to \omega_{b+}$; $\omega \to \omega_{b-}$	$\omega \to \infty$ Free Space
a	$(\gamma^2 + k_0^2 K_{\parallel}) \frac{K_{\parallel}}{K_{\perp}}$	$(\gamma^2 + k_0^2 K_{\parallel}) \frac{K_{\parallel}}{K_{\perp}} \approx p_1^2$	$k_0^2 K_{\parallel} = p_1^2$	$\gamma^2 \frac{K_{\parallel}}{K_{\perp}} \to p_1^2$	$\gamma^2 \frac{K_{\parallel}}{K_{\perp}} + 2k_0^2 K_{\parallel} \to p_1^2$	$\gamma^2 + k_0^2 = p_1^2$
b	$-j\omega\mu_0\gamma \frac{K_{\times}}{K_{\perp}}$	0	0	$p_2^2 \to \gamma^2$	$p_2^2 \to \gamma^2$	0
c	$\gamma^2 + k_0^2 \frac{K_r K_{\ell}}{K_{\perp}}$	$\gamma^2 + k_0^2 K_{\perp} \approx p_2^2$	$k_0^2 \frac{K_r K_{\ell}}{K_{\perp}} = p_2^2$	$H_z \to 0$	$H_z \to 0$	$\gamma^2 + k_0^2 = p_2^2$
d	$j\omega\epsilon_0\gamma \frac{K_{\times} K_{\parallel}}{K_{\perp}}$	0	0	0		0

For guided propagation along $\underset{\sim}{B}_0$, it is convenient to write this equation as shown at the top of Table 9.3 and to regard it as relating the propagation constant γ along $\underset{\sim}{B}_0$ to the transverse wave numbers p and the frequency. (See Section 9.5.1.) Another functional equation relating γ, p, and ω is obtained from the requirement that the field solutions for a particular geometry must satisfy the boundary conditions at the waveguide walls. (See Chapter 10 for examples.) This equation, usually called the determinantal equation from the boundary conditions, together with the dispersion relation, determines γ and p as functions of frequency, plasma parameters, and geometry. The field solutions are obtained from Equations 9.162 through 9.166. Some characteristic features of the solution are the following: (1) For a given frequency and propagation constant there are two transverse wave numbers; (2) the transverse wave numbers are usually functions of frequency and geometry; hence there are no field patterns that are independent of frequency as in a free-space waveguide bounded by perfectly conducting walls; (3) the propagation constant and transverse wave numbers may be complex even though the plasma and boundaries are lossless; (4) the fields are in general mixed E and H waves.

For the restricted frequency ranges shown at the top of columns 3 through 7 in Table 9.3, the wave equations are approximately decoupled, and the dispersion relation takes on the simplified form shown in the respective columns. In the MHD regime (column 3) we can distinguish E waves with dispersion relation given by p_1^2 and H waves with dispersion relation given by p_2^2. The E waves cut off at zero frequency, while the H-wave cutoff frequencies are finite. Hence, these H-wave solutions are valid only if their cutoff frequency is much below both ω_p and $\omega_{b\pm}$. The fields associated with these solutions and perfectly conducting boundary conditions lead to values of p_1 and p_2 that are dependent on geometry only (see Chapter 10). At cutoff the wave equations split exactly into E and H waves, given respectively by p_1^2 and p_2^2 in column 4. These are recognized to be the dispersion relations for the ordinary and extraordinary waves, respectively, propagating across the magnetic field $\underset{\sim}{B}_0$. For the E-wave cutoffs, perfectly conducting electric boundary conditions can be satisfied with constant values of p_1 corresponding to the resonant wave numbers for the cross section of a free-space waveguide. For the H-wave cutoffs, the electric field of the extraordinary wave is elliptically polarized in the transverse plane, and p_2 must in general be a function of frequency. The resonances in propagation occur at the plasma and cyclotron frequencies. Near the resonances the fields are mixed E and H, with the dispersion relation given by p_1^2 and p_2^2 in columns 5 and 6. Solutions with both E and H waves are necessary for properly matching boundary conditions. The character of the propagation near resonance is essentially given by the equation for p_1^2 constant (see Figure 9.8). Both traveling waves and backward waves are possible. Finally, for extremely high frequencies, the plasma model reduces to free space, and the guided-wave solutions are the well-known E and H waves, with dispersion relations given respectively by p_1^2 and p_2^2 in column 7. The field solutions are summarized in Section 9.2, and the perfectly conducting boundary conditions on the waveguide walls determine p_1 and p_2 as functions of geometry and independent of frequency (see Chapter 10).

The energy and power relations of Chapter 7 can now be summarized for the particular case of the LMG, temperate plasma waveguide. For the temperate plasma tensor (Equation 2.13) we have $\underset{\approx}{K}_{Tz} = 0$, and hence $U_{\epsilon zT} = 0$ (Equation 7.133). The propagation constant and frequency are related to the real and reactive power flow (P and Q of Equation 7.112), and to $U_{\epsilon z}$ and $U_{\epsilon T}$ (Equations 7.134 and 7.135), by Equations 7.137 through 7.140. Using the tensor of Equation 2.13, we find that Equations 7.134 and 7.135 become

$$U_{\epsilon z} = \tfrac{1}{4}\epsilon_0 \int_A K_{\parallel} |E_z|^2 \, da \tag{9.266}$$

$$U_{\epsilon T} = \tfrac{1}{4}\epsilon_0 \int_A \left[K_{\perp}\left(|E_{T_1}|^2 + |E_{T_2}|^2\right) + 2\,\mathrm{Re}\left(E_{T_1}E_{T_2}{}^{*}K_{\times}\right)\right] da \tag{9.267}$$

where E_{T_1} and E_{T_2} are the two orthogonal components of the electric field in the transverse plane (for example, E_x and E_y or E_r and E_ϕ). Equation 9.266 is positive when $K_{\parallel} > 0$, $\omega > \omega_p$, and negative when $K_{\parallel} < 0$, $\omega < \omega_p$. The sign of $U_{\epsilon T}$ can be determined only if the field solutions are known. The time-averaged energy per unit length of guide U is given by Equation 7.119 and can be shown to be positive definite.

For a propagating wave ($\alpha = 0$, $\gamma = j\beta$), Equations 7.149a through 7.150b, 7.152, and 7.153a give

$$Q = 0 \tag{9.268}$$

$$U_{mT} + U_{mz} = U_{\epsilon T} + U_{\epsilon z} \tag{9.269}$$

$$\frac{\partial\omega}{\partial\beta} = \frac{\omega}{\beta}\,\frac{2(U_{mT} - U_{\epsilon z})}{U} = \frac{\omega}{\beta}\,\frac{2(U_{\epsilon T} - U_{mz})}{U} \tag{9.270}$$

Since U_{mT} (Equation 7.113 with $\underset{\sim}{H} \equiv \underset{\sim}{H}_T$) is positive definite and $U_{\epsilon z} > 0$ for $\omega < \omega_p$, Equation 9.270 predicts that for $\omega < \omega_p$ the group velocity and phase velocity must have the same sign. At the cutoff frequencies, for which $\gamma = 0$, Equations 7.146b and 7.149b must be satisfied simultaneously, and hence

$$U_{\epsilon z} = U_{mT} \tag{9.271}$$

$$U_{\epsilon T} = U_{mz} \tag{9.272}$$

Equations 9.271 and 9.272 apply to the E-wave and H-wave cutoff fields, respectively. Complex waves ($\gamma = \alpha + j\beta$, Equations 7.141 through 7.144) must carry neither real nor reactive power and must satisfy Equations 9.271 and 9.272. Hence, if $U_{\epsilon z} \neq 0$, complex waves cannot exist for $\omega < \omega_p$. Cutoff waves ($\gamma = \alpha$) have zero real power flow and must satisfy Equation 7.146b.

9.5.6 Infinite Magnetic Fields. When the externally applied magnetic field B_0 is infinite, the charged particles of the plasma are constrained to move along the direction of $\underset{\sim}{B}_0$. Under these conditions, $K_\times = 0$, $K_\perp = 1$, and the plasma tensor becomes diagonal $(1, 1, K_\parallel)$. The wave equations for E_z and H_z, Equations 9.157 and 9.158, become independent, and we can identify E- and H-wave solutions for all frequencies.

The E-wave solutions are

$$H_z = 0 \tag{9.273}$$

$$\nabla_T^2 E_z + p_e^2 E_z = 0 \tag{9.274}$$

$$p_e^2 = (\gamma^2 + k_0^2)K_\parallel \tag{9.275}$$

$$\underset{\sim}{E}_T = -\frac{\gamma}{\gamma^2 + k_0^2} \nabla_T E_z \tag{9.276}$$

$$\underset{\sim}{H}_T = -\frac{j\omega\epsilon_0}{\gamma^2 + k_0^2} \underset{\sim}{i}_z \times \nabla_T E_z \tag{9.277}$$

The charged particles interact with the electromagnetic fields only through E_z. Thus Equations 9.276 and 9.277 for the transverse fields are identical in form with those of a free-space waveguide, Equations 9.44 and 9.45. For a plasma-filled waveguide with perfectly conducting walls, the transverse wave numbers p_e^2 of Equation 9.275 are real, positive, and independent of frequency. Hence, for each value of p_e, the dispersion relation for the E waves, Equation 9.275, can be written as

$$\gamma^2 = -k_0^2 \, \frac{\omega^2 - (\omega_p^2 + p_e^2 c^2)}{\omega^2 - \omega_p^2} \tag{9.278}$$

Equation 9.278 shows one resonance at $\omega = \omega_p$ and one cutoff at $\omega = (\omega_p^2 + p_e^2 c^2)^{\frac{1}{2}}$ for each distinct value of p_e. Figure 9.10 illustrates the dispersion characteristics for one particular value of p_e.

The H-wave solutions are identical to those in a free-space waveguide, Equations 9.38 through 9.41. This is understandable, because in the z direction, where the charged particles are free to move, the H waves have only a magnetic field, and hence, there is no interaction between these fields and the plasma.

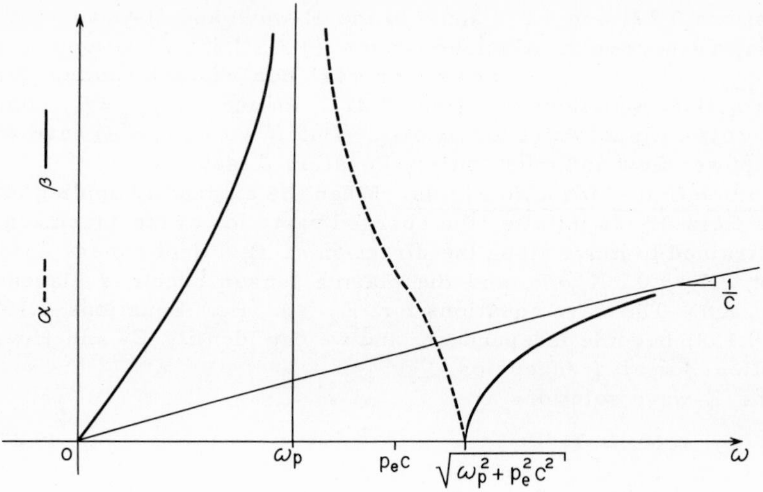

Figure 9.10. Dispersion characteristics for E waves in an LMG plasma waveguide with infinite magnetic field.

9.6 Transversely Magnetized, Inhomogeneous Plasma Waveguides

We now turn our attention to the field analysis of TMG, temperate plasma waveguides. These systems are uniform in a direction (we choose it to be along the z axis) that is orthogonal to the applied magnetic field. Two such plasma waveguide systems are shown in Figure 9.11. Figure 9.11a is a rectangular waveguide containing plasma with the applied magnetic field in the y direction, and Figure 9.11b is a coaxial waveguide containing plasma with the

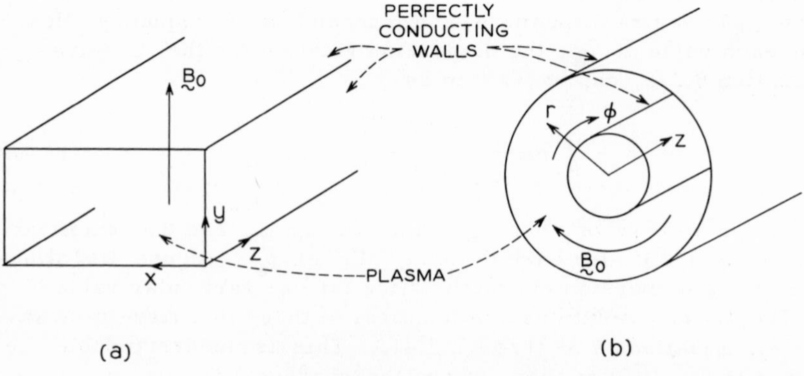

Figure 9.11. TMG waveguides: (a) rectangular waveguide; (b) coaxial waveguide.

applied magnetic field in the ϕ direction. With the z axis orthogonal to $\underset{\sim}{B}_0$, the normalized dielectric tensor for the temperate plasmas, Equation 2.13, takes on the form

$$\underset{\approx}{K} = \begin{bmatrix} K_\perp & 0 & K_\times \\ 0 & K_\parallel & 0 \\ -K_\times & 0 & K_\perp \end{bmatrix} \tag{9.279}$$

where all elements are as defined in Part I. We shall restrict our analysis to the case when the plasma parameters are also uniform in the direction of the applied magnetic field $\underset{\sim}{B}_0$. Thus we shall allow the plasma parameters to be functions of only the transverse coordinate that is orthogonal to $\underset{\sim}{B}_0$. For example, the plasma density and magnetic field in Figure 9.11a may be functions of x, and hence $\underset{\approx}{K}(x)$, and in Figure 9.11b they may be functions of r, and hence $\underset{\approx}{K}(r)$. Under these conditions the analysis of Section 9.1 can be transformed to apply directly to the TMG, inhomogeneous plasma waveguide.

9.6.1 Transversely Magnetized, Rectangular Plasma Waveguide. Consider the system of Figure 9.11a to have a plasma and an applied magnetic field that are uniform in the y and z directions but that may vary in the x direction. Equation 9.279 shows that the electric displacement vector in the y direction is independent of the x and z components of the electric field and that $K_\parallel$ is independent of y. Therefore, we may construct solutions to Maxwell's equations with dependence $\exp(j\omega t)\exp(\mp\gamma z)\exp(\mp jp_y y)$ with γ and p_y as constants independent of the coordinates. Because of the inhomogeneity in the x direction, there is no corresponding propagation constant for this direction. The analysis of Section 9.1 can be made to apply to the present system by making the identifications shown in Table 9.4. With these transformations of Equations 9.21 through 9.37, we obtain the system of equations applicable to TMG, rectangular plasma waveguides, inhomogeneous in the x direction. Equations 9.30 and 9.31 become a set of coupled, second-order differential equations in x for E_y and H_y, and Equation 9.21 gives the remaining field components in terms of E_y and H_y. In general, both solutions, $\exp(-jp_y y)$ and $\exp(jp_y y)$, are necessary for matching boundary conditions. These are readily obtained by noting the bidirectional character of Maxwell's equations, Equations 9.14 through 9.17, for the temperate plasma: If the fields E_z, H_z, $\underset{\sim}{E}_T$, and $\underset{\sim}{H}_T$ are solutions with $\exp(-\gamma z)$ dependence, then the field solutions with $\exp(\gamma z)$ dependence are $\pm E_z$, $\mp H_z$, $\mp\underset{\sim}{E}_T$, and $\pm\underset{\sim}{H}_T$, all with either the upper or the lower signs. Using Table 9.4, we note that if the field solutions E_y, H_y, $E_{x,z}$, and $H_{x,z}$ have the y dependence $\exp(-jp_y y)$, the field solutions with $\exp(+jp_y y)$ dependence, for the same γ, are $\pm E_y$, $\mp H_y$, $\mp E_{x,z}$, and $\pm H_{x,z}$.

Table 9.4. Transformations for Equations 9.21 through 9.37, Which Make Them Applicable to the TMG, Rectangular Plasma Waveguide, Inhomogeneous in the x Direction.

LMG (Section 9.1)		TMG (Section 9.6.1)
z direction	⟶	y direction
γ	⟶	$-jp_y$
$\nabla_T \underset{\approx}{K}$	⟶	$\frac{\partial}{\partial x} \underset{\approx}{K}$
∇_T	⟶	$\underset{\sim}{i}_x \frac{\partial}{\partial x} - \underset{\sim}{i}_z \gamma$
∇_T^2	⟶	$\frac{\partial^2}{\partial x^2} + \gamma^2$
E_z	⟶	E_y
H_z	⟶	H_y
$\underset{\sim}{E}_T$	⟶	$\underset{\sim}{i}_x E_x + \underset{\sim}{i}_z E_z$
$\underset{\sim}{H}_T$	⟶	$\underset{\sim}{i}_x H_x + \underset{\sim}{i}_z H_z$

9.6.2 Transversely Magnetized, Coaxial Plasma Waveguide. For the coaxial waveguide system shown in Figure 9.11b, the plasma and the applied magnetic field are uniform in the ϕ and z directions and can vary with radial position. Here we can construct solutions to Maxwell's equations with dependence $\exp(j\omega t) \exp(\mp\gamma z) \exp(\mp jm\phi$ with γ and m independent of the coordinates. Equations 9.21 through 9.37, with the transformations shown in Table 9.5, may again be made applicable. Thus Equations 9.30 and 9.31 become a set of coupled, second-order differential equations in r for E_ϕ and H_ϕ, and Equation 9.21 gives the remaining fields in terms of E_ϕ and H_ϕ. Again both solutions with $\exp(-jm\phi)$ and $\exp(jm\phi)$ can be easily obtained. If the field solutions E_ϕ, H_ϕ, $E_{r,z}$, and $H_{r,z}$ have a ϕ dependence $\exp(-jm\phi)$, then the field solutions with ϕ dependence $\exp(jm\phi)$, are $\pm E_\phi$, $\mp H_\phi$, $\mp E_{r,z}$, and $\pm H_{r,z}$.

The solution of the TMG, inhomogeneous plasma waveguide problems formulated in Sections 9.6.1 and 9.6.2 is thus very similar to the LMG, inhomogeneous plasma waveguide problem of Section 9.1, and follows steps similar to those outlined in Section 9.1 following Equation 9.37. The major difficulty lies in the simultaneous solution of the coupled, variable-coefficient differential equations in the electric and magnetic fields along $\underset{\approx}{B}_0$. Unlike the LMG case, in the TMG case these equations are coupled even at cutoff.

Table 9.5. Transformations for Equations 9.21 through 9.37, Which Make Them Applicable to the TMG, Coaxial Plasma Waveguide, Inhomogeneous in the r Direction

LMG (Section 9.1)		TMG (Section 9.6.2)
z direction	$\longrightarrow$	ϕ direction
γ	$\longrightarrow$	$-jm$
$\nabla_T \underset{\approx}{K}$	$\longrightarrow$	$\nabla_r \underset{\approx}{K}$
∇_T	$\longrightarrow$	$\nabla_r - \underset{\sim}{i}_z \gamma$
∇_T^2	$\longrightarrow$	$\nabla_r^2 + \gamma^2$
E_z	$\longrightarrow$	E_ϕ
H_z	$\longrightarrow$	H_ϕ
$\underset{\sim}{E}_T$	$\longrightarrow$	$\underset{\sim}{i}_r E_r + \underset{\sim}{i}_z E_z$
$\underset{\sim}{H}_T$	$\longrightarrow$	$\underset{\sim}{i}_r H_r + \underset{\sim}{i}_z H_z$

9.7 Transversely Magnetized, Homogeneous Plasma Waveguides

When the plasma is homogeneous and the applied magnetic field is uniform, the elements of the tensor of Equation 9.279 are independent of coordinates. For the coaxial plasma waveguide of Figure 9.11b, the assumption of an applied magnetic field that is independent of r can be a reasonable approximation only when the waveguide is thin compared with the radius of the inner cylinder. In a manner similar to the treatment in Section 9.6, we can adapt the the analysis of Section 9.5 to the present problem. We shall use the notation shown in Table 9.6 so that the analysis can apply to both the rectangular and coaxial waveguides. We shall summarize the resulting field equations and solutions with dependence $\exp(-jp_B r_B)$ $\exp(-\gamma z) \exp(j\omega t)$, where r_B is the coordinate along $\underset{\sim}{B}_0$. All the results are obtainable through the use of the transformations shown in Tables 9.4 and 9.5 on the equations of Sections 9.1 and 9.5.

The coupled wave equations for the electric and magnetic field components along $\underset{\sim}{B}_0$ are

$$\nabla_{T'}^2 E_B + (a + \gamma^2) E_B = bH_B \tag{9.280}$$

$$\nabla_{T'}^2 H_B + (c + \gamma^2) H_B = dE_B \tag{9.281}$$

Table 9.6. Unified Vector Operator and Field Notation for the TMG, Homogeneous Plasma Waveguide

The subscript B refers to the coordinate along $\underset{\sim}{B}_0$; the subscript T' refers to the coordinate orthogonal to $\underset{\sim}{B}_0$ and z; the subscript TB refers to the two coordinates orthogonal to $\underset{\sim}{B}_0$, that is, T' and z.

Notation	Rectangular Waveguide	Coaxial Waveguide
$\nabla_{T'}$	∇_x	∇_r
$\nabla_{T'}^2$	∇_x^2	∇_r^2
$\nabla_{TB} = \nabla_{T'} - \underset{\sim}{i}_z\gamma$	$\nabla_x - \underset{\sim}{i}_z\gamma$	$\nabla_r - \underset{\sim}{i}_z\gamma$
$\underset{\sim}{i}_B$	$\underset{\sim}{i}_y$	$\underset{\sim}{i}_\phi$
E_B	E_y	E_ϕ
$\underset{\sim}{E}_{TB}$	$\underset{\sim}{i}_x E_x + \underset{\sim}{i}_z E_z$	$\underset{\sim}{i}_r E_r + \underset{\sim}{i}_z E_z$
p_B	p_y	m
$p_{1,2}$	$p_{x1,2}$	$p_{1,2}$

where

$$a = (-p_B^2 + k_0^2 K_\perp)\frac{K_\parallel}{K_\perp} \tag{9.282}$$

$$b = \omega\mu_0 p_B \frac{K_\times}{K_\perp} \tag{9.283}$$

$$c = -p_B^2 + k_0^2 \frac{K_r K_\ell}{K_\perp} \tag{9.284}$$

$$d = -\omega\epsilon_0 p_B \frac{K_\times K_\parallel}{K_\perp} \tag{9.285}$$

The solutions to Equations 9.280 and 9.281 can be written as

$$E_B = E_{B1} + E_{B2} \tag{9.286}$$

$$H_B = h_1 E_{B1} + h_2 E_{B2} \tag{9.287}$$

where E_{B1} and E_{B2} are solutions of

$$\nabla_{T'}^2 E_{B1,2} + p_{1,2}^2 E_{B1,2} = 0 \tag{9.288}$$

and

$$h_{1,2} = \frac{a + \gamma^2 - p_{1,2}^2}{b} = \frac{d}{c + \gamma^2 - p_{1,2}^2} \tag{9.289}$$

The transverse wave numbers p_1^2 and p_2^2 are related to the propagation constant through the dispersion relation:

$$p^4 - (2\gamma + a + c)p^2 + [\gamma^4 + (a + c)\gamma^2 + ac - bd] = 0 \tag{9.290}$$

The remaining field components can be expressed in terms of E_B and H_B:

$$\begin{bmatrix} \underset{\sim}{E}_{TB} \\ \underset{\sim}{H}_{TB} \\ \underset{\sim}{i}_B \times \underset{\sim}{E}_{TB} \\ \underset{\sim}{i}_B \times \underset{\sim}{H}_{TB} \end{bmatrix} = \begin{bmatrix} P & R & Q & S \\ T & P & U & Q \\ -Q & -S & P & R \\ -U & -Q & T & P \end{bmatrix} \begin{bmatrix} \nabla_{TB} E_B \\ \nabla_{TB} H_B \\ \underset{\sim}{i}_B \times \nabla_T E_B \\ \underset{\sim}{i}_B \times \nabla_T H_B \end{bmatrix} \tag{9.291}$$

where

$$P = \frac{-jp_B(-p_B^2 + k_0^2 K_\perp)}{D} \tag{9.292}$$

$$R = \frac{j\omega\mu_0 k_0^2 K_\times}{D} \tag{9.293}$$

$$Q = \frac{jp_B k_0^2 K_\times}{D} \tag{9.294}$$

$$S = \frac{j\omega\mu_0(-p_B^2 + k_0^2 K_\perp)}{D} \tag{9.295}$$

$$T = \frac{-p_B^2 j\omega\epsilon_0 K_\times}{D} \tag{9.296}$$

$$U = \frac{-j\omega\epsilon_0(-p_B^2 K_\perp + k_0^2 K_r K_\ell)}{D} \tag{9.297}$$

and

$$D = (-p_B^2 + k_0^2 K_r)(-p_B^2 + k_0^2 K_\ell) \tag{9.298}$$

The field solutions with $\exp(jp_B r_B)$ dependence for the same γ are $\pm E_B$, $\mp H_B$, $\mp \underset{\sim}{E}_{TB}$, and $\pm \underset{\sim}{H}_{TB}$, all with either the upper sign or the lower sign. Thus, for each wave solution, $\exp(-\gamma z)$, the fields have a standing-wave variation along $\underset{\sim}{B}_0$, corresponding to a particular value of p_B^2, and a variation in a direction orthogonal to both $\underset{\sim}{B}_0$ and z given by the two p^2 solutions to Equation 9.288. The complete field solutions, when used to satisfy the boundary conditions on the perfectly conducting walls of the waveguide, lead to a determinantal equation, which is a functional equation in γ and p, that together with the dispersion relation, Equation 9.290, can be solved for γ and p.

From Equations 9.280 and 9.281 it is evident that at cutoff the wave equations are still coupled, and two p^2 solutions are usually needed. If either $p_B = 0$ or $p = 0$, corresponding to uniformity either along or across $\underset{\sim}{B}_0$, the wave equations separate. These solutions are possible only if the boundaries are also uniform, respectively, along or across $\underset{\sim}{B}_0$.

The case of $p = 0$ corresponds to a waveguide system that is uniform in all directions at right angles to the applied magnetic field $\underset{\sim}{B}_0$. A parallel-plate plasma waveguide with the magnetic field perpendicular to the plates is such a system (Figure 9.11a with side walls removed and upper and lower plates extending to $x = \pm\infty$). The field solutions for this system are made up merely of the free waves propagating at an angle θ with respect to the magnetic field (Equation 2.27 and Section 3.4). Each of these waves can satisfy the boundary conditions at the perfectly conducting walls, as can be ascertained readily from their polarization properties (Equation 4.1). The fields can be also deduced from Equations 9.280, 9.281, and 9.291, with $\nabla_{T'}$ set equal to zero. We note that all field components are in general coupled. The values of $p_B = p_y$ are determined by the boundary conditions and are independent of frequency. For each value of p_B^2 we obtain two values of γ^2 from the dispersion relation, Equation 9.290 with $p = 0$. The ratio of $|p_B|$ to $|\gamma|$ is the tangent of the angle for the free wave that makes up this solution. The cutoffs in propagation correspond to waves at $\theta = 0$, which are the right and left circularly polarized waves forming standing waves in between the plates. Resonance in propagation occur for frequencies corresponding to the resonances of the free waves at $\theta = \pi/2$, that is, of the extraordinary wave. The cutoffs and resonances will be discussed more fully in Section 9.7.2.

9.7.1 Transversely Magnetized, Homogeneous Plasma Waveguides with Uniformity along $\underset{\sim}{B}_0$. The simplest field solutions for the TMG homogeneous plasma waveguide are the ones that have no variations along the applied magnetic field $\underset{\sim}{B}_0$.[36,37] Two systems for which this is of interest are the coaxial (Figure 9.11b) plasma waveguide solutions with $(\partial/\partial\phi) = 0$, and the parallel-plate waveguide with $\underset{\sim}{B}_0$ along the plates (Figure 9.11a with upper and lower plates removed, and side plates extending to $\pm y = \infty$) and $(\partial/\partial y) = 0$. For these solutions p_B, the wave number for the direction along $\underset{\sim}{B}_0$, is zero, and the wave equations for E_B and H_B, Equations 9.280 and 9.281, are uncoupled. Clearly, these solutions bear a close relationship to the cutoff solutions for the LMG plasma waveguide treated in Section 9.5.2. Furthermore, the cutoff fields of these solutions are related to the free waves that propagate across $\underset{\sim}{B}_0$. Since Equations 9.280 and 9.281 are independent of each other, we can consider separately the solutions with $H_B = 0$ and with $E_B = 0$.

Solutions with $H_B = 0$

From Equations 9.280 through 9.298 with p_B and H_B equal to zero, we find:

$$\nabla_{T'}^2 E_B + p_1^2 E_B = 0 \tag{9.299}$$

$$p_1^2 = k_0^2 K_{\parallel} + \gamma^2 \tag{9.300}$$

$$\underset{\sim}{E}_{TB} = 0 \tag{9.301}$$

$$\underset{\sim}{H}_{TB} = \frac{1}{j\omega\mu_0} \underset{\sim}{i}_B \times \nabla_{TB} E_B \tag{9.302}$$

(At cutoff, $\gamma = 0$, these fields are precisely the E-wave cutoff solutions for the LMG plasma waveguide of Section 9.5.2.) For perfectly conducting electric boundary conditions on the waveguide walls, p_1^2 is a positive real number that depends upon geometry only (see footnote on page 139 in Section 9.2). These field solutions are H waves that have the electric field entirely along $\underset{\sim}{B}_0$ and hence apply to the rectangular waveguide. The dispersion equation, Equation 9.300, is identical with that of an isotropic plasma waveguide, Equation 9.69, and is as illustrated in Figure 9.3. For these field solutions the waveguide can essentially be regarded as filled with a medium of effective dielectric constant $\epsilon_0 K_{\parallel}$. The value of p_1 is the same as for an H wave of the same symmetry and in a free-space waveguide of the same geometry. For a fixed value of p_1, Equation 9.300 shows that these solutions are merely the ordinary wave propagating across the magnetic field at an angle with respect to the boundary determined by the relative magnitude of p_1 to γ.

If the plasma boundary is free, the solutions outside the plasma must satisfy Equations 9.148 through 9.154. However, within the first-order fields there is no surface current, and hence, surface waves cannot exist for these solutions.

Solutions with $E_B = 0$

In the absence of an electric field component along $\underset{\sim}{B}_0$, and with $p_B = 0$, Equations 9.280 through 9.298 become

$$\nabla_{T'}^2 H_B + p_2^2 H_B = 0 \tag{9.303}$$

$$p_2^2 = k_0^2 \frac{K_r K_\ell}{K_\perp} + \gamma^2 \tag{9.304}$$

$$\underset{\sim}{H}_{TB} = 0 \tag{9.305}$$

$$\underset{\sim}{E}_{TB} = \frac{j\omega\mu_0}{(p^2 - \gamma^2)} \left(\frac{K_\times}{K_\perp} \nabla_{TB} H_B + \underset{\sim}{i}_B \times \nabla_{TB} H_B \right) \tag{9.306}$$

(At cutoff, $\gamma = 0$, these fields are precisely the H-wave cutoff solutions for the LMG plasma waveguide treated in Section 9.5.2). For perfectly conducting electric boundary conditions on the waveguide walls, the tangential electric field at the boundary must vanish, and Equation 9.306 gives

$$\frac{K_\times}{K_\perp} \gamma H_B + \underset{\sim}{i}_{T'} \cdot \nabla_{T'} H_B = 0 \tag{9.307}$$

where H_B is to be evaluated at the waveguide walls. For the coaxial geometry, Equation 9.307 leads to a rather complicated determinantal equation. We therefore restrict ourselves to the parallel-plate geometry.

For the parallel-plate geometry, Equation 9.307 gives the following determinantal equation:

$$\left(p_2^2 + \gamma^2 \frac{K_\times^2}{K_\perp^2} \right) \sin p_2 2d = 0 \tag{9.308}$$

where $2d$ is the plate separation. Hence, the boundary conditions can be satisfied in two distinct ways that lead to distinct field solutions.

Boundary condition I:

$$\sin p_2 2d = 0 \tag{9.309}$$

The values of p_2^2 are therefore real, positive, and entirely determined by the geometry of the waveguide cross section. The field solutions are E waves, and the values of p_2 are the E-wave eigenvalues for solutions with the same symmetry in a free-space waveguide of the same geometry. Equation 9.304 shows that these solutions are simply the extraordinary wave propagating across the magnetic field and at an angle with respect to the boundary that is determined by the relative magnitudes of p_2 and γ. The dispersion characteristic of γ can be deduced from Equation 9.304. Resonances in the propagation $(\gamma \to \infty)$ occur at the resonant frequency of the extraordinary wave and are independent of p_2. The cutoffs in propagation $(\gamma = 0)$ occur at the frequencies for which $p_2 c = \omega n_\times$. For each value of p_2 these can be obtained from a graphical solution, as shown in Figure 9.7 with p_h equal to p_2. Since for finite fields p_2 must be finite, these solutions have a low-frequency cutoff that occurs above $\omega = 0$. For each value of p_2 there are three cutoff frequencies. Hence, there are three frequency ranges in which there are an infinite number of cutoff frequencies corresponding to the infinite number of distinct p_2 values. A typical dispersion characteristic for a particular value of p_2 is shown in Figure 9.12.

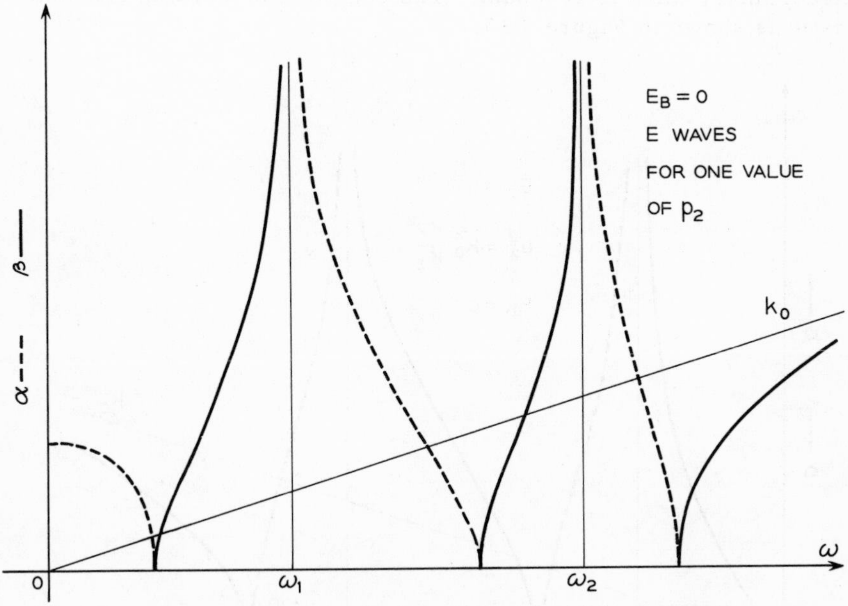

Figure 9.12. Dispersion characteristics for a TMG, homogeneous, parallel-plate plasma waveguide with $\underset{\sim}{B}_0$ along the plates and no variation along $\underset{\sim}{B}_0$; ω_1 and ω_2 are the resonances of the extraordinary wave; p_2 is constant, Equation 9.309.

Boundary condition II:

$$p_2^2 = -\gamma^2 \frac{K_\times^2}{K_\perp^2} \qquad (9.310)$$

When Equation 9.310 is taken as the determinantal equation, the longitudinal electric field E_z vanishes everywhere, and the fields are entirely contained in the transverse plane of the waveguide. Both γ and p_2 are functions of frequency but independent of the separation distance between the plates. The values of γ and p_2^2 are found from a simultaneous solution of Equations 9.304 and 9.310,

$$\gamma^2 = -k_0^2 K_\perp = -k_0^2 \tfrac{1}{2}(K_\ell + K_r) \qquad (9.311)$$

$$p_2^2 = k_0^2 \frac{K_\times^2}{K_\perp} \tag{9.312}$$

The propagation constant of Equation 9.311 has resonances at the cyclotron frequencies and cutoffs at the frequencies for which the extraordinary wave is resonant. The complete dispersion characteristic is shown in Figure 9.13.

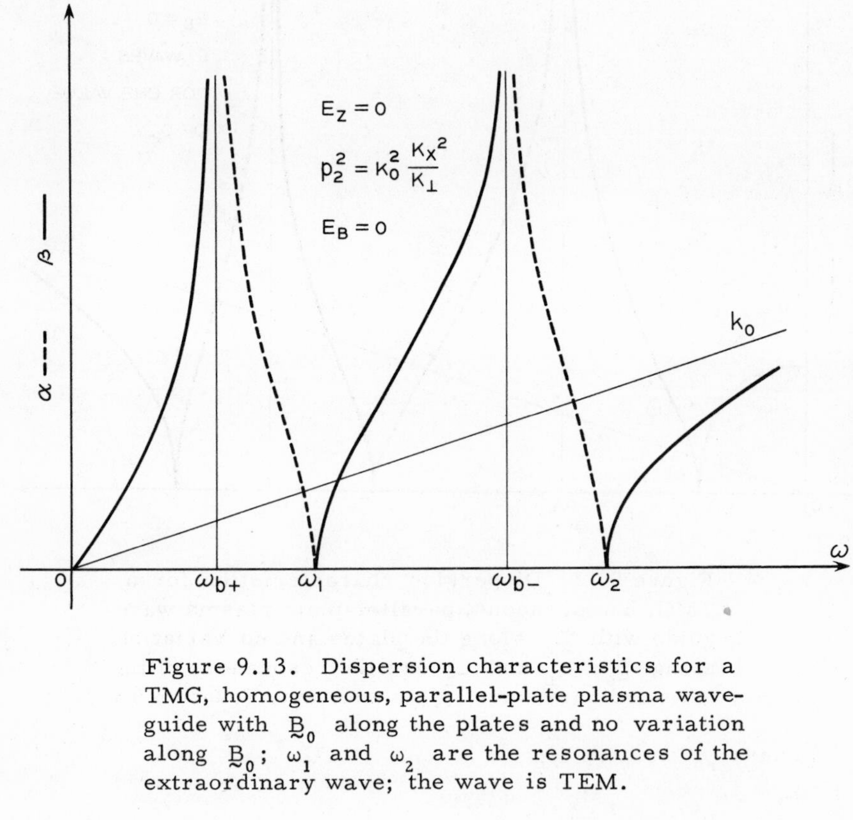

Figure 9.13. Dispersion characteristics for a TMG, homogeneous, parallel-plate plasma waveguide with $\underset{\approx}{B}_0$ along the plates and no variation along $\underset{\approx}{B}_0$; ω_1 and ω_2 are the resonances of the extraordinary wave; the wave is TEM.

In the low-frequency regime, as $\omega \to 0$, Equations 9.311 and 9.312 give

$$\gamma^2 \to -\frac{\omega^2}{u_a^2}\left(1 + \frac{u_a^2}{c^2}\right) \tag{9.313}$$

$$p_2^2 \to 0 \tag{9.314}$$

where u_a is the Alfvén speed as given by Equation 9.241. This is the dispersionless TEM Alfvén wave propagating across the applied magnetic field $\underset{\sim}{B}_0$. The wave impedance is readily found from Equation 9.306:

$$\frac{E_x}{H_y} = \sqrt{\frac{\mu_0}{\epsilon_0}} \frac{\frac{u_a}{c}}{\left(1 + \frac{u_a^2}{c^2}\right)^{\frac{1}{2}}} \tag{9.315}$$

which is approximately the free-space impedance reduced by the factor u_a/c.

9.7.2 Transversely Magnetized, Homogeneous, Rectangular Plasma Waveguide. In order to obtain a better understanding of the field solutions, we shall develop a set of equations that are different from those in the previous section. In Section 9.7 we developed the wave equations for the field components along $\underset{\sim}{B}_0$ and expressed the remaining field components in terms of these. Instead, here we shall develop the differential equations for the longitudinal field components and express the transverse fields in terms of them. This will allow us to see more readily the coupling of the free-space waveguide E and H waves and will prove useful in other respects too.

Following the analysis of Section 9.1, we separate Maxwell's equations into longitudinal and transverse parts, express the transverse fields in terms of the longitudinal fields, and derive the differential equations for the longitudinal fields. Since this is analogous to what was done in Section 9.1, we omit the details. The important difference arises from the new form of the tensor, Equation 9.279, for which $\underset{\approx}{K}_{Tz}$ is finite, that is, from the fact that the plane transverse to the applied magnetic field contains the uniform direction of propagation. We obtain the following results:

$$\begin{bmatrix} E_x \\ H_y \\ E_y \\ H_x \\ E_z \end{bmatrix} = \begin{bmatrix} -\frac{\gamma}{\Gamma_\perp} & \frac{-j\omega\mu_0}{\gamma} & 0 & 0 & -\frac{k_0^2 K_\times}{\Gamma_\perp} \\ \frac{-j\omega\epsilon_0 K_\perp}{\Gamma_\perp} & -\frac{\gamma}{\Gamma_\perp} & 0 & 0 & \frac{j\omega\epsilon_0 K_\times \gamma}{\Gamma_\perp} \\ 0 & 0 & -\frac{\gamma}{\Gamma_\parallel} & \frac{j\omega\mu_0}{\Gamma_\parallel} & 0 \\ 0 & 0 & \frac{j\omega\epsilon_0 K_\parallel}{\Gamma_\parallel} & -\frac{\gamma}{\Gamma_\parallel} & 0 \\ 0 & 0 & 0 & 0 & 1 \end{bmatrix} \begin{bmatrix} \frac{\partial E_z}{\partial x} \\ \frac{\partial H_z}{\partial y} \\ \frac{\partial E_z}{\partial y} \\ \frac{\partial H_z}{\partial x} \\ E_z \end{bmatrix} \tag{9.316}$$

$$K_\perp \Gamma_\parallel \frac{\partial^2 E_z}{\partial x^2} + K_\parallel \Gamma_\perp \frac{\partial^2 E_z}{\partial y^2} + \Gamma_\parallel (K_\perp \Gamma_\perp + k_0^2 K_\times^2) E_z$$

$$= j\omega\mu_0 \left[\gamma (K_\parallel - K_\perp) \frac{\partial^2 H_z}{\partial x \, \partial y} - \Gamma_\parallel K_\times \frac{\partial H_z}{\partial y} \right] \tag{9.317}$$

$$\Gamma_\perp \frac{\partial^2 H_z}{\partial x^2} + \Gamma_\parallel \frac{\partial^2 H_z}{\partial y^2} + \Gamma_\perp \Gamma_\parallel H_z$$

$$= j\omega\epsilon_0 \left[\gamma (K_\parallel - K_\perp) \frac{\partial^2 E_z}{\partial x \, \partial y} + \Gamma_\parallel K_\times \frac{\partial E_z}{\partial y} \right] \tag{9.318}$$

where the time and z dependencies of the fields were assumed of the form $\exp(j\omega t - \gamma z)$, and

$$\Gamma_\parallel = \gamma^2 + k_0^2 K_\parallel \tag{9.319}$$

$$\Gamma_\perp = \gamma^2 + k_0^2 K_\perp \tag{9.320}$$

Equation 9.316 gives the transverse fields in terms of the longitudinal fields. Equations 9.317 and 9.318 are a coupled set of linear partial differential equations of second order for the longitudinal fields E_z and H_z. If we assume solutions of the form $\exp(-jp_x x)$ $\exp(-jp_y y)$, then for nonzero values of E_z and H_z Equations 9.317 and 9.318 yield the following dispersion equation:

$$(\gamma^2 - p_x^2)^2 - (\gamma^2 - p_x^2) \left[p_y^2 \left(\frac{K_\parallel}{K_\perp} + 1 \right) - k_0^2 \left(K_\parallel + \frac{K_r K_\ell}{K_\perp} \right) \right]$$

$$+ \frac{K_\parallel}{K_\perp} \left(p_y^2 - k_0^2 K_r \right) \left(p_y^2 - k_0^2 K_\ell \right) = 0 \tag{9.321}$$

Equation 9.321 is of course the usual dispersion relation (Equations 2.24 or 9.161 or 9.290) recast into a form that is convenient for the solutions we are seeking. It is directly obtainable also from Equation 9.167 with γ^2 replaced by $-p_y^2$ and p^2 replaced by $p_x^2 - \gamma^2$. We note that for a given value of γ Equation 9.321 is of fourth order in both p_x and p_y. As we have shown earlier (see Sections 9.6 and 9.7), standing-wave solutions may be set up in the direction of $\underset{\sim}{B}_0$, and thus the boundary conditions can be met with one positive value of p_y^2. Hence, for a given value of p_y^2 and γ, Equation 9.321 gives two values of p_x^2. Solutions with two values of p_x^2 are necessary for matching boundary conditions. Equation 9.321 can be solved for $\gamma^2 - p_x^2$:

$$\gamma^2 - p_x^2 = \tfrac{1}{2} B \pm \tfrac{1}{2} \sqrt{B^2 - 4C} \tag{9.322}$$

where

$$B = p_y^2 \left(\frac{K_\parallel}{K_\perp} - 1 \right) - k_0^2 \left(K_\parallel + \frac{K_r K_\ell}{K_\perp} \right) \tag{9.323}$$

$$C = \frac{K_\parallel}{K_\perp} \left(p_y^2 - k_0^2 K_r \right) \left(p_y^2 - k_0^2 K_\ell \right) \tag{9.324}$$

It is clear that both γ and p_x can take on arbitrary complex values. At a particular frequency their values are determined from the simultaneous solution of the dispersion equation, Equation 9.321, and the determinantal equation from the boundary conditions. This is usually very complicated. Instead, we shall turn our attention to the solutions at the critical frequencies of cutoff and resonance.

9.7.3 Cutoff Conditions and Fields. Unlike the LMG, homogeneous plasma waveguide, the differential equations for E_z and H_z for the TMG, homogeneous plasma waveguide at cutoff, Equations 9.317 and 9.318 with $\gamma = 0$, do not decouple for arbitrary field variations. Hence, the field solutions at cutoff are in general made up of mixed E- and H-wave fields. The dispersion equation is of fourth order in p_x, and the boundary conditions may require that p_x be a function of frequency.

We should note that Equation 9.321 for $\gamma = 0$ is identical with Equation 2.27 with $k_0^2 n^2 = p_x^2 + p_y^2$. The cutoffs in the waveguide correspond to resonances in the transverse plane of the waveguide of free waves that propagate at various angles with respect to the magnetic field $\underset{\sim}{B}_0$. If the waveguide cross section is closed, free waves at $\theta = 0$ or $\theta = 90°$ cannot be made to satisfy the boundary conditions on all walls. Hence, the free waves that resonate in the waveguide cross section at cutoff and satisfy the boundary conditions on the walls must be traveling neither along nor across $\underset{\sim}{B}_0$. In fact, Equation 9.322 shows that for each value of p_y we require two independent solutions of p_x^2 and, hence, waves at two independent angles θ with respect to $\underset{\sim}{B}_0$. The necessity of two such waves (that is, two n^2 solutions of Equation 2.27) can also be seen from the polarization properties of free waves as given by the set of Equations 4.1. We note that when the magnetic field $\underset{\sim}{B}_0$ is perpendicular to the perfectly conducting walls (top and bottom walls of waveguide) a free wave incident on the wall at an angle θ with respect to $\underset{\sim}{B}_0$ is reflected at an angle $\pi - \theta$ with respect to $\underset{\sim}{B}_0$. For this case, Equations 4.1 show that the requirement of zero tangential electric field at the wall can be satisfied. Consider now that this same wave is incident on the side wall of the waveguide where $\underset{\sim}{B}_0$ is tangent to the wall; the angle of incidence is again θ with respect to $\underset{\sim}{B}_0$, but the reflected wave is at an angle $-\theta$ with respect to $\underset{\sim}{B}_0$. For these incident and reflected waves, Equations 4.1 show that the boundary conditions cannot be satisfied. Hence, for a closed waveguide the boundary conditions cannot be met by a free wave at one angle θ with respect to $\underset{\sim}{B}_0$ (one n^2 value) resonating in the transverse plane of the waveguide. When the waveguide has a uniform parallel-plane boundary, so that the magnetic field is either along or across the walls, the cutoff solutions can be easily determined.

Parallel-plate waveguides. Consider the parallel-plate waveguides, shown in Figure 9.14, for which we seek the cutoff solutions for propagation in the z direction. We shall consider only the lowest-order cutoff solutions, namely, those for which there is no variation in the uniform direction perpendicular to the z direction. For the case when the magnetic field is tangent to the walls (Figure 9.14a), the cutoff solutions can be formed from the fields of the extraordinary wave or the fields of the ordinary wave propagating between the plates (See also Section 9.7.1). These waves travel

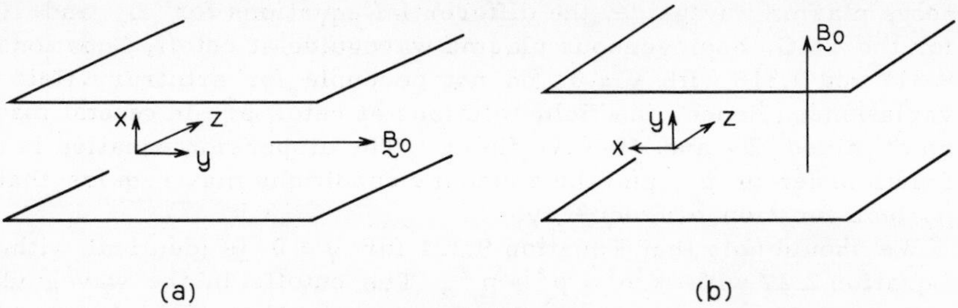

Figure 9.14. TMG, homogeneous, parallel-plate plasma waveguides: (a) $\underset{\sim}{B}_0$ in the y direction along the plates; (b) $\underset{\sim}{B}_0$ in the y direction perpendicular to the plates.

across $\underset{\sim}{B}_0$, have no variation along $\underset{\sim}{B}_0$, and can be made to satisfy the boundary conditions at the walls. For the case when the magnetic field $\underset{\sim}{B}_0$ is perpendicular to the plates (Figure 9.14b), the cutoff solutions are formed by the right and left circularly polarized waves propagating along $\underset{\sim}{B}_0$ in between the plates. These cutoff solutions for the parallel-plate waveguide follow directly from Equations 9.316 through 9.321:

(a) $\underset{\sim}{B}_0$ parallel to plates: $p_y = 0$

This corresponds to the system of Figure 9.14a, already discussed in Section 9.7.1, for which there is no variation with respect to y. The wave equations, Equations 9.317 and 9.318, are decoupled, and we can identify separately E- and H-wave cutoff fields:

(1) Cutoff E wave:

$$\frac{\partial^2 E_z}{\partial x^2} + p_{xe}{}^2 E_z = 0 \tag{9.325}$$

$$p_{xe}{}^2 = k_0^2 \frac{K_r K_\ell}{K_\perp} \tag{9.326}$$

$$E_x = -\frac{K_\times}{K_\perp} E_z \tag{9.327}$$

$$H_y = \frac{1}{j\omega\mu_0} \frac{\partial E_z}{\partial x} \tag{9.328}$$

$$H_z = H_x = E_y = 0 \tag{9.329}$$

These equations describe the extraordinary wave that is elliptically polarized in the (x, z) plane and propagates in the x direction. The value of p_{xe} depends upon the plate separation

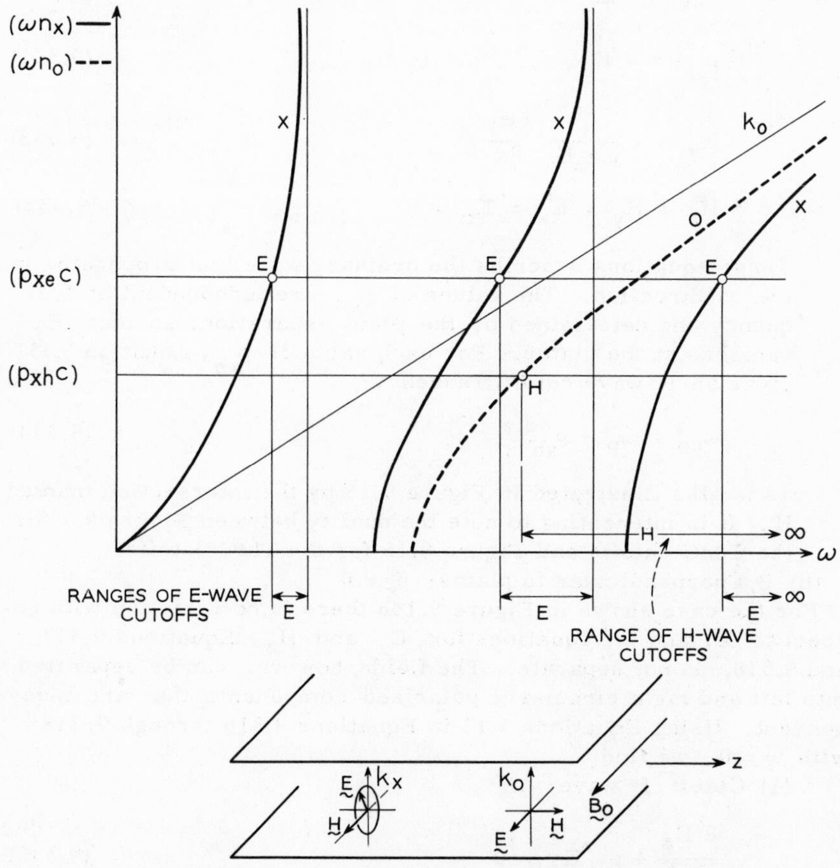

Figure 9.15. Graphical determination of the cutoff frequencies for a TMG, homogeneous, parallel-plate plasma waveguide with $\underset{\sim}{B}_0$ along the plates and no variation along $\underset{\sim}{B}_0$; shown are the solutions for the lowest values of p_{xe} and p_{xh}, and the ranges where the cutoffs occur for all values of p_x.

only and is determined by the condition that E_z vanish at the plates. The solution for the cutoff frequencies given by Equation 9.326 is shown schematically in Figure 9.15. For each value of p_{xe} we have three E-wave cutoff frequencies, as indicated in Figure 9.15 by the intersections marked E. We note from Equations 9.325 through 9.328 that solutions are also possible with $K_\perp$ and E_z equal to zero. These were already discussed from the determinantal equation, Equation 9.310.

(2) Cutoff H wave:

$$\frac{\partial^2 H_z}{\partial x^2} + p_{xh}{}^2 H_z = 0 \tag{9.330}$$

$$p_{xh}{}^2 = k_0{}^2 K_{\parallel} \tag{9.331}$$

$$E_y = -\frac{1}{j\omega\epsilon_0 K_{\parallel}} \frac{\partial H_z}{\partial x} \tag{9.332}$$

$$H_x = H_y = E_x = E_z = 0 \tag{9.333}$$

These equations describe the ordinary wave that propagates in the x direction. The values of p_{xh} are independent of frequency and determined by the plate separation, so that E_y vanishes at the plates. For each value of p_{xh}, Equation 9.331 gives an H-wave cutoff frequency:

$$\omega_{co}{}^2 = \omega_p{}^2 + p_{xh}{}^2 c^2 \tag{9.334}$$

as is also illustrated in Figure 9.15 by the intersection marked H. It is interesting to note the duality between Figure 9.7 for the LMG cutoffs and Figure 9.15 for the TMG cutoffs.

(b) $\underset{\sim}{B}_0$ perpendicular to plates: $p_x = 0$

For the case shown in Figure 9.14b there is no variation with respect to x, and the equations for E_z and H_z, Equations 9.317 and 9.318, do not separate. The fields, however, can be separated into left and right circularly polarized components that are independent. Using Equations 1.11 in Equations 9.316 through 9.318 with $\gamma = 0$, we find

(1) Cutoff ℓ wave:

$$\frac{\partial^2 E_\ell}{\partial y^2} + p_{y\ell}{}^2 E_\ell = 0 \tag{9.335}$$

$$p_{y\ell}{}^2 = k_0{}^2 K_\ell \tag{9.336}$$

$$E_z = \frac{1}{\sqrt{2}} E_\ell \tag{9.337}$$

$$H_x = -\frac{1}{j\omega\mu_0}\frac{\partial E_z}{\partial y} \tag{9.338}$$

$$E_x = jE_z \tag{9.339}$$

$$H_z = -jH_x \tag{9.340}$$

(2) Cutoff r wave:

$$\frac{\partial^2 E_r}{\partial y^2} + p_{yr}{}^2 E_r = 0 \tag{9.341}$$

$$p_{yr}{}^2 = k_0^2 K_r \tag{9.342}$$

$$E_z = \frac{1}{\sqrt{2}} E_r \tag{9.343}$$

$$H_x = -\frac{1}{j\omega\mu_0}\frac{\partial E_z}{\partial y} \tag{9.344}$$

$$E_x = -jE_z \tag{9.345}$$

$$H_z = jH_x \tag{9.346}$$

The fields in Equations 9.335 through 9.340 and 9.341 through 9.346 are, respectively, those of the left and right circularly polarized waves propagating along $\underset{\sim}{B}_0$. The values of p_y are independent of frequency and functions of the plate separation only. Their values are determined from the boundary-condition requirement that both E_x and E_z must vanish at the plates. The solutions for the cutoff frequencies is obtained from Equations 9.336 and 9.342. Figure 9.16 illustrates how these cutoff frequencies can be determined graphically. A plot of ωn_ℓ and ωn_r can be obtained from Figure 3.3 for a given ratio of m_+/m_- and $\omega_p^2/\omega_{b+}\omega_{b-}$ as shown in Figure 3.2. The pole of n_ℓ is at $\omega = \omega_{b+}$, and the pole of n_r is at $\omega = \omega_{b-}$. The finite frequency zero of n_ℓ is at the positive real root of

$$\omega^2 + (\omega_{b-} - \omega_{b+})\omega - (\omega_p{}^2 + \omega_{b+}\omega_{b-}) = 0 \tag{9.347}$$

and the finite frequency zero of n_r is at the positive real root of

$$\omega^2 - (\omega_{b-} - \omega_{b+})\omega - (\omega_p{}^2 + \omega_{b+}\omega_{b-}) = 0 \tag{9.348}$$

For each value of p_y, we find two ℓ-wave cutoff frequencies and two r-wave cutoff frequencies. The lowest cutoff frequency is one corresponding to ℓ-wave fields.

9.7.4 Resonance Conditions and Fields. From the dispersion relation, Equation 9.321, or Equations 9.322 through 9.324, we note that for p_y finite, resonances in the propagation $(\gamma \to \infty)$ can occur

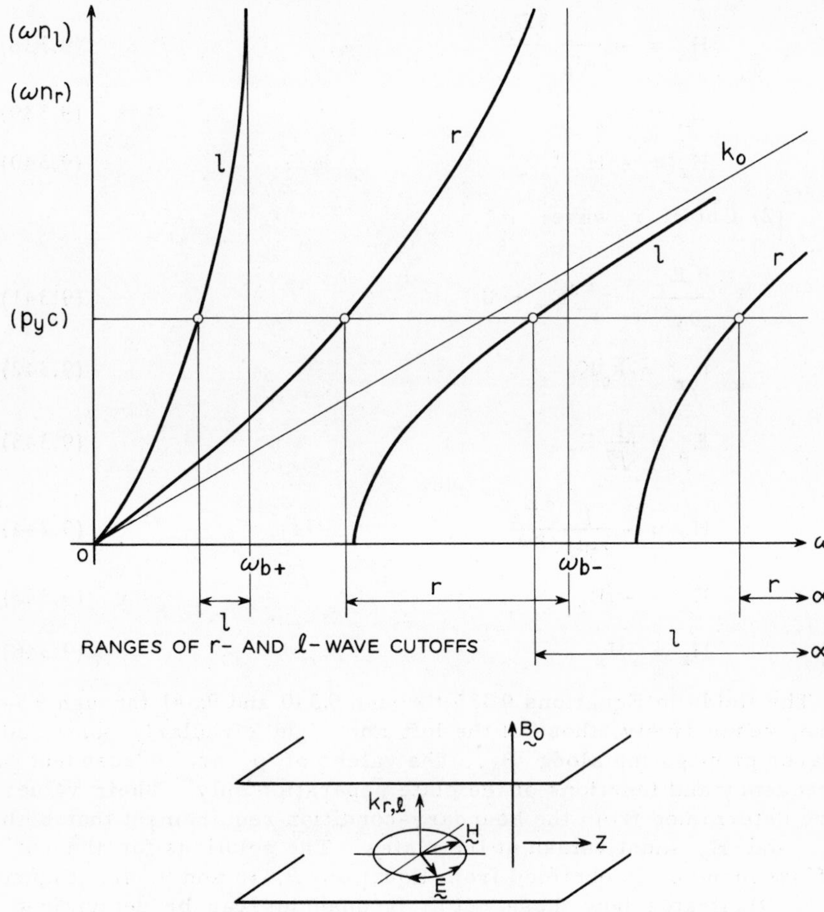

Figure 9.16. Graphical determination of the cutoff frequencies for a TMG, homogeneous, parallel-plate plasma waveguide with $\underset{\sim}{B}_0$ normal to the plates and no variation across $\underset{\sim}{B}_0$; shown are the solutions for the lowest value of p_y, and the ranges where the cutoffs occur for all values of p_y.

at frequencies for which $K_\perp \to 0$ and at the frequency for which $K_\parallel \to -\infty$. (The resonances for $p_y = 0$ were discussed in Section 9.7.1 and will be omitted here.) The first set of frequencies correspond to the resonances of the extraordinary wave, and the other resonance is at zero frequency. Equation 9.322 shows that near the resonances one value of p_x^2 is independent of γ and p_y is therefore determined by the geometry only, and the other value of p^2 approaches γ^2, with the difference between p_x^2 and γ^2 depending in general upon frequency and p_y. Near these resonances the field solutions and the matching of boundary conditions are therefore simplified. The resonances associated with surface waves on the boundary between a plasma and free space will be discussed in Chapter 10.

Resonances for $K_\perp \to 0$

The frequencies at which these resonances occur are readily determined from either Figure 3.5 or Figure 3.6 (along the parabolas marked plasma resonance), or from Equation 3.10, or Equation 9.197. For a given set of plasma parameters there are two such resonances. The first one lies in the range $\omega_{b+} < \omega < \sqrt{\omega_{b+}\omega_{b-}}$, and the second one is always above ω_{b+} and below ω_p. These lower and upper frequency limits of the resonances correspond to densities that are, respectively, very low $(\omega_p < \omega_{b+})$ and very high $(\omega_p > \omega_{b-})$.

The dispersion relation near these resonances is obtained from Equation 9.322:

$$\gamma^2 \to p_y^2 \frac{K_\parallel}{K_\perp} - k_0^2 \frac{K_r K_\ell}{K_\perp} \tag{9.349}$$

One value of p_x^2 is arbitrary and determined from boundary conditions, and the other value approaches γ^2. Equation 9.349 shows that the frequency variation of γ near these resonances depends upon both the plasma frequency ω_p and the waveguide dimensions along the magnetic field, which determine p_y. The results are summarized in Table 9.7. We note that at the high-frequency resonance the propagation character is that of a traveling wave, regardless of the values of the plasma frequency or p_y. This is because ω_p is always below the high-frequency resonance. The low-frequency resonance may be either traveling or backward wave in character. If the plasma frequency is below this resonance, we have a traveling wave, regardless of the value of p_y. If the plasma frequency is higher than this resonance, the wave may be either traveling or backward depending upon the value of p_y; large values of p_y give rise to the backward-wave characteristic. These characteristics are illustrated in Figure 9.17 at the frequencies marked ω_1 and ω_2.

The field equations near the resonances ω_1 and ω_2 can be obtained from Equations 9.316 through 9.318. For p_x and p_y finite, $\gamma \to \infty$, and keeping only terms to order $1/\gamma$ or larger, we obtain

$$H_z \to 0 \tag{9.350}$$

$$\frac{\partial^2 E_z}{\partial y^2} + p_y^2 E_z = 0 \tag{9.351}$$

$$E_x \to -\frac{1}{\gamma}\frac{\partial E_z}{\partial x} \tag{9.352}$$

$$H_y \to -\frac{\omega\epsilon_0 K_\ell}{\gamma} E_z \qquad (9.353)$$

$$H_x \to 0 \qquad (9.354)$$

The solutions with $p_x^2 \to \gamma^2$ correspond to H_z finite. These fields have a monotonic variation with x, and taken alone cannot satisfy perfectly conducting boundary conditions.

Table 9.7. Characteristics of the Propagation Constant in the Vicinity of the Resonances ω_1 and ω_2 (of the Extraordinary Wave) for a TMG Homogeneous Waveguide

	γ	$\frac{\partial\omega}{\partial\beta}$	$\frac{\partial\omega}{\partial\alpha}$
I. $\omega_p < \omega_1$ and p_y arbitrary			
$\omega < \omega_{1,2}$	$j\beta$	> 0	
$\omega > \omega_{1,2}$	α		< 0
II. $\omega_p > \omega_1$ and $\frac{(p_y c)^2}{\omega_1^2} < \frac{K_r^2}{\lvert K_\parallel \rvert}$			
$\omega < \omega_{1,2}$	$j\beta$	> 0	
$\omega > \omega_{1,2}$	α		< 0
III. $\omega_p > \omega_1$ and $\frac{(p_y c)^2}{\omega_1^2} > \frac{K_r^2}{\lvert K_\parallel \rvert}$			
$\omega < \omega_1$	α		> 0
$\omega > \omega_1$	$j\beta$	< 0	
$\omega < \omega_2$	$j\beta$	> 0	
$\omega > \omega_2$	α		< 0

Resonance for $K_\parallel \to -\infty$

This resonance occurs at zero frequency. In the absence of boundaries, $K_\parallel \to -\infty$ corresponds to a degenerate set of solutions for which $\theta = \pi/2$ and n is arbitrary (see Equation 2.27). In the presence of boundaries, the degeneracy disappears, and we have a true resonance. This situation is very similar to the degenerate solutions at $\omega = \omega_p$ for $\theta = 0$ (see Equation 2.27 and Section 9.5.3).

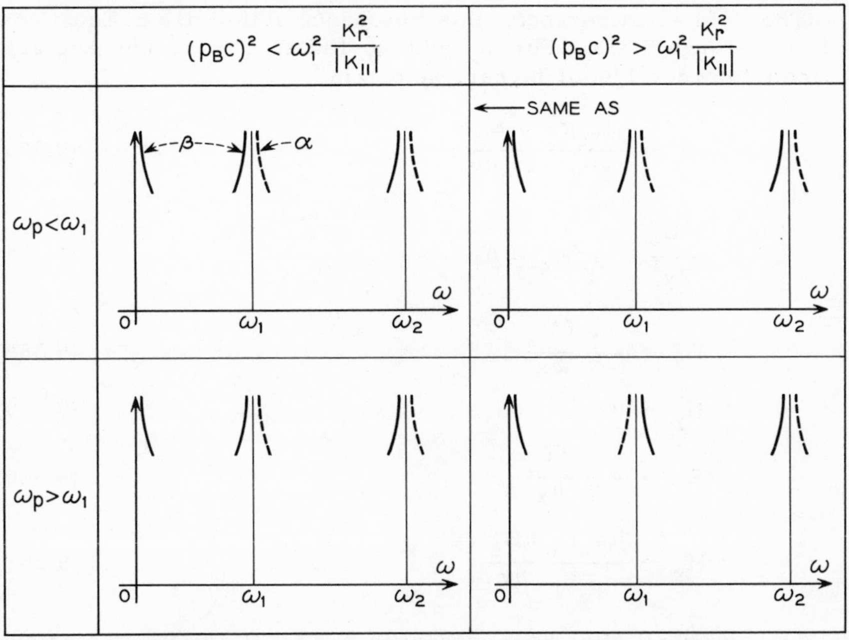

Figure 9.17. Dispersion of the propagation constant $\gamma = \alpha + j\beta$ near the resonances of a TMG, homogeneous plasma waveguide; p_B is the eigenvalue for the direction of $\underset{\sim}{B}_0$; ω_1 and ω_2 are the resonances of the extraordinary wave.

In the low-frequency regime collisions may play a dominant role, and any resonance would be highly damped. Collisions may be accounted for by introducing a damping term into the force equation (see Sections 2.2 and 2.3). If the collisions are between the ions and molecules and the damping term is as in Equation 2.4, we find that there is no resonance near zero frequency. On the other hand, if the collisions are only between the electrons and ions and the damping term is proportional to the relative velocities between ions and electrons, the resonance persists and is accompanied by a resonant damping ($\alpha \to \infty$ and $\beta \to \infty$). We shall consider only the case where a resonance occurs.

From Equation 9.322 with $\gamma \to \infty$ near zero frequency, we obtain

$$\gamma^2 \to \frac{K_{\parallel}}{K_{\perp}} p_y^2 \tag{9.355}$$

The frequency dependence of the propagation constant is that of a backward wave, as shown in Figure 9.17. We note that Equation 9.355 is independent of p_x. As with the other resonances at $K_{\perp} = 0$, we find that one value of p_x^2 is arbitrary and the other value approaches γ^2.

The field equations near this resonance follow from Equations 9.316 through 9.318. For p_x and p_y finite, $\gamma \to \infty$, and keeping terms to order $1/\gamma$ or higher, we obtain

$$H_z \to \frac{j\omega\epsilon_0 K_\perp}{p_y^2 \gamma} \frac{\partial^2 E_z}{\partial x \, \partial y} \tag{9.356}$$

$$\frac{\partial^2 E_z}{\partial y^2} + p_y^2 E_z = 0 \tag{9.357}$$

$$E_x \to -\frac{1}{\gamma} \frac{\partial E_z}{\partial x} \tag{9.358}$$

$$H_y \to 0 \tag{9.359}$$

$$E_y \to -\frac{1}{\gamma} \frac{\partial E_z}{\partial y} \tag{9.360}$$

$$H_x \to \frac{j\omega\epsilon_0 K_\perp}{p_y^2} \frac{\partial E_z}{\partial y} \tag{9.361}$$

The solutions with $p_x^2 \to \gamma^2$ correspond to H_z finite. These fields have a monotonic variation with x, and taken alone cannot satisfy the boundary conditions on perfectly conducting walls.

9.7.5 Summary and Energy and Power Relations (TMG). Table 9.8 summarizes some of the important dispersion characteristics for the TMG, homogeneous plasma waveguide. In the general case, with $\underset{\sim}{B}_0$ along one of the orthogonal coordinate axes, p_B is a constant determined from boundary conditions, and two values of p^2 are required. The field solutions are therefore mixed E and H waves. The field solutions are outlined in Equations 9.280 through 9.298. The values of p^2 and γ must be obtained from a simultaneous solution of the dispersion equation and the determinantal equation that results from satisfying the boundary conditions. At cutoff, $\gamma = 0$, the wave equations are still coupled, and solutions with two independent values of p^2 are required. Hence, the fields are still mixed E and H waves. Resonances, $\gamma \to \infty$, occur at zero frequency and at the frequencies for which the extraordinary wave is resonant. The zero-frequency resonance, if it is undamped, exhibits a backward-wave character. The high-frequency extraordinary wave resonance is always a traveling wave, and the low-frequency extraordinary-wave resonance may be either a traveling or a backward wave (see Figure 9.17). The field solutions near the resonances are essentially E waves with one value of p^2 constant. The second value of p^2 approaches γ^2 and belongs to the coupled H-wave solutions that vanish at resonance.

When the waveguide system is uniform in the direction of $\underset{\sim}{B}_0$, the

Table 9.8 Summary of Dispersion Characteristics for the TMG, Homogeneous Plasma Waveguide with $\underset{\sim}{B}_0$ along One Orthogonal Coordinate Axis

		$p_B = 0$			$p = 0$
			$E_B = 0$		Parallel-plate
	General	$H_B = 0$	$E_z \neq 0$	$E_z = 0$ Parallel-plate	
Dispersion Relation	Equation 9.290 $\gamma, p_B^2, p_1^2, p_2^2$ p_B = constant	$\gamma^2 = p^2 - k_0^2 K_\parallel$ p = constant	$\gamma^2 = p^2 - k_0^2 \frac{K_r K_\ell}{K_\perp}$ parallel plate: p = constant coaxial: $p(\omega)$	$\gamma^2 = -k_0^2 K_\perp$ $p^2 = k_0^2 \frac{K_\times^2}{K_\perp}$	p_y = constant γ_1^2, γ_2^2 Free waves at θ
Cutoffs $\gamma = 0$	p_B^2, p_1^2, p_2^2	$\omega_{co}^2 = \omega_p^2 + p^2c^2$	$p^2c^2 = \omega^2 n_\times$ coaxial: $p(\omega)$ parallel plate: three for each p_x = constant	$\omega_{co} = \omega\vert_{n_\times = \infty}$ (two) $\omega \to 0$ (MHD)	$p_y^2 = k_0^2 K_{r,\ell}$ (four for each p_y)
Resonances $\gamma \to \infty$	1. $\omega = \omega\vert_{n_\times = \infty}$ (two) $\gamma^2 \to p_B^2 \frac{K_\parallel}{K_\perp} - 2k_0^2 \frac{K_r K_\ell}{K_\perp}$ $p_1^2 \to$ constant; $p_2^2 \to \gamma^2$ 2. $\omega = 0$ $\gamma^2 \to p^2 \frac{K_\parallel}{K_\perp}$ $p_1^2 \to$ constant; $p_2^2 \to \gamma^2$	None	$\omega = \omega\vert_{n_\times = \infty}$ (two)	$\omega = \omega_{b\pm}$	Same as general case

field solutions with no variations along this direction ($p_B = 0$) can be separated into two categories. The first set of solutions has zero magnetic field along $\underset{\sim}{B}_0$; these solutions are identical to H waves of the same symmetry in an isotropic plasma waveguide and represent the ordinary wave propagating across $\underset{\sim}{B}_0$ at various angles to the boundary. The second set of solutions has zero electric field along $\underset{\sim}{B}_0$. In general, these E waves are made up of the extraordinary wave propagating across $\underset{\sim}{B}_0$ at various angles to the boundary. For the parallel-plate boundary the angles of the waves (and hence p) are independent of frequency and determined by the plate separation. For the circular boundary the angles (and hence p) are functions of both frequency and propagation constant and must be determined from boundary conditions. All these E waves have resonances at the frequencies for which $n_\times \to \infty$. For the parallel-plate system the cutoffs are determined as in Figure 9.15, which gives three cutoff frequencies for each value of p; for the coaxial system the cutoff frequencies are determined from the simultaneous solution of the determinantal equation from the boundary conditions and the dispersion relation. In the case of the parallel-plate boundaries a solution is also possible with $E_z = 0$ and p frequency dependent. The fields are then entirely transverse to the z direction, with resonances at the cyclotron frequencies and cutoffs at the frequencies for which $n_\times = \infty$. These waves propagate down to zero frequency, where they become MHD Alfvén-type waves across $\underset{\sim}{B}_0$.

For the parallel-plate waveguide with $\underset{\sim}{B}_0$ normal to the plates, the field solutions that have no variation across $\underset{\sim}{B}_0$ are made up of the free waves that propagate at an angle θ with respect to $\underset{\sim}{B}_0$. These have cutoff solutions for $\theta = 0$, which are the right and left circularly polarized waves. The resonances correspond to $\theta = \pi/2$ and occur at $\omega = 0$ and at the resonant frequencies of the extraordinary wave. Finally, in all cases, as $\omega \to \infty$, the plasma model reduces to free space, and all solutions must approach those of a free-space waveguide given in Section 9.2.

The results of Chapter 7 can be applied to the TMG temperate-plasma waveguide. For the dielectric tensor of Equation 9.279, $\underset{\approx}{K}_{Tz} \neq 0$, and hence $U_{\epsilon zT} \neq 0$. Therefore, the propagation constant and frequency are related to the real and reactive power flow (P and Q of Equation 7.112) and to $U_{\epsilon zT}$, $U_{\epsilon z}$, and $U_{\epsilon T}$ (Equations 7.133 through 7.135), as given by Equations 7.137 through 7.140. Equations 7.133 through 7.135 become

$$U_{\epsilon zT} = \tfrac{1}{4}\epsilon_0 \int_A E_{T'}^{*} E_z K_\times \, da \tag{9.362}$$

$$U_{\epsilon z} = \tfrac{1}{4}\epsilon_0 \int_A |E_z|^2 K_\perp \, da \tag{9.363}$$

$$U_{\epsilon T} = \tfrac{1}{4}\epsilon_0 \int_A \left(|E_{T'}|^2 K_\perp + |E_B|^2 K_\parallel \right) da \tag{9.364}$$

where the subscript T' denotes the transverse component that is orthogonal to $\underset{\sim}{B}_0$ (for example, x or r), and the subscript B denotes the transverse component that is along $\underset{\sim}{B}_0$ (for example, y or ϕ). Equation 9.362 is in general complex. Equation 9.363 is real, and its sign depends upon the sign of $K_\perp$. We recall from Part I that $K_\perp$ has poles at the cyclotron frequencies (ω_{b+} and ω_{b-}) and zeros at the resonances of the extraordinary wave (ω_1 and ω_2). Hence, $K_\perp > 0$ for $0 < \omega < \omega_{b+}$, $\omega_1 < \omega < \omega_{b-}$, and $\omega > \omega_2$, and $K_\perp < 0$ for $\omega_{b+} < \omega < \omega_1$ and $\omega_{b-} < \omega < \omega_2$. These regions are shown schematically in Figure 9.18 as a function of ω_p. Equation

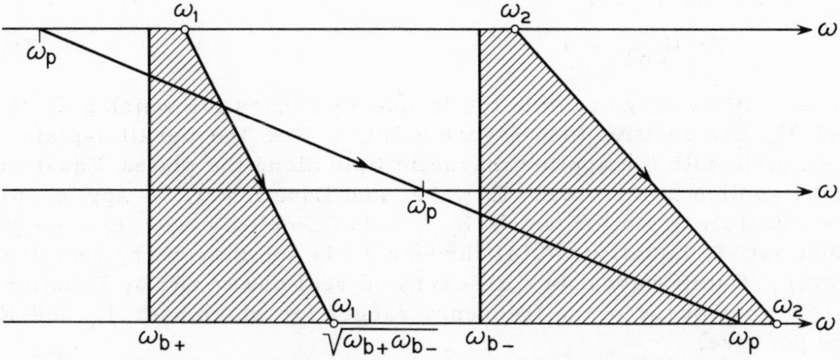

Figure 9.18. Schematic illustration of the loci of the resonances ω_1 and ω_2 with increasing plasma frequency. These resonances are not linear functions of ω_p (see Figure 3.6). In the shaded region, $K_\perp < 0$.

9.364 is also real, but its sign depends upon both $K_\perp$ and $K_\parallel$ and the relative excitations of the transverse fields. In the frequency ranges where both $K_\perp$ and $K_\parallel$ are either positive or negative, $U_{\epsilon T}$ is respectively positive or negative, regardless of the relative excitation of the transverse fields; since for $\omega > \omega_p$, $K_\parallel > 0$, and for $\omega < \omega_p$ $K_\parallel < 0$, these frequency ranges are easily deduced from Figure 9.18. The time-averaged energy density per unit length of guide U, given by Equation 7.119, is positive definite.

For a propagating wave ($\alpha = 0$, $\gamma = j\beta$), Equations 7.149a through 7.150b, 7.152, and 7.153a give

$$Q = -\frac{\omega}{\beta}\, 2 \,\mathrm{Im}\, U_{\epsilon zT} \tag{9.365}$$

$$U_{mT} + U_{mz} = U_{\epsilon T} + U_{\epsilon z} + 2\,\mathrm{Re}\, U_{\epsilon zT} \tag{9.366}$$

$$\frac{\partial \omega}{\partial \beta} = \frac{\omega}{\beta} \frac{2(U_{mT} - U_{\epsilon z} - \text{Re } U_{\epsilon zT})}{U} \tag{9.367}$$

$$= \frac{\omega}{\beta} \frac{2(U_{\epsilon T} + \text{Re } U_{\epsilon zT} - U_{mz})}{U} \tag{9.368}$$

We note that a propagating wave may carry reactive power. (See, for example, the solutions corresponding to Equations 9.311 and 9.312). At the cutoff frequencies, for which $\gamma = 0$, Equations 7.146b, 7.147b, and 7.149b must be satisfied simultaneously, and hence

$$U_{mT} = U_{\epsilon z} + \text{Re } U_{\epsilon zT} \tag{9.369}$$

$$U_{\epsilon T} = U_{mz} - \text{Re } U_{\epsilon zT} \tag{9.370}$$

$$\text{Im } U_{\epsilon zT} = 0 \tag{9.371}$$

These cutoffs may occur in the frequency ranges for which both $K_\perp$ and $K_\parallel$ are positive (see Figure 9.18). For the parallel-plate waveguide with the applied magnetic field along the plates, Equation 9.369 applies to the cutoff E wave, and Equation 9.370 applies to the cutoff H wave for which $U_{\epsilon zT} = 0$. Complex waves $(\gamma = \alpha + j\beta)$ must satisfy Equations 7.141 through 7.144 and may carry reactive power. Cutoff waves $(\gamma = \alpha)$ carry no real power and by Equation 7.146b may occur in the frequency ranges for which both $K_\perp$ and $K_\parallel$ are positive.

9.7.6 Infinite Magnetic Fields. For an externally applied magnetic field $\underset{\sim}{B}_0$ that is infinite in strength, $K_\times = 0$, $K_\perp = 1$, and Equation 9.279 is diagonal $(1, K_\parallel, 1)$. However, the longitudinal fields are still coupled (see Equations 9.317 and 9.318). The electric and magnetic field components along $\underset{\sim}{B}_0$ are decoupled (see Equations 9.280 and 9.281), and the field solutions are conveniently found through the equations of Section 9.7. The two types of waves are

(a) $H_B = 0$:

$$\nabla_{T'}^2 E_B + p_1^2 E_B = 0 \tag{9.372}$$

$$p_1^2 = (-p_B^2 + k_0^2)K_\parallel + \gamma^2 \tag{9.373}$$

$$\underset{\sim}{E}_{TB} = \frac{-jp_B}{-p_B^2 + k_0^2} \nabla_{TB} E_B \tag{9.374}$$

$$\underset{\sim}{H}_{TB} = \frac{-j\omega\epsilon_0}{-p_B^2 + k_0^2} \underset{\sim}{i}_B \times \nabla_{TB} E_B \tag{9.375}$$

(b) $E_B = 0$

$$\nabla_{T'}^2 H_B + p_2^2 H_B = 0 \tag{9.376}$$

$$p_2^2 = -p_B^2 + k_0^2 + \gamma^2 \tag{9.377}$$

$$\underset{\sim}{H}_{TB} = \frac{-jp_B}{-p_B^2 + k_0^2} \nabla_{TB} H_B \tag{9.378}$$

$$\underset{\sim}{E}_{TB} = \frac{-j\omega\mu_0}{-p_B^2 + k_0^2} \underset{\sim}{i}_B \times \nabla_{TB} H_B \tag{9.379}$$

Since the charged particles are constrained to move along $\underset{\sim}{B}_0$, the solutions with $E_B = 0$, Equations 9.376 through 9.379, are identical to those for a free-space waveguide.

The solutions with $H_B = 0$ and E_B finite, Equations 9.372 through 9.375, are dependent upon the plasma density. For a homogeneously filled plasma waveguide with perfectly conducting boundaries, the transverse wave numbers p_1^2 and p_B^2 are positive real numbers dependent upon geometry only. The dispersion relation, Equation 9.373, can be written as

$$\gamma^2 = -\frac{\omega^4 - \omega^2(p^2c^2 + \omega_p^2) + \omega_p^2 p_B^2 c^2}{\omega^2 c^2} \tag{9.380}$$

where

$$p^2 = p_1^2 + p_B^2 \tag{9.381}$$

is the transverse wave number of the empty waveguide. For $\omega \to 0$, Equation 9.380 has a resonance in propagation with $\gamma^2 \to -\omega_p^2 p_B^2/\omega^2$. For $\omega \to \infty$, $\gamma^2 \to -\omega^2/c^2$, as expected. For finite and nonzero frequencies, we have two cutoff frequencies, for which $\gamma = 0$. These occur at

$$\omega_{co}^2 = \frac{p^2c^2 + \omega_p^2}{2}\left[1 \pm \sqrt{1 - \frac{4\dfrac{p_B^2 c^2}{\omega_p^2}}{\left(\dfrac{p^2c^2}{\omega_p^2} + 1\right)^2}}\,\right] \tag{9.382}$$

and are dependent upon both the plasma frequency and the free-space waveguide cutoff frequencies p^2c^2 and $p_B^2c^2$. A typical dispersion characteristic for some particular values of p_1^2 and p_B^2 is shown in Figure 9.19.

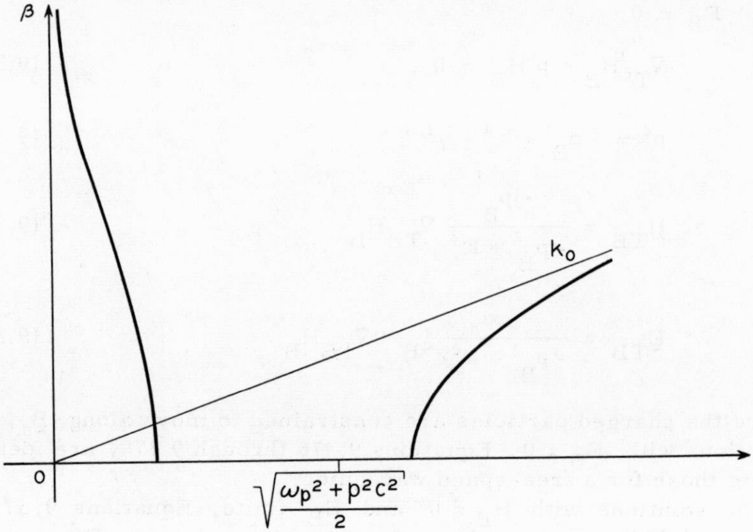

Figure 9.19. Dispersion characteristic for a TMG, homogeneous plasma waveguide with $\underset{\sim}{B}_0 = \infty$ and $H_B = 0$; the figure is for one set of values of p_1 and p_B; the cutoffs are given by Equation 9.382.

Chapter 10

BOUNDARY-VALUE PROBLEMS AND APPROXIMATE TECHNIQUES

The field analysis of Chapter 9 has consisted mainly in the formulation of the solution for the electromagnetic fields in temperate plasma waveguides and the discussion of the associated dispersion relations (for example, Equations 9.144 and 9.149, 9.161, and 9.321). The complete solution of the plasma waveguide problem involves the simultaneous solution of the dispersion equation and the determinantal equation which results from imposing particular boundary conditions that the field solutions must satisfy. In Sections 10.1 through 10.4 we discuss some aspects of this determinantal equation for the various plasma waveguides considered in Chapter 9. We restrict ourselves to temperate plasmas and simple geometries (circular and parallel plane) so as to keep the mathematical complexity to a minimum. We treat both filled and partially filled plasma waveguides, and for the latter we discuss the possible surface waves and their character at resonance. In the last section, Section 10.5, the quasi-static analysis of guided waves is developed and illustrated for both LMG and TMG plasmas.

10.1 Homogeneous, Isotropic Plasma Waveguides

In this section we shall briefly give the field solutions to the free-space waveguide (Section 9.2) of some simple geometries with perfectly conducting electric boundaries.[28] It follows from the analysis of Section 9.3.1 that these solutions will also apply to such waveguides that are filled with a homogeneous, isotropic, temperate plasma.

We choose the geometries of the waveguide cross section to be either rectangular or circular, as shown in Figure 10.1. We shall

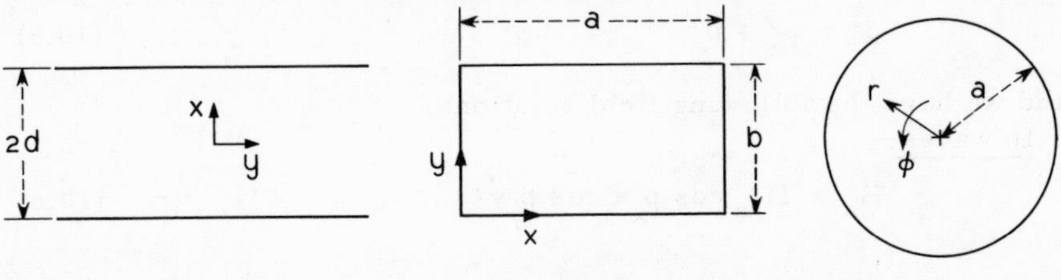

Figure 10.1. Cross sections of common waveguides: (a) parallel-plate, (b) rectangular, and (c) circular.

be interested in the source-free field solutions to Equations 9.38 through 9.45 and, hence, require that the fields be finite and single-valued. The boundary condition on the perfectly conducting electric walls of the waveguide is as given by Equation 9.6. We now give the solutions of Equation 9.39 for H waves and of Equation 9.43 for E waves, subject to the boundary conditions just outlined.

For the parallel-plate waveguide of Figure 10.1a we give the solutions that are independent of y, $(\partial/\partial y) \equiv 0$. We have solutions that are symmetric and antisymmetric with respect to $x = 0$.

H waves:

$$H_z^s = H_{z0}^s \cos p_h^s x \tag{10.1a}$$

$$p_h^s d = m\pi; \quad m = 0, 1, 2 \ldots \tag{10.2a}$$

$$H_z^a = H_{z0}^a \sin p_h^a x \tag{10.1b}$$

$$p_h^a d = (2m+1)\frac{\pi}{2}; \quad m = 0, 1, 2 \ldots \tag{10.2b}$$

E waves:

$$E_z^s = E_{z0}^s \cos p_e^s x \tag{10.3a}$$

$$p_e^s d = (2m+1)\frac{\pi}{2}; \quad m = 0, 1, 2 \ldots \tag{10.4a}$$

$$E_z^a = E_{z0}^a \sin p_e^a x \tag{10.3b}$$

$$p_e^a d = m\pi; \quad m = 1, 2 \ldots \tag{10.4b}$$

The parallel-plate waveguide can of course also propagate TEM waves, having $E_z = 0$ and $H_z = 0$.

For the rectangular waveguide of Figure 10.1b the transverse wave number is

$$p^2 = p_x^2 + p_y^2 \tag{10.5}$$

and we have the following field solutions.

H waves:

$$H_z = H_{z0} \cos p_x x \cos p_y y \tag{10.6}$$

$$p_x a = m\pi; \qquad m = \text{integer} \tag{10.7}$$

$$p_y b = n\pi; \qquad n = \text{integer} \tag{10.8}$$

$$m + n \neq 0 \tag{10.9}$$

E waves:

$$E_z = E_{z0} \sin p_x x \sin p_y y \tag{10.10}$$

$$p_x a = m\pi; \qquad m = \text{integer} \tag{10.11}$$

$$p_y b = n\pi; \qquad n = \text{integer} \tag{10.12}$$

$$mn \neq 0 \tag{10.13}$$

For the circular waveguide of Figure 10.1c we obtain the following solutions.

H waves:

$$H_z = H_{z0} J_m(p_h r) e^{jm\phi} \tag{10.14}$$

$$J_m'(p_h a) = 0 \tag{10.15}$$

E waves:

$$E_z = E_{z0} J_m(p_e r) e^{jm\phi} \tag{10.16}$$

$$J_m(p_e a) = 0 \tag{10.17}$$

where J_m is the Bessel function of order m, and J_m' is the derivative of J_m with respect to its argument.

In all of these solutions, H_{z0} and E_{z0} are arbitrary complex amplitudes, and the z dependence, $\exp(-\gamma z)$, has been omitted. The determinantal equations that follow from the boundary conditions in each case are the following: Equations 10.2 and 10.4 for the parallel-plate waveguide; Equations 10.5, 10.7 through 10.9, and 10.11 through 10.13 for the rectangular waveguide; and Equations 10.15 and 10.17 for the circular waveguide. These functional equations together with the dispersion equations, Equations 9.39 and 9.43, completely determine the dependence of the propagation constant γ upon frequency ω.

In the case where these waveguides are completely filled with a homogeneous and isotropic plasma (Section 9.3), the dispersion relation is given by Equation 9.69, which together with the determinantal equations gives the dependence of γ upon ω and ω_p. We note that in all cases the determinantal equations are independent of both frequency and propagation constant, and hence the transverse wave numbers are completely specified by the geometry of the waveguide cross section. It then also follows that the field patterns for a fixed value of p are also independent of frequency. When the plasma is either inhomogeneous or anisotropic, the determinantal equation from boundary conditions is in general dependent upon both the propagation constant and the frequency.

10.2 Inhomogeneous, Isotropic Plasma Waveguides

The field analysis for the inhomogeneous, isotropic plasma waveguides was given in Section 9.4. Here we shall choose some simple examples for which the boundary-value problem is readily solved. We shall consider a parallel-plate and circular waveguide filled with a plasma whose density varies over the cross section, and we solve for the H-wave fields having the electric field perpendicular to the density gradient (Section 9.4.1). In addition, we shall discuss the boundary-value problem for a homogeneous plasma that either does not fill the waveguide or is in free space (Section 9.4.2).

10.2.1 Inhomogeneous, Plasma-Filled Waveguides. Consider the parallel-plate, plasma-filled waveguide of Figure 10.2a. In accord-

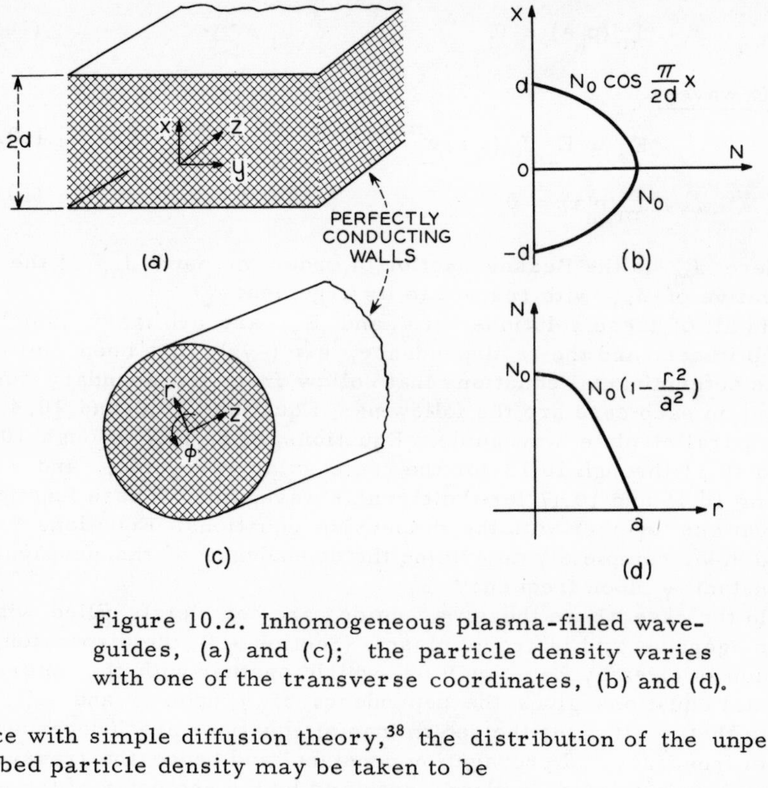

Figure 10.2. Inhomogeneous plasma-filled waveguides, (a) and (c); the particle density varies with one of the transverse coordinates, (b) and (d).

ance with simple diffusion theory,[38] the distribution of the unperturbed particle density may be taken to be

$$N = N_0 \cos \frac{\pi}{2d} x \tag{10.18}$$

and is shown in Figure 10.2b. This holds for the case when the mean free path is short compared with the separation between the plates, 2d. The normalized dielectric constant of the plasma is then

$$K_{\parallel} = 1 - \frac{\omega_{p0}^2}{\omega^2} \cos^2 \frac{\pi}{2d} x \tag{10.19}$$

where ω_{p0} is the plasma frequency corresponding to the maximum density N_0 at $x = 0$. We choose to study H-wave field solutions that are independent of y, $(\partial/\partial y) \equiv 0$. For these fields and the chosen density variation, Equation 9.134 is satisfied, and the electric field in the y direction is as given by Equation 9.135:

$$\frac{d^2E_y}{dx^2} + p^2Ey = 0 \tag{10.20}$$

where, from Equations 9.116 and 10.19, p^2 is

$$p^2 = \gamma^2 + k_0^2 - k_{p0}^2 \cos^2 \frac{\pi}{2d} x \tag{10.21}$$

with

$$k_{p0} = \frac{\omega_{p0}}{c} \tag{10.22}$$

Hence, Equation 10.20 for E_y is the Mathieu differential equation. Equations 10.20 and 10.21 can be put in standard form[39] by the following transformation of variables:

$$z = \frac{\pi}{2d} x \tag{10.23}$$

$$\delta = \left(\frac{d}{\pi}\right)^2 \left(\gamma^2 + k_0^2 - \tfrac{1}{2} k_{p0}^2\right) \tag{10.24}$$

$$q = \left(\frac{d}{\pi}\right)^2 \tfrac{1}{8} k_{p0}^2 \tag{10.25}$$

and we obtain

$$\frac{1}{4} \frac{d^2E_y}{dz^2} + (\delta - 4q \cos 2z)E_y = 0 \tag{10.26}$$

Equation 10.26 has as solutions the Mathieu functions[39]

$$E_y \sim ce_m(z, q) \qquad \text{and} \qquad se_n(z, q) \tag{10.27}$$

Since q is real, we have an infinite sequence of δ eigenvalues with periodic solutions that will allow the boundary conditions to be satisfied at $x = \pm d$. From Equations 10.24 and 10.25 the propagation constant is

$$\gamma^2 = (\delta + 4q) \left(\frac{\pi}{d}\right)^2 - k_0^2 \tag{10.28}$$

and hence the cutoff frequency is given by

$$\omega_{co}^2 = (\delta + 4q)\left(\frac{\pi}{d}\right)^2 c^2 \tag{10.29}$$

Since q is independent of frequency, the dispersion characteristic of γ has a frequency dependence similar to that of the homogeneously filled waveguide (Figure 9.3), except that the cutoff frequency must be determined from Equation 10.29. Unlike in the homogeneously filled guide, the field pattern is now a function of the plasma density ($q \sim k_{po}^2 d^2$). The remaining field components (H_z and H_x) can be determined from E_y and Equations 9.132 and 9.133.

As an example, we consider the lowest-order Mathieu function for which the boundary conditions can be satisfied,

$$E_y = E_{y0}\, ce_1(z, q) \tag{10.30}$$

For this mode, Figure 10.3a shows the amplitude distribution of E_y as a function of x ($0 \le x \le d$) for four values of q corresponding to four values of $k_{po}d$ (Equation 10.25). For $q = 0$, corresponding to empty space, E_y has a cosine variation with x, as would be expected from Equations 10.1 and 9.41. As the maximum density N_0 increases, the x position of the maximum of E_y moves out toward the plates where the density is lower. Thus the electric field is "expelled" from the regions of high plasma density, and may even vanish at $x = 0$. The eigenvalues δ corresponding to different values of q for this mode are shown in Figure 10.3b. For small values of q ($q < \frac{1}{2}$),

$$4\delta \approx 1 + 8q - 8q^2 - 8q^3 - \frac{8}{3}q^4 + \cdots \tag{10.31}$$

For large values of q ($q > \frac{1}{2}$),

$$\delta \approx -4q + 3\sqrt{2q} + \frac{10}{32} - \frac{3}{2^{10}}\frac{12}{\sqrt{2q}} - \cdots \tag{10.32}$$

Comparing Equation 10.29 with Equation 9.70, we find that the cutoff frequency is lower than that of a homogeneously filled waveguide with particle density N_0. From Equations 9.132 and 9.133 we see that H_z has an x dependence which is the derivative of E_y with respect to x, while H_x has an x dependence similar to that of E_y. Hence, as the density N_0 increases, the power density becomes concentrated in the regions where E_y is appreciable.

We next consider briefly a similar plasma in a circular waveguide, Figure 10.2c. Again, under the conditions of a diffusion-controlled plasma with mean free path small compared with the waveguide radius a, the unperturbed density distribution is[38]

$$N = N_0 J_0\left(2.405\frac{r}{a}\right) \tag{10.33}$$

where J_0 is the Bessel function of order zero. The field equations can now be solved by series expansions; this general proce-

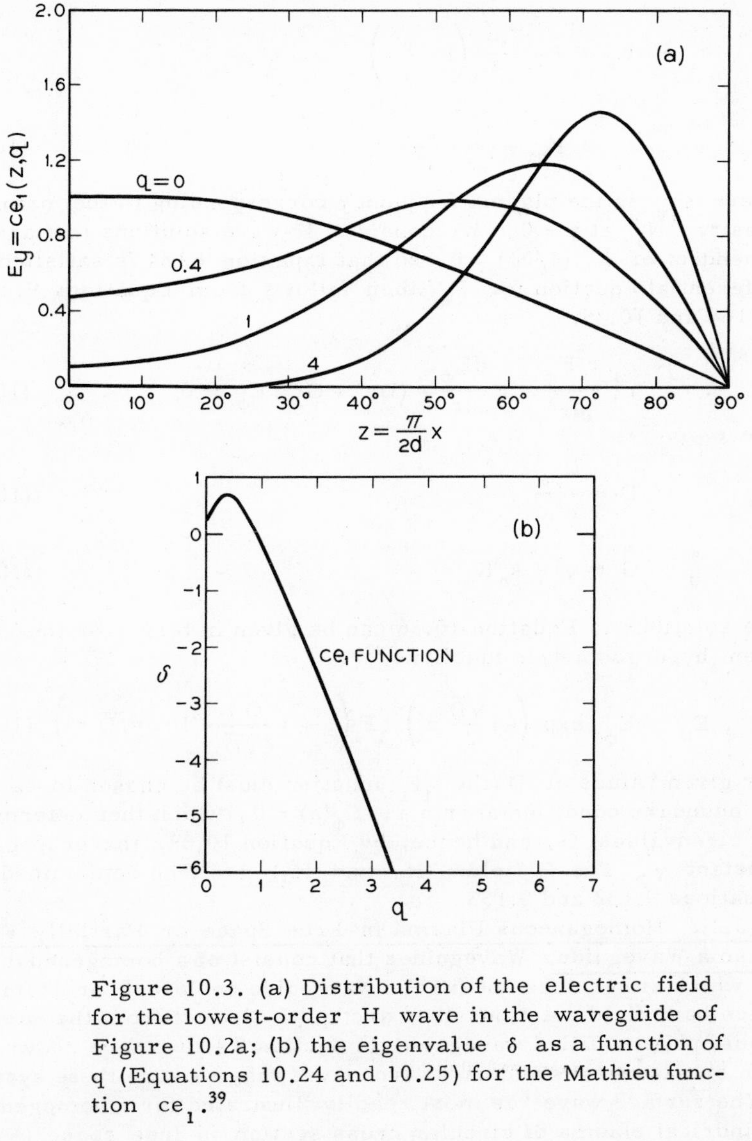

Figure 10.3. (a) Distribution of the electric field for the lowest-order H wave in the waveguide of Figure 10.2a; (b) the eigenvalue δ as a function of q (Equations 10.24 and 10.25) for the Mathieu function ce_1.[39]

dure although straightforward is quite involved and will not be given here. Instead we shall use a parabolic approximation to the density distribution of Equation 10.33 (see Figure 10.2d):

$$N \approx N_0 \left(1 - \frac{r^2}{a^2}\right) \tag{10.34}$$

for which

$$K_{\parallel} = 1 - \frac{\omega_{p0}^2}{\omega^2}\left(1 - \frac{r^2}{a^2}\right)$$

$$= K_{\parallel 0} + \frac{\omega_{p0}^2}{\omega^2}\frac{r^2}{a^2} \qquad (10.35)$$

where ω_{p0} is the plasma frequency corresponding to the maximum density N_0 at $r = 0$. We consider H-wave solutions that are independent of ϕ, $(\partial/\partial\phi) \equiv 0$, so that Equation 9.134 is satisfied. The differential equation for E_ϕ then follows from Equations 9.135, 9.116, and 10.35,

$$r^2 \frac{d^2E_\phi}{dr^2} + r\frac{dE_\phi}{dr} + (Dr^4 + Gr^2)E_\phi = 0 \qquad (10.36)$$

where

$$D = \frac{\omega_{p0}^2}{c^2a^2} \qquad (10.37)$$

$$G = \gamma^2 + k_0^2K_{\parallel 0} \qquad (10.38)$$

The solutions to Equation 10.36 can be given in terms of the confluent hypergeometric functions:[40]

$$E_\phi = E_{\phi 0}\exp\left(-j\frac{\sqrt{D}}{2}r^2\right)\ {}_1F_1\left(\frac{1}{2} + j\frac{G}{4\sqrt{D}},\ 1;\ j\sqrt{D}\,r^2\right) \qquad (10.39)$$

For given values of D the ${}_1F_1$ function must be chosen to satisfy the boundary condition at $r = a$, $E_\phi(a) = 0$, which then determines the eigenvalues G, and hence, by Equation 10.38, the propagation constant γ. The H fields (H_r and H_z) are then determined by Equations 9.132 and 9.133.

10.2.2 Homogeneous Plasma in Free Space or Partially Filled Plasma Waveguide. Waveguides that consist of a homogeneous plasma which either does not fill the waveguide enclosure or is in free space have field solutions that are quite distinct from the homogeneous, plasma-filled waveguide. In Section 9.4.2 we have shown that both surface waves and backward waves may exist in these systems.

The surface wave[41] is most readily illustrated for a homogeneous, cylindrical plasma of circular cross section in free space (Figure 10.4a). Fast waves ($q^2 > 0$, Equation 9.149) will radiate away from the plasma. Our interest is in waves that are guided along the plasma and which must be slow ($\beta > k$). For these slow waves both q^2 (Equation 9.149) and p^2 (Equation 9.144) must be negative, and we write

$$-p^2 = \beta^2 - k_0^2K_{\parallel} = \Gamma^2 \qquad (10.40)$$

$$-q^2 = \beta^2 - k_0^2 = T^2 \qquad (10.41)$$

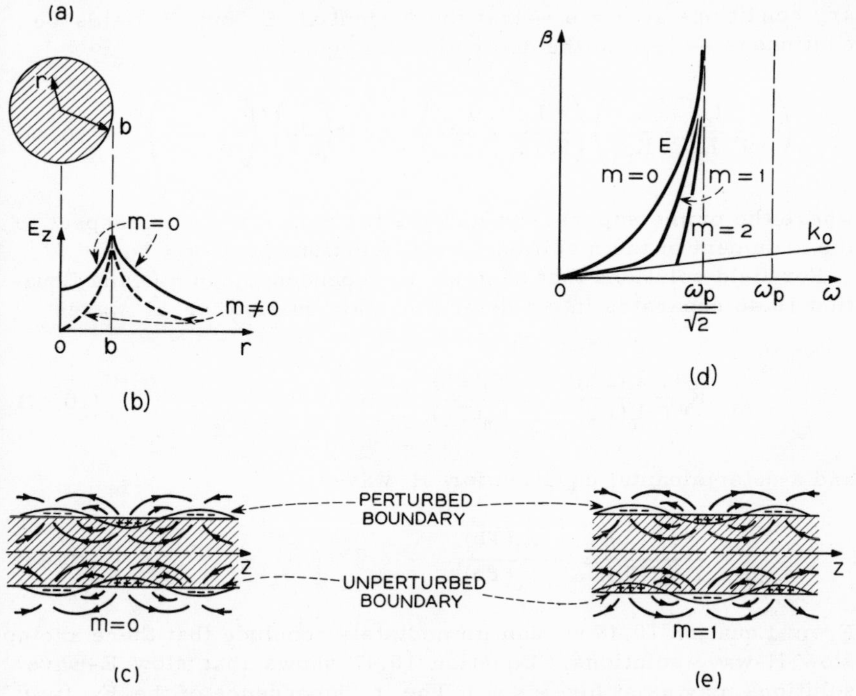

Figure 10.4. (a) Homogeneous, isotropic plasma in free space; (b) radial variation of E_z for slow surface waves; (c) and (e) longitudinal variation of electric field for slow surface waves; (d) dispersion characteristic.

The solutions for the longitudinal fields (Equations 9.142, 9.143, 9.147, and 9.148), subject to the requirements of finiteness at $r = 0$ and vanishing amplitude at $r = \infty$, are for $r < b$:

$$E_z^i = AI_m(\Gamma r)e^{jm\phi} \tag{10.42}$$

$$H_z^i = BI_m(\Gamma r)e^{jm\phi} \tag{10.43}$$

and for $r > b$:

$$E_z^o = CK_m(Tr)e^{jm\phi} \tag{10.44}$$

$$H_z^o = DK_m(Tr)e^{jm\phi} \tag{10.45}$$

where I_m and K_m are the modified Bessel functions,[39] and A, B, C, and D are arbitrary complex constants. The transverse fields follow from Equations 9.145, 9.146, 9.150, and 9.151. The boundary conditions at $r = b$ — that the tangential E and H fields be continuous — lead to the determinantal equation

$$\left(K_{\parallel}\frac{T}{\Gamma}\frac{I_m'}{I_m} - \frac{K_m'}{K_m}\right)\left(\frac{T}{\Gamma}\frac{I_m'}{I_m} - \frac{K_m'}{K_m}\right) = m^2\beta^2\left(\frac{T}{k_0 b}\right)^2\left(\frac{1}{\Gamma^2} - \frac{1}{T^2}\right)^2 \qquad (10.46)$$

where the prime superscript denotes the derivative with respect to the argument of the modified Bessel functions, and $r = b$.

For field solutions that have no ϕ dependence, $m = 0$ and Equation 10.46 separates into a determinantal equation for E waves:

$$K_{\parallel}\frac{T}{\Gamma}\frac{I_1(\Gamma b)}{I_0(\Gamma b)} = -\frac{K_1(Tb)}{K_0(Tb)} \qquad (10.47)$$

and a determinantal equation for H waves:

$$\frac{T}{\Gamma}\frac{I_1(\Gamma b)}{I_0(\Gamma b)} = -\frac{K_1(Tb)}{K_0(Tb)} \qquad (10.48)$$

From Equation 10.48 we can immediately conclude that there are no slow H-wave solutions. Equation 10.47 shows that slow E-wave solutions may exist for $\omega < \omega_p$. The r dependence of the E_z field (Equations 10.42 and 10.44) is sketched in Figure 10.4b, and the longitudinal variation of the electric field is shown in Figure 10.4c. As $K_{\parallel} \to -1$, Equation 10.47 admits the solution $T \to \infty$ and $\Gamma \to \infty$, that is, $\beta \to \infty$. Hence, these E waves have a resonance, $\beta = \infty$ for $K_{\parallel} = -1$, that is, at

$$\omega = \frac{\omega_p}{\sqrt{2}} \qquad (10.49)$$

When the medium outside the plasma rod is a homogeneous isotropic dielectric with dielectric constant ϵ, the resonance frequency at which $\beta \to \infty$ is

$$\omega = \frac{\omega_p}{\sqrt{1 + \frac{\epsilon}{\epsilon_0}}} \qquad (10.50)$$

At this resonance the fields are highly concentrated at the plasma surface, and hence this resonance persists even when the plasma is inside of a cylindrical metallic enclosure whose radius is larger than the outer radius of the plasma. At low frequencies $(\omega \to 0)$ Equation 10.47 for the $m = 0$ E waves admits slow-wave solutions. The complete dispersion curve is shown in Figure 10.4d. Near $\omega = 0$ we find approximately

$$-(\beta^2 - k_0^2)b^2 \ln[(\beta^2 - k_0^2)^{\frac{1}{2}} b] \approx k_p b \frac{I_0(k_p b)}{I_1(k_p b)} \frac{\omega^2}{\omega_p^2} \tag{10.51}$$

where $k_p = \omega_p/c$.

When the plasma is inside of a cylindrical metallic tube of radius a $(a > b)$, the electric field E_z inside the plasma is of the same form as in Equation 10.42. However, the electric field E_z outside the plasma must vanish at $r = a$ and is given by

$$E_z^o = CG_{mm}(Tr, Ta) \tag{10.52}$$

where

$$G_{mm}(Tr, Ta) = I_m(Tr)K_m(Ta) - K_m(Tr)I_m(Ta) \tag{10.53}$$

For the ϕ-independent E waves the determinantal equation becomes

$$K_{\parallel} \frac{T}{\Gamma} \frac{I_0'}{I_0} = \frac{G_{00}'}{G_{00}} \tag{10.54}$$

where the prime denotes the derivative with respect to the argument containing r, and $r = b$. The low-frequency limit of Equation 10.54 gives

$$\left(\frac{\beta}{k_0}\right)^2 \approx 1 + \left[k_p b \frac{I_1(k_p b)}{I_0(k_p b)} \ln \frac{a}{b}\right]^{-1} \tag{10.55}$$

It should be noted that both of the low-frequency approximations of Equations 10.51 and 10.55 show that $(k_0/\beta) < 1$ as $\omega \to 0$.

Waves with arbitrary ϕ dependence $(m \neq 0)$ also exhibit a resonance at the frequency given by Equation 10.49 or Equation 10.50. This can be shown directly from the form of Equation 10.46 in the limit of $\beta \to \infty$:

$$\frac{|\beta|}{k_0} \to \frac{1}{4k_0 b} \frac{\omega_p^2}{\left(\frac{\omega_p^2}{2} - \omega^2\right)} \tag{10.56}$$

The same limiting form is also obtained for a plasma that is concentric with a cylindrical metallic enclosure which it does not fill. We note that these solutions correspond to pure propagation for frequencies that are below the resonance, and hence they are traveling waves. The fields associated with the surface waves for $m \neq 0$ are mixed E and H waves, and at $r = 0$ they have $E_z = 0$ (see Figure 10.4b). A more complete idea of the dispersion characteristics of these waves can be obtained by determining their cutoffs.

For the plasma cylinder in free space the cutoffs occur for $\beta \to k_0$.

In this limit the determinantal equation, Equation 10.46, gives for $m = 1$

$$\frac{\omega^2}{\omega_p^2} \to - \frac{(k_p b)\dfrac{I_1'(k_p b)}{I_1(k_p b)} + 1}{(k_p b)^2 \ln(\beta^2 - k_0^2)} \tag{10.57}$$

and for $m \geq 2$

$$\frac{\omega_p^2}{\omega^2} \to 2 + \frac{(k_p b)^2}{(m-1)\left[(k_p b)\dfrac{I_m'(k_p b)}{I_m(k_p b)} + m\right]} \tag{10.58}$$

Hence, for $m = 1$ the cutoff is at $\omega = 0$, and for $m \geq 2$ the cutoff $(\beta \to k_0)$ is at a finite frequency below the resonant frequency $\omega_p/\sqrt{2}$. A sketch of a possible dispersion curve is shown in Figure 10.4d.

When the plasma partially fills a concentric metallic cylinder, the cutoffs occur for frequencies at which $\beta \to 0$. Then it is necessary to know the behavior of the propagation for fast waves, $0 \leq \beta \leq k_0$. The radial variation of the fields is now oscillatory in nature, and the modified Bessel functions of Equations 10.42, 10.43, 10.52, and a similar equation for H_z^a, become replaced by the ordinary Bessel and Neumann functions. The determinantal equation becomes

$$\left(K_{\parallel}\frac{q}{p}\frac{J_m'}{J_m} - \frac{H'_{mm}}{H_{mm}}\right)\left(\frac{q}{p}\frac{J_m'}{J_m} - \frac{H'_{mm'}}{H_{mm'}}\right) = m^2\beta^2\left(\frac{q}{k_0 b}\right)^2\left(\frac{1}{p^2} - \frac{1}{q^2}\right)^2 \tag{10.59}$$

$$H_{mm}(qr, qa) = J_m(qr)\,N_m(qa) - N_m(qr)\,J_m(qa) \tag{10.60}$$

$$H_{mm'}(qr, qa) = J_m(qr)\,N_m'(qa) - N_m(qr)\,J_m'(qa) \tag{10.61}$$

where J_m and N_m are the Bessel and Neumann functions of order m, the prime superscript denotes the derivative with respect to the argument, and all functions in Equation 10.59 are evaluated at $r = b$. For simplicity we shall restrict ourselves to the case of $k_p b \ll 1$ and determine the frequencies for which $\beta = k_0$ and $\beta = 0$. For $\beta \to k_0$, Equation 10.59 gives for $m = 1$

$$\omega \to \frac{\omega_p}{\sqrt{2}}\left[1 + \left(\frac{k_p b}{2}\right)^2 \ln\frac{a}{b}\right]^{-\frac{1}{2}} \tag{10.62}$$

which is always below the resonance frequency $\omega_p/\sqrt{2}$. For cutoff, $\beta \to 0$, Equation 10.60 separates into a determinantal equation for E waves and a determinantal equation for H waves. The E-limit waves have their cutoff frequency above both ω_p and the cutoff frequency for E waves in the free-space waveguide. The determinantal equation for the H-limit waves from Equation 10.59 with $k_p b << 1$ is[42]

$$\frac{N_m'(k_0 a)}{J_m'(k_0 a)} \approx \frac{m!(m-1)!2^{2m}}{\pi(k_0 b)^{2m}} \frac{(K_{\parallel}+1)}{(K_{\parallel}-1)} \tag{10.63}$$

Equation 10.63 shows that the lowest cutoff frequency is always below both $\omega_p/\sqrt{2}$ and the lowest cutoff frequency for H waves in a free-space waveguide. For $k_0 a << 1$, Equation 10.63 gives

$$\omega_{co} < \frac{\omega_p}{\sqrt{2}} \left[1 - \left(\frac{b}{a}\right)^{2m}\right]^{\frac{1}{2}} \tag{10.64}$$

which is below the frequency for which $\beta = k_0$. The dispersion curves for the ϕ-dependent waves in a partially filled plasma waveguide with $k_p b << 1$ and $k_0 a << 1$ are shown in Figure 10.5b. A backward-wave region may exist near $\beta = k_0$. For $m = 1$, $k_p b << 1$, and $a/b >> 1$, the condition for backward waves can be deduced from Equations 10.62 and 10.64,

$$\left(\frac{a}{b}\right)^2 \ln \frac{a}{b} > \left(\frac{2}{k_p b}\right)^2 \tag{10.65}$$

Backward-wave characteristics are possible with various geometries and boundary conditions. Some general criteria can be established from the group velocity-phase velocity relations of Chapter 7. Since the plasma is isotropic, Equation 7.153 reduces to the form of Equation 9.79. However, because of the inhomogeneity of the waveguide cross section, the integrations that enter into the evaluation of the U's must be carried out separately inside and outside the plasma. Thus, for example,

$$U_\epsilon = \int_{A_i} \tfrac{1}{4}\epsilon_0 K_{\parallel} |\underset{\sim}{E}^i|^2\, da + \int_{A_o} \tfrac{1}{4}\epsilon |E^o|^2\, da$$

$$= U_\epsilon^i + U_\epsilon^o \tag{10.66}$$

$$U_{ek} = \int_{A_i} \tfrac{1}{4}\epsilon_0 \frac{\partial(\omega K)}{\partial\omega} |E^i|^2\, da + \int_{A_o} \tfrac{1}{4}\epsilon |E^o|^2\, da \tag{10.67}$$

Two equivalent forms of the condition for backward waves follow from Equation 9.79:

$$U_{mT} - U_{\epsilon z} < 0 \tag{10.68}$$

$$U_{\epsilon T} - U_{mz} < 0 \tag{10.69}$$

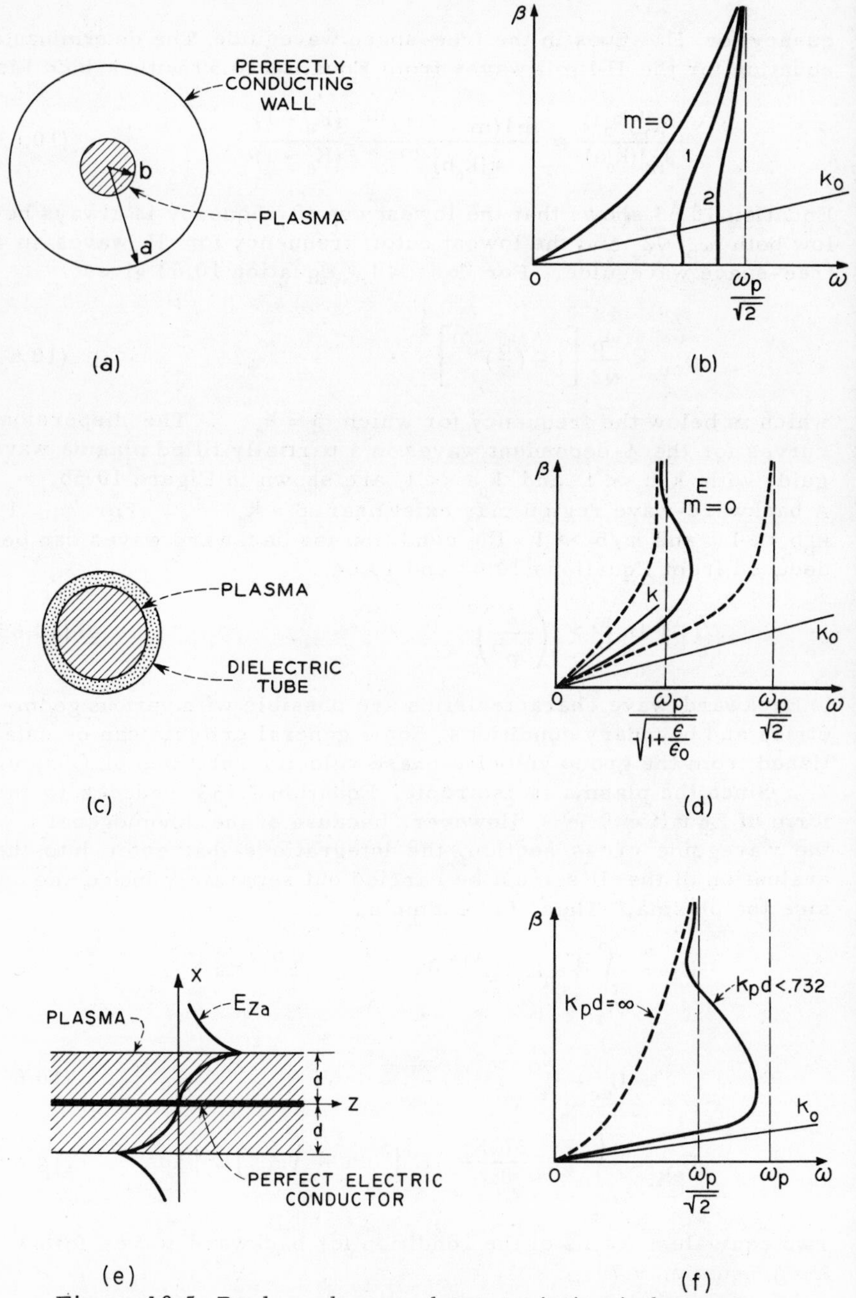

Figure 10.5. Backward-wave characteristics in homogeneous, isotropic plasma waveguides; (a) and (b) partially filled plasma waveguide for $k_0 b << 1$ and $k_0 b |K_{\parallel}|^{\frac{1}{2}} << 1$; (c) and (d) plasma rod surrounded by a thin dielectric tube; (e) and (f) plasma slab in free space or above a perfect electric conductor.

As examples, consider the ϕ-independent waves for which E and H waves can be identified separately. The H waves have $E_z = 0$, and hence, $U_{\epsilon z} = 0$; since U_{mT} is positive definite, Equation 10.68 shows that ϕ-independent H waves cannot be backward waves. On the other hand, E waves have $H_z = 0$. Hence, $U_{mz} = 0$, and Equation 10.69 predicts that backward waves will exist if $U_{\epsilon T} < 0$; this may occur if

$$\omega < \omega_p \tag{10.70}$$

and

$$|U_{\epsilon T}{}^{i}| > |U_{\epsilon T}{}^{o}| \tag{10.71}$$

A possible configuration for ϕ-independent backward E waves, shown in Figure 10.5a, consists of a plasma surrounded by a dielectric cylinder whose thickness is small compared with the plasma radius.[43] At resonance the fields are concentrated near the plasma surface, and the resonance occurs at the frequency corresponding to Equation 10.50. At low frequencies the fields extend out into the free-space region, and the effect of the thin dielectric cylinder is negligible. The frequency dependence of the propagation constant may then contain a backward-wave region, as shown in Figure 10.5d.

The simplest isotropic plasma configuration that exhibits backward surface waves is the parallel-plane, homogeneous plasma slab or the perfect conductor covered with a homogeneous plasma layer,[44] sketched in Figure 10.5e. Equations 9.142 through 9.151 readily show that y-independent wave solutions exist as independent E and H waves. The H-wave solutions cannot satisfy the boundary condition Equation 9.141. The E-wave solutions can satisfy the boundary conditions. By symmetry, two distinct types of E waves can exist: One has E_z antisymmetric about $x = 0$, corresponding to a perfect electric conductor in the $x = 0$ plane; the second has E_z symmetric about $x = 0$, corresponding to a perfect magnetic conductor in the $x = 0$ plane. We shall discuss the antisymmetric solutions because only they may have backward-wave characteristics.

The antisymmetric E-wave solution, illustrated in Figure 10.5e, is

$$E_{za}{}^{i} = A \sinh \Gamma x \tag{10.72}$$

for $|x| < d$, and

$$E_{za}{}^{o} = A \sinh \Gamma d \exp[-T(x - d)] \tag{10.73}$$

for $x > d$. Γ and T are given by Equations 10.40 and 10.41, and A is an arbitrary complex amplitude. The transverse fields are then obtained from Equations 9.145, 9.146, 9.150, and 9.151. The boundary condition of Equation 9.141 gives the determinantal equation

$$\Gamma \tanh \Gamma d = -TK_{\parallel} \tag{10.74}$$

For Γ and T real and positive, the solutions of Equation 10.74 must occur for $K_{\parallel} < 0$; that is, $\omega < \omega_p$. Since $\Gamma > T$, we note that solutions can occur for both $K_{\parallel} < -1$ and $K_{\parallel} > -1$, that is, for $\omega < \omega_p/\sqrt{2}$ and $\omega > \omega_p/\sqrt{2}$. The condition for backward waves, Equation 10.71, becomes

$$\tanh \Gamma d \left(\tanh \Gamma d + \frac{\Gamma d}{\cosh^2 \Gamma d} \right) > |K_{\parallel}|^2 \tag{10.75}$$

From Equations 9.145 and 9.146 for an E wave, we note that the sign of the time-averaged power flow in the plasma is the same as the sign of $\beta K_{\parallel}$, while outside the plasma the time-averaged power flow has the sign of β. Hence, the E-wave solution with a positive phase velocity has its time-averaged power flow inside the plasma in a direction opposite to the phase velocity, and outside the plasma in the direction of the phase velocity. It can be readily shown that the backward wave condition of Equations 10.71 and 10.75 is equivalent to the condition that the magnitude of the time-averaged power flow inside the plasma be larger than outside the plasma. Detailed computations (Oliner and Tamir[44]) show that backward waves exist for $k_p d < 0.732$, and in the frequency range from about $0.91\omega_p/\sqrt{2}$ to just below ω_p. A typical dispersion characteristic is shown in Figure 10.5f. Near resonance, $\beta \to \infty$, Equation 10.74 requires $K_{\parallel} < -1$, and the characteristic is that of a traveling wave. This is to be expected because near resonance the plasma is effectively semi-infinite in the x direction. Near cutoff, $\beta \to k_0$, we find that Equation 10.74 requires $\omega \to 0$, and we obtain

$$\frac{\beta}{k_0} \to 1 + 2\frac{\omega^2}{\omega_p^2} \tanh^2 k_p d \tag{10.76}$$

As $\omega \to 0$, the power flow is predominantly in free space and increasing, while the power flow in the plasma is about constant. As β increases, the power flow in the plasma increases. However, if the plasma is thin, $k_p d \ll 1$, the power flow is predominantly outside the plasma even up to frequencies near ω_p. Hence, for such cases a backward wave characteristic must exist.

Finally, it should be noted that complex wave solutions can also exist. In the dispersion characteristic of the propagation constant, the complex solutions usually start at the frequencies for which the group velocity vanishes for finite values of ω and β (see Figures 10.5d and f).

10.3 Homogeneous, Anisotropic Plasma Waveguides

Here we shall consider the boundary-value problem for a waveguide filled with a homogeneous plasma in the presence of an applied uniform magnetic field either parallel with or transverse to the direction of propagation.[36,37,45] In Sections 9.5 and 9.6 we have shown that for four special frequencies (ω for $\gamma \to 0$, ω for $\gamma \to \infty$, $\omega \to 0$, and $\omega \to \infty$) the solution of this problem could be obtained from approximate, uncoupled forms of Equations 9.157 and 9.158. For arbitrary frequencies the solution is obtained as outlined by Equations 9.159 through 9.166. In this section we shall derive the determinantal equations from the perfectly conducting electric boundary conditions for parallel-plate and circular waveguide cross sections. These functional equations must then be solved simultaneously with Equations 9.168 for γ, p_1, and p_2.

10.3.1 Longitudinally Magnetized, Parallel-Plate Waveguide. The geometry of the LMG parallel-plate waveguide is as shown in Figure 10.1a.[37,46] The applied unperturbed, uniform magnetic field is in the z direction. The solutions to Equation 9.162 that are independent of y ($\partial/\partial y = 0$) are

$$E_{z1} = A_1 \sin p_1 x + B_1 \cos p_1 x \tag{10.77}$$

$$E_{z2} = A_2 \sin p_2 x + B_2 \cos p_2 x \tag{10.78}$$

where A_1, A_2, B_1, and B_2 are arbitrary constants. The total E_z field is given by the sum of Equations 10.77 and 10.78. The transverse fields follow from Equation 9.166, and the boundary conditions are

$$\left.\begin{aligned} E_z &= 0 \\ E_y &= 0 \end{aligned}\right\} \quad \text{at } x = \pm d \text{ for all } z \tag{10.79}$$

The boundary conditions of Equation 10.79 must be satisfied for arbitrary values of the constants A_1, A_2, B_1, and B_2. Hence, the determinant of the coefficients of A_1, A_2, B_1, and B_2 in Equation 10.79 must vanish. This gives the determinantal equation, and establishes the relative values of A_1, A_2, B_1, and B_2. It is convenient to separate the E_z field into antisymmetric and symmetric components, and to obtain the results in the following form:

$$E_z^a = A_1 \sin p_1 x + A_2 \sin p_2 x \tag{10.80}$$

$$f_1 p_2 \sin p_1 d \cos p_2 d - f_2 p_1 \sin p_2 d \cos p_1 d = 0 \tag{10.81}$$

$$\frac{A_2}{A_1} = -\frac{\sin p_1 d}{\sin p_2 d} = -\frac{f_2 p_1}{f_1 p_2}\,\frac{\cos p_1 d}{\cos p_2 d} \tag{10.82}$$

and

$$E_z^s = B_1 \cos p_1 x + B_2 \cos p_2 x \tag{10.83}$$

$$-f_1 p_2 \cos p_1 d \sin p_2 d + f_2 p_1 \cos p_2 d \sin p_1 d = 0 \tag{10.84}$$

$$\frac{B_2}{B_1} = -\frac{\cos p_1 d}{\cos p_2 d} = -\frac{f_2 p_1}{f_1 p_2}\,\frac{\sin p_1 d}{\sin p_2 d} \tag{10.85}$$

where

$$f_{1,2} = p_{1,2}^{\;2}\frac{S}{b} - \frac{1}{\gamma}\frac{K_\perp}{K_\times} \tag{10.86}$$

with S and b as given by Equations 9.25 and 9.33, respectively, and $b \neq 0$. For a given plate separation d and plasma parameters ω_p, ω_{b+}, and ω_{b-}, Equations 10.81, 10.84, and 9.168 can in principle be solved simultaneously for the propagation constant γ as a function of frequency ω. Equations 10.81 and 10.84 can be combined into a single equation for both antisymmetric and symmetric E_z solutions:

$$\sin p_1 2d \sin p_2 2d = \frac{2f_1 f_2 p_1 p_2}{(f_1^2 p_2^2 + f_2^2 p_1^2)}(1 - \cos p_1 2d \cos p_2 2d) \tag{10.87}$$

For the special frequencies of cutoff, resonance, zero, and infinity, the dispersion relations and fields are determined by the methods of Section 9.5.

Cutoff. The E-wave cutoff fields are given by Equations 9.184 through 9.187, with E_z and p_e as given by Equations 10.3a through 10.4b. The cutoff frequencies are determined by Equation 9.194:

$$\omega_{co}^{\;s} = \sqrt{\omega_p^2 + \left[\frac{(2m+1)\pi c}{2d}\right]^2}\;; \quad m = 0, 1, 2, \cdots \tag{10.88}$$

$$\omega_{co}^{\;a} = \sqrt{\omega_p^2 + \left(\frac{m\pi c}{d}\right)^2}\;; \quad m = 1, 2, \cdots \tag{10.89}$$

as is also shown in Figure 9.7 at the intersection marked E. The H-wave cutoff fields are given by Equations 9.188 through 9.191 with H_z given by Equation 10.1a and 10.2a. For the parallel-plate geometry with solutions independent of y, a standing wave can be established with the extraordinary waves propagating between the plates in the $\pm x$ directions. The boundary condition on the perfectly conducting electric plates is then given by Equation 9.196. Hence, the eigenvalues p_h are a function of the geometry only and are given by Equations 10.1b and 10.2b. The cutoff frequencies are determined by Equations 9.183:

$$(\omega n_\times)_s = \frac{m\pi c}{d}\;; \quad m = 0, 1, 2, \cdots \tag{10.90}$$

$$(\omega n_\times)_a = \frac{(2m+1)\pi c}{2d}\;; \quad m = 0, 1, 2, \cdots \tag{10.91}$$

as illustrated graphically in Figure 9.7 by the intersections marked H_0.

Resonance. Near plasma resonance the transverse wave numbers are related to γ by Equations 9.206 and 9.207, and near the cyclotron resonances by Equations 9.221 and 9.222. For all resonances, Equation 10.81 shows that $f_1 p_2 \to 0$ and $f_2 p_1$ remains finite. The p_1 eigenvalues belong to E-wave solutions and are given approximately by Equations 10.3b and 10.4b. The p_2 eigenvalues approach infinity as γ. Hence, for a propagating wave $(\gamma = j\beta)$ near resonance, the E_z fields are approximately

$$E_z^{\,a} \approx A_1 \left(\sin p_1^{\,a} x - \frac{\sin p_1^{\,a} d}{\sinh \beta d} \sinh \beta x \right) \tag{10.92}$$

$$E_z^{\,s} \approx B_1 \left(\cos p_1^{\,s} x - \frac{\cos p_1^{\,s} d}{\cosh \beta d} \cosh \beta x \right) \tag{10.93}$$

and are sketched in Figure 10.6 for the lowest value of p_1. It is clear that the values of p_1 are actually somewhat smaller than those given by Equations 10.3b and 10.4b. The approximate form of

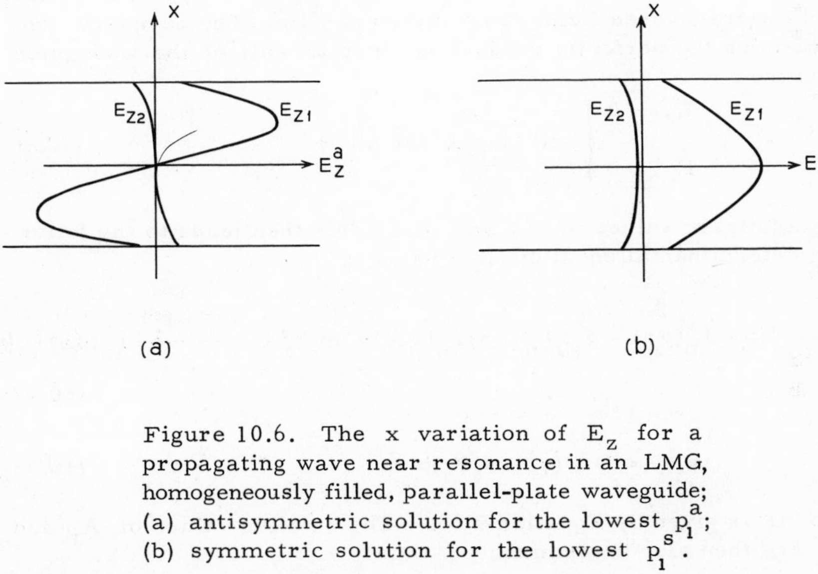

Figure 10.6. The x variation of E_z for a propagating wave near resonance in an LMG, homogeneously filled, parallel-plate waveguide; (a) antisymmetric solution for the lowest p_1^a; (b) symmetric solution for the lowest p_1^s.

the transverse fields of Equations 9.212, 9.213 through 9.216, and 9.228 through 9.231 cannot be used for satisfying the boundary conditions, especially the fields near cyclotron resonances (Equations 9.212 and 9.228 through 9.231) that have finite transverse magnetic fields. The proper transverse fields are obtained from Equations 9.21 through 9.28, evaluated near the resonances, with E_z as in Equations 10.92 and 10.93 and H_z from Equations 9.164 and 9.165.

Magnetohydrodynamic Regime. In the low-frequency limit described in Section 9.5.4, we find that the determinantal equation can be satisfied with either $p_1 = 0$ and p_2 finite ($f_2 p_1$ finite and $f_1 p_2 \to \infty$), or p_1 finite and $p_2 = 0$ ($f_1 p_2$ finite and $f_2 p_1 \to \infty$). These limits lead to the E and H waves described in Section 9.5.4.

10.3.2 Longitudinally Magnetized, Circular Waveguide. The LMG circular waveguide, as shown in Figure 10.1c, is filled with a homogeneous plasma. The applied unperturbed, uniform magnetic field is along the z direction. The general solutions to Equation 9.162, regular at $r = 0$, are

$$E_{z1} = A_1 J_m(p_1 r) e^{jm\phi} \tag{10.94}$$

$$E_{z2} = A_2 J_m(p_2 r) e^{jm\phi} \tag{10.95}$$

where A_1 and A_2 are arbitrary constants. The boundary conditions require that the values of m be the same for both solutions. The E_z field in the waveguide is the sum of Equations 10.94 and 10.95. The H_z field is then found from Equations 9.164 and 9.165, and the transverse fields from Equation 9.166. The boundary conditions on the perfectly conducting electric wall of the waveguide require

$$\left.\begin{aligned} E_z &= 0 \\ E_\phi &= 0 \end{aligned}\right\} \quad \text{at } r = a \text{ for all } \phi \tag{10.96}$$

for arbitrary values of A_1 and A_2. This then leads to the following determinantal equation ($b \neq 0$):

$$g_2 p_1 J_m'(p_1 a) J_m(p_2 a) - g_1 p_2 J_m'(p_2 a) J_m(p_1 a) = mjk_0^2 K_\times \frac{(p_2^2 - p_1^2)}{a} J_m(p_1 a) J_m(p_2 a) \tag{10.97}$$

where

$$g_{1,2} = D - (\gamma^2 + k_0^2 K_\perp) p_{1,2}^{\ 2} \tag{10.98}$$

and D is given by Equation 9.28. The relative values of A_1 and A_2 are then also determined;

$$\frac{A_2}{A_1} = -\frac{J_m(p_1 a)}{J_m(p_2 a)} \tag{10.99}$$

The simultaneous solution of Equations 10.97 and 9.168 gives the propagation constant γ and the transverse wave numbers $p_{1,2}$ for given plasma parameters ω_p, $\omega_{b\pm}$, waveguide radius a, and frequency ω.

Cutoff. For the E waves the fields are given by Equations 9.184 through 9.187, with E_z and p_e given by Equations 10.16 and 10.17, respectively. If we denote the solutions of Equation 10.17 by

$$p_e a \equiv \epsilon_{mn} \tag{10.100}$$

where ϵ_{mn} is the nth zero of the mth-order Bessel function, then the cutoff frequencies are (Equation 9.194)

$$\omega_{co} = \sqrt{\omega_p^2 + \left(\frac{\epsilon_{mn} c}{a}\right)^2} \tag{10.101}$$

This solution is also illustrated graphically in Figure 9.7, for the lowest value of ϵ_{mn}, by the intersection marked E.

The H-wave fields at cutoff are given by Equations 9.188 through 9.191. The H_z field is given by Equation 10.14, and

$$p_h^2 = k_0^2 \frac{K_r K_\ell}{K_\perp} \tag{10.102}$$

is in general not a constant. The boundary condition (Equation 9.195), requiring that the tangential electric field vanish at the waveguide wall, gives

$$(p_h a) \frac{J_m'(p_h a)}{J_m(p_h a)} = -m \frac{jK_\times}{K_\perp} \tag{10.103}$$

The cutoff frequency and p_h are obtained from a simultaneous solution of Equations 10.102 and 10.103. For the special case of cutoff H waves with fields independent of ϕ, $m = 0$ and Equation 10.103 reduces to Equation 10.15. This is also evident from the boundary condition that reduces to Equation 9.196 and allows the extraordinary waves to exist in the transverse plane of the waveguide as a cylindrical wave. Thus p_h becomes a constant determined by the geometry only (Equation 10.15 for $m = 0$), and the cutoff frequencies are found from Equation 10.102. For this case, let

$$p_h a = \chi_{on} \tag{10.104}$$

where χ_{on} is the nth zero of the <u>first</u>-order Bessel function. Equation 10.9 can then be written as

$$\left(\frac{\chi_{on} c}{a}\right) = (\omega n_\times) \tag{10.105}$$

and the solutions for the cutoff frequencies for a particular n are illustrated in Figure 9.7 by the intersections marked H_0.

Resonances. From the dispersion equation (Equations 9.167 through 9.170), the transverse wave numbers can be related to the propagation constant. Thus, near plasma resonance we obtain Equations 9.206 and 9.207, and near the cyclotron resonances Equations 9.221 and 9.222. For all resonances the determinantal equation,

Equation 10.97, is approximately reduced to only the term containing $g_1 p_2$. The p_1 eigenvalues are then approximately given by Equation 10.17, and the fields are approximately only E waves, as discussed in Section 9.5.3. To obtain fields that properly meet the boundary conditions, those associated with the p_2 eigenvalue that approaches γ must be retained. Thus, for a propagating wave ($\gamma = j\beta$) near resonance, the E_z field is approximately

$$E_z \approx A_1 \left[J_m(p_1 r) - \frac{J_m(p_1 a)}{I_m(\beta a)} I_m(\beta r) \right] e^{jm\phi} \tag{10.106}$$

where I_m is the modified Bessel function of order m. The r dependence of Equation 10.106 for $m = 0$ and $m = 1$ is sketched in Figure 10.7. We note that the values of p_1 are actually smaller than those given by Equation 10.17. The approximate form of the transverse fields is obtained by evaluating Equations 9.21 through 9.28 near the resonances and by using E_z of Equation 10.106 and H_z from Equations 9.164 and 9.165.

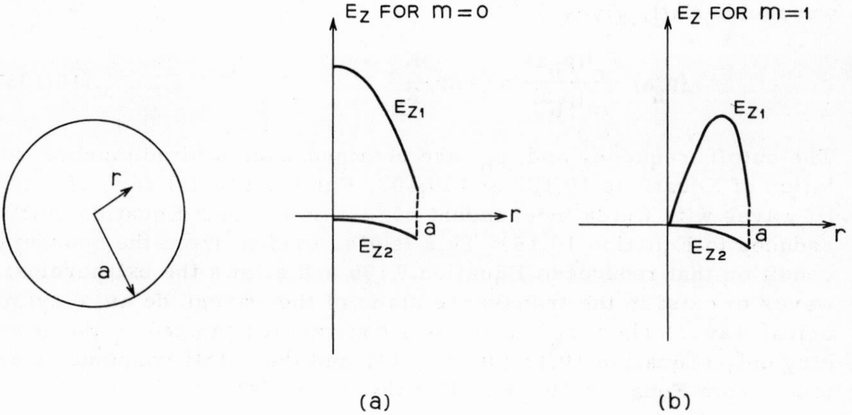

Figure 10.7. The r variation of E_z for a propagating wave near resonance in an LMG homogeneously filled, circular plasma waveguide; (a) lowest-order variation for $m = 0$; (b) lowest-order variation for $m = 1$.

Magnetohydrodynamic Regime. In the low-frequency region the dispersion equation, Equation 9.161, separates into Equations 9.252 and 9.258. The determinantal equation, Equation 10.97 reduces to

$$J_m(p_1 a) \left[(\gamma^2 + k_0^2 K_\perp) p_2 a J_m'(p_2 a) + jmk_0^2 K_\times J_m(p_2 a) \right] = 0 \tag{10.107}$$

The coupled wave equations, Equations 9.157 and 9.158, separate into E and H waves as discussed in Section 9.5.4. For the E waves, p_e of Equation 9.252 is the same as p_1 in Equation 10.107, and the determinantal equation is

$$J_m(p_1 a) = 0 \tag{10.108}$$

For the H waves, p_h of Equation 9.258 is the same as p_2 in Equation 10.107, and the determinantal equation is

$$p_2 a \frac{J_m{}'(p_2 a)}{J_m(p_2 a)} = -jm \frac{k_0^2 K_\times}{(\gamma^2 + k_0^2 K_\perp)} \tag{10.109}$$

We note again that the H waves have cutoffs $(\gamma = 0)$ at finite frequencies (Equation 9.260). For these cutoffs Equation 10.109 becomes exactly Equation 10.103, as it should. Thus the H waves belong to the MHD waves only if their cutoff frequency is much lower than ω_{b+} and ω_p. Under these conditions the determinantal equation, Equation 10.109, reduces to

$$J_m{}'(p_2 a) = 0 \tag{10.110}$$

and p_2 is a function of only the waveguide radius a.

10.3.3 Transversely Magnetized Waveguides. We shall restrict our attention to a few interesting examples for which the determinantal equation is not too complicated.

As a first example, consider the fields in a rectangular, TMG plasma waveguide near resonance (Section 9.7.4). Near the resonances of the extraordinary wave $(K_\perp \to 0)$, the fields are given by the solutions of Equations 9.351 through 9.353:

$$E_z \to A \sin p_y y \sin p_x x \tag{10.111}$$

$$E_x \to -\frac{p_x}{\gamma} A \sin p_y y \cos p_x x \tag{10.112}$$

$$H_y \to -\frac{\omega\epsilon_0 K_\ell}{\gamma} A \sin p_y y \sin p_x x \tag{10.113}$$

$$E_y \to -\frac{p_y}{\gamma} A \cos p_y y \sin p_x x \tag{10.114}$$

$$H_x \to 0 \tag{10.115}$$

$$H_z \to 0 \tag{10.116}$$

The boundary conditions can be satisfied with $p_y = m\pi/b$ and $p_x = n\pi/a$. The propagation constants γ and p_y are related by the dispersion relation, Equation 9.349. Away from resonance the solutions associated with the second p_x^2 value $(p_x^2 \to \gamma^2)$ must also be included. These have a monotonic variation in x and y, and the fulfillment of the boundary conditions requires that the first p_x value be smaller than $n\pi/a$. Similar considerations apply to the fields near the resonance at zero frequency, Equations 9.355 through 9.361.

As another example, we consider the fields in a TMG plasma waveguide that are uniform along $\underset{\sim}{B}_0$ (Section 9.7.1). Since the solutions with zero magnetic field along $\underset{\sim}{B}_0$ are the same as for an isotropic waveguide, we need not discuss them in detail. The solutions with zero electric field along $\underset{\sim}{B}_0$ are more interesting. Let us consider the parallel-plate geometry for which the determinantal equation is Equation 9.308. The first set of solutions corresponding to $p_2 = n\pi/2d$ and Equation 9.309 leads to the following fields (Equations 9.303 through 9.306):

$$H_y = A\left[\cos p_2 x - \frac{\gamma}{p_2}\frac{K_\times}{K_\perp}\sin p_2 x\right] \tag{10.117}$$

$$E_x = -A\frac{j\omega\mu_0}{p_2^2 - \gamma^2}\left[\gamma\left(1 + \frac{K_\times}{K_\perp}\right)\cos p_2 x + \frac{(p_2^2 - \gamma^2)}{p_2}\frac{K_\times}{K_\perp}\sin p_2 x\right] \tag{10.118}$$

$$E_z = A\frac{j\omega\mu_0}{(p_2^2 - \gamma^2)p_2}\left(p_2^2 + \gamma^2\frac{K_\times^2}{K_\perp^2}\right)\sin p_2 x \tag{10.119}$$

It is interesting to note that the field distributions of H_y and E_x in x depend upon γ, and hence are different for forward and backward traveling waves. A similar situation exists for the corresponding fields H_ϕ and E_r in the coaxial plasma waveguide. The second set of solutions, corresponding to Equations 9.311 and 9.312, has only transverse fields:

$$E_z = 0 \tag{10.120}$$

$$\frac{E_x}{H_y} = \frac{j\omega\mu_0}{\gamma} \tag{10.121}$$

where γ is given by Equation 9.311. Since these solutions extend down to zero frequency (see Equations 9.313 through 9.315), it is of interest to investigate the motion of the particles. It can be shown readily that the particles have a velocity component in the x direction that is out of phase with E_x, and a velocity component in the z direction that is in phase with E_x. For a propagating wave the balance required by Equation 7.149a must be satisfied. For low frequencies, $\omega \to 0$, the x component of the velocity becomes small, and U_k is essentially the longitudinal kinetic energy of the particles. However, since the velocity is elliptically polarized about $\underset{\sim}{B}_0$, U_d remains finite and contributes to the necessary balance of Equation 7.149a.

10.4 Waveguides Partially Filled with a Homogeneous, Anisotropic Plasma

The simplest inhomogeneous plasma waveguide is one containing a homogeneous plasma that partially fills the cross section of the waveguide. The boundary-value problem for the fields is now more complicated. At the surface between a neutral plasma and free space the variation of the first-order electric field may be appreciable, so that adjacent particles crossing the surface may experience widely different forces. (See also Section 9.4.2.) In the following section we shall simplify the boundary conditions. We assume that under the influence of the first-order fields, consistent with the force and continuity equations, a surface charge density is present at the unperturbed position of the plasma surface. Under these conditions the discontinuity in the normal electric field is given by this surface charge, and the other field components are continuous across the boundary. We shall not enter into the mathematical details of this boundary-value problem, but consider only some of the new features of the field solutions. These are the surface waves that may propagate and their resonances. At resonance the surface waves have their fields concentrated at the boundary of the plasma. Hence, these resonances may be determined from the analysis of surface waves on a semi-infinite plasma.

10.4.1 Surface Waves. The two cases to be considered are illustrated in Figure 10.8. In the free space outside the plasma the fields

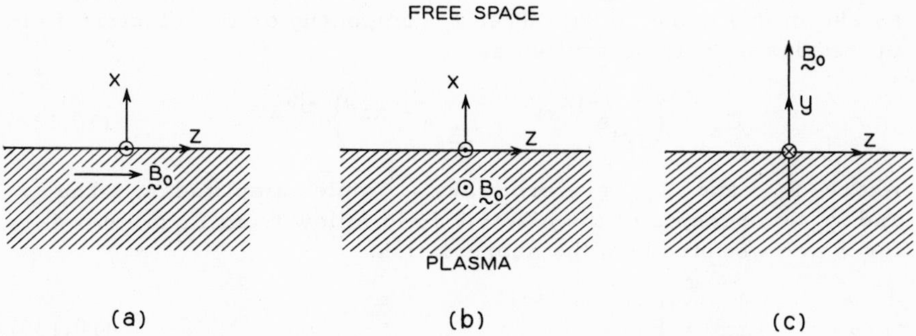

Figure 10.8. The surface between plasma and free space; (a) LMG plasma; (b) TMG plasma with $\underset{\sim}{B}_0$ parallel to surface; (c) TMG plasma with $\underset{\sim}{B}_0$ perpendicular to surface.

shall be taken as a superposition of E and H nonuniform plane waves[24] having the propagation constant β in the z direction and the attenuation constant α in the x direction. These fields, with space dependence $\exp(-\alpha x)\exp(-j\beta z)$, are

E wave:

$$E_z = E_{z0} \tag{10.122}$$

$$E_x = -\frac{j\beta}{\alpha} E_{z0} \tag{10.123}$$

$$H_y = -\frac{j\omega\epsilon_0}{\alpha} E_{z0} \tag{10.124}$$

H wave:

$$H_z = H_{z0} \tag{10.125}$$

$$H_x = -\frac{j\beta}{\alpha} H_{z0} \tag{10.126}$$

$$E_y = \frac{j\omega\mu_0}{\alpha} H_{z0} \tag{10.127}$$

where

$$\beta^2 = \alpha^2 + k_0^2 \tag{10.128}$$

Here E_{z0} and H_{z0} are arbitrary complex amplitudes.

In the plasma we shall use the plane-wave solutions of Chapter 3. We shall treat separately the LMG and TMG plasmas.

Longitudinally Magnetized Plasma. For this case the applied magnetic field is along the boundary in the direction of propagation, as shown in Figure 10.8a. The z component of the electric field in the plasma may be written as

$$E_z = \left(E_{z1} e^{-jk_{x1}x} + E_{z2} e^{-jk_{x2}x}\right) e^{-jk_z z} \tag{10.129}$$

where E_{z1} and E_{z2} are arbitrary, complex amplitudes. The x and y components of the electric field follow from Equations 4.1, for which we introduce the abbreviations

$$\frac{E_x}{E_z} = P_x \tag{10.130}$$

$$\frac{E_y}{E_z} = P_y \tag{10.131}$$

The magnetic field can then be evaluated from Equation 7.52 and Equations 10.129 through 10.131. The dispersion equation for these waves is Equation 2.24 or its alternate form given by Equation 9.167 with $p^2 = k_x^2$ and $\gamma^2 = -k_z^2$. Matching the tangential E and H at the surface between the plasma and free space gives the following determinantal equation:

$$P_{y2}\left(1 - \frac{jk_{x2}}{a}\right)\left(\beta P_{x1} - k_{x1} + \frac{jk_0^2}{a}\right) - P_{y1}\left(1 - \frac{jk_{x1}}{a}\right)\left(\beta P_{x2} - k_{x2} + \frac{jk_0^2}{a}\right) = 0 \tag{10.132}$$

Equation 10.132 together with the dispersion relations of Equations 10.128 and 9.167 determines the wave numbers k_{x1}, k_{x2}, β, and a. We shall look only for the resonance ($\beta \to \infty$) of the surface wave for which both k_{x1} and k_{x2} have positive, imaginary values. From Equation 9.167 it is readily shown that for $\beta \to \infty$ we may have

$$k_{x1} \to j\beta\sqrt{\frac{K_{\parallel}}{K_{\perp}}} \tag{10.133}$$

$$k_{x2} \to j\beta \tag{10.134}$$

The other solutions for which one of the k_x values remains finite have already been discussed in Section 9.5.3. For the limit of the surface-wave resonance of Equations 10.133 and 10.134, Equation 10.132 gives

$$K_{\parallel}K_{\perp} = 1 \tag{10.135}$$

with

$$K_{\perp} < 0 \text{ and } K_{\parallel} < 0 \tag{10.136}$$

For a given set of plasma parameters (ω_p, ω_{b+}, and ω_{b-}), Equations 10.135 and 10.136 determine the frequency at which this surface-wave resonance occurs.

For the temperate, neutral plasma described by Equations 2.13, 2.20, and 2.21, Equation 10.135 has two solutions. For $\omega_{b-} >> \omega_{b+}$, these solutions are approximately

$$\omega_{r1}^2 \approx \frac{\omega_{b+}\omega_{b-}(\omega_p^2 - \omega_{b+}\omega_{b-})}{\omega_p^2 + \omega_{b-}^2\left(1 + \dfrac{\omega_{b+}}{\omega_{b-}}\right)} \tag{10.137}$$

$$\omega_{r2}^2 \approx \frac{1}{2}\left[\omega_p^2 + \omega_{b-}^2\left(1 + \frac{\omega_{b+}}{\omega_{b-}}\right)\right] \tag{10.138}$$

The zeros of $K_{\perp}$ (Equation 9.197) are approximately at

$$\omega_{01}^2 \approx \frac{\omega_{b+}\omega_{b-}(\omega_p^2 + \omega_{b+}\omega_{b-})}{\omega_p^2 + \omega_{b-}^2} \tag{10.139}$$

$$\omega_{02}^2 \approx \omega_p^2 + \omega_{b-}^2 \tag{10.140}$$

and $K_{\perp}$ is negative for $\omega_{b+} < \omega < \omega_{01}$ and $\omega_{b-} < \omega < \omega_{02}$. On the other hand, $K_{\parallel}$ is negative for $\omega < \omega_p$ and positive for $\omega > \omega_p$.

Hence, the surface-wave resonances can occur only in the restricted frequency ranges illustrated in Figure 10.9. We note that for $\omega_p < \omega_{b+}$ no surface-wave resonances exist. For $\omega_{b+} < \omega_p < \omega_{b-}$, one such resonance appears between ω_{b+} and ω_{01}. For $\omega_p > \omega_{b-}$,

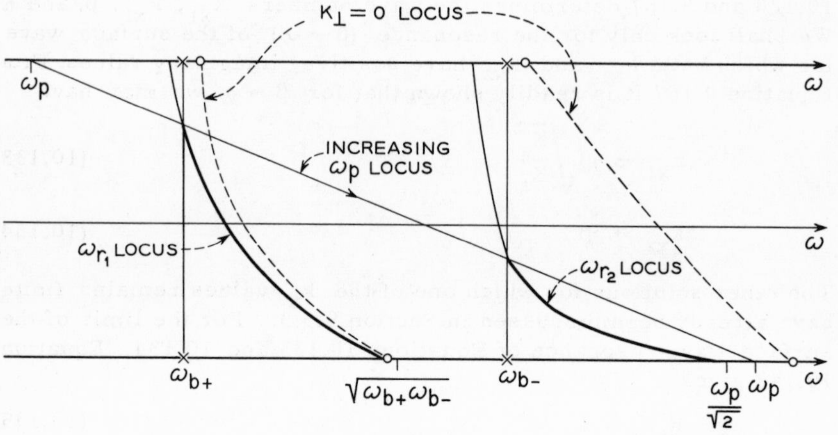

Figure 10.9. Schematic illustration of the frequency ranges where the LMG plasma surface-wave resonances ω_{r1} and ω_{r2} occur as a function of ω_p; lower horizontal lines correspond to higher ω_p.

a second surface-wave resonance is also present between ω_{b-} and ω_p. In the high-density limit, $\omega_p >> \omega_{b-}$, two surface-wave resonances occur at

$$\omega_{r1} \approx \sqrt{\omega_{b+}\omega_{b-}} \tag{10.141}$$

and

$$\omega_{r2} \approx \frac{\omega_p}{\sqrt{2}} \tag{10.142}$$

In this limit, the low-frequency resonance ω_{r1} is the electron-ion "hybrid" resonance, and the high-frequency resonance ω_{r2} corresponds to the surface-wave resonance of the isotropic plasma (Equation 10.49).

Near the surface-wave resonance the fields are highly concentrated at the surface and rapidly decay away from the surface. Further details of the dispersion of these slow waves are given in Section 10.5.

Transversely Magnetized Plasma. Next we consider the transversely magnetized plasma of Figure 10.8b. Since we are interested in possible surface-wave resonances, we shall omit field variations in the y direction. The fields in the TMG plasma with uniformity along $\underset{\sim}{B}_0$ were discussed in Section 9.7.1. The H-wave solutions (H_z, H_x, E_y) cannot support surface waves. The fields associated with the E wave can be obtained from Equations 9.303 through 9.306. Assuming a space dependence of $\exp(-jpx)\exp(-j\beta z)$, we obtain

$$H_y = H_{yp} \tag{10.143}$$

$$E_x = \frac{\omega\mu_0}{(p^2 + \beta^2)}\left(\frac{K_\times}{K_\perp}p_2 + \beta\right)H_{yp} \tag{10.144}$$

$$E_z = \frac{\omega\mu_0}{(p^2 + \beta^2)}\left(\frac{K_\times}{K_\perp}\beta - p_2\right)H_{yp} \tag{10.145}$$

where H_{yp} is an arbitrary complex amplitude, and the dispersion relation is

$$p^2 = k_0^2\frac{K_r K_\ell}{K_\perp} - \beta^2 \tag{10.146}$$

Matching at the surface the tangential electric and magnetic fields of Equations 10.122, 10.123, 10.143, and 10.145, we obtain the following determinantal equation:

$$-jp = \frac{-jK_\times}{K_\perp}\beta - \frac{K_r K_\ell}{K_\perp}\alpha \tag{10.147}$$

Equations 10.128, 10.146, and 10.147 determine p, β, and α as functions of frequency and plasma parameters. The conditions for a surface wave are obtained when the right-hand side of Equation 10.147 is positive and real with α positive and real and β real.

The frequency for which we have a surface-wave resonance is readily determined in the limit $\beta \to \infty$ and $k_0^2 K_r K_\ell / K_\perp$ finite. This is found to be at

$$K_\ell = -1 \tag{10.148}$$

which for the temperate, neutral plasma occurs at the solutions of

$$(\omega + \omega_{b-})(\omega - \omega_{b+}) = \tfrac{1}{2}\omega_p^2 \tag{10.149}$$

For low densities $(\omega_p^2 \ll \omega_{b+}\omega_{b-})$, this frequency is just above ω_{b+}, and for high densities $(\omega_p^2 \gg \omega_{b-}^2)$, it is just above $\omega_p/\sqrt{2}$. The dispersion characteristic near this resonance is

$$\beta^2 \to \frac{-\frac{3}{2}k_0^2}{(K_\ell + 1)} \tag{10.150}$$

Since $K_\ell < -1$ for frequencies below the resonance, the dispersion is that of a traveling wave (that is, group and phase velocities have the same sign). Near this resonance both α and $-jp$ approach β and are very large, and the fields are concentrated near the surface and decay rapidly away from the surface. The cutoff of this surface wave occurs at a frequency for which $\beta = k_0$. From Equations 10.128, 10.146, and 10.147, we find that this frequency corresponds to $K_\perp = \frac{1}{4}$.

The surface-wave resonance for the TMG plasma of Figure 10.8c is similar to the LMG plasma resonance and will be treated in Section 10.5.2.

10.4.2 Longitudinally Magnetized, Cylindrical Plasma Waveguide. We now turn to the specific problem of an LMG homogeneous plasma cylinder that is either in free space or inside of a concentric waveguide with perfectly conducting walls (Figure 10.10). We shall set up the field solutions and the determinantal equation and look at some of the solutions.[47]

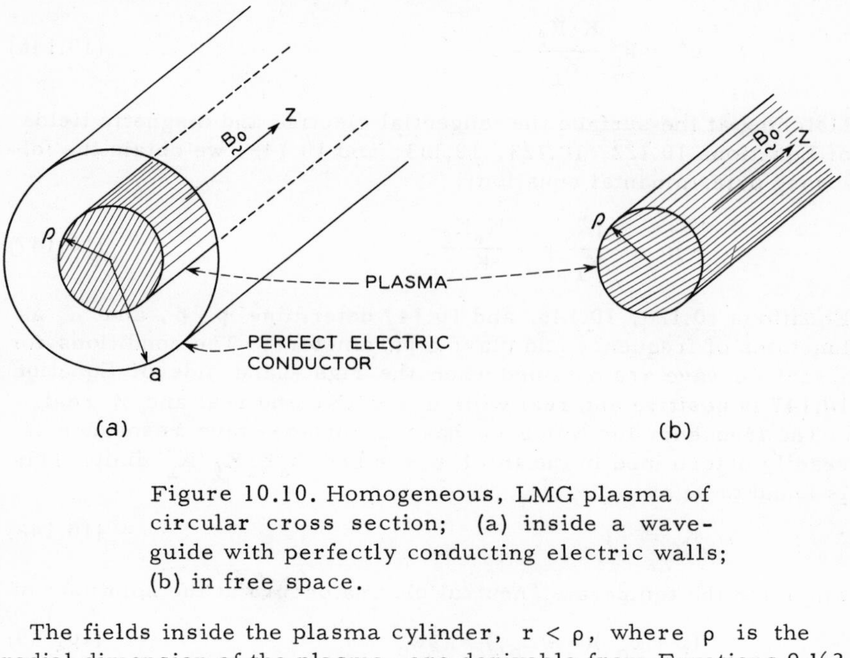

Figure 10.10. Homogeneous, LMG plasma of circular cross section; (a) inside a waveguide with perfectly conducting electric walls; (b) in free space.

The fields inside the plasma cylinder, $r < \rho$, where ρ is the radial dimension of the plasma, are derivable from Equations 9.162 through 9.166. The longitudinal electric field is

$$E_z = \left[AJ_m(p_1 r) + BJ_m(p_2 r) \right] e^{jm\phi} \tag{10.151}$$

where J_m is the ordinary Bessel function of order m, A and B are arbitrary complex constants, and the z dependence $\exp(-\gamma z)$ will be omitted for brevity. The propagation constants γ, p_1, and

p_2 are related by the dispersion equation, Equation 9.161. The longitudinal magnetic field is obtained from Equation 9.164, and the transverse fields from Equation 9.166. We shall write down the components that are tangential to the plasma boundary:

$$H_z = \left[Ah_1J_m(p_1r) + Bh_2J_m(p_2r)\right] e^{jm\phi} \quad (10.152)$$

where h_1 and h_2 are given by Equation 9.165;

$$E_\phi = A\left[\frac{jm}{r} \ell_2 J_m(p_1 r) + L_2 p_1 J_m'(p_1 r)\right] e^{jm\phi}$$
$$+ B\left[\frac{jm}{r} \ell_1 J_m(p_2 r) + L_1 p_2 J_m'(p_2 r)\right] e^{jm\phi} \quad (10.153)$$

$$H_\phi = A\left[\frac{jm}{r} y_2 J_m(p_1 r) + Y_2 p_1 J_m'(p_1 r)\right] e^{jm\phi}$$
$$+ B\left[\frac{jm}{r} y_1 J_m(p_2 r) + Y_1 p_2 J_m'(p_2 r)\right] e^{jm\phi} \quad (10.154)$$

where

$$\ell_{1,2} = p_{1,2}^2 \frac{R}{b} - \frac{1}{\gamma} \quad (10.155)$$

$$L_{1,2} = p_{1,2}^2 \frac{S}{b} - \frac{1}{\gamma}\frac{K_\perp}{K_\times} \quad (10.156)$$

$$y_{1,2} = \frac{p_{1,2}^2 P + \gamma}{b} \quad (10.157)$$

$$Y_{1,2} = p_{1,2}^2 \frac{Q}{b} \quad (10.158)$$

with R, S, P, Q, and b given by Equations 9.23, 9.25, 9.22, 9.24, and 9.33. (Note that the dimensions of ℓ and L are length and the dimensions of y and Y are admittance times length.)

The fields outside the plasma $(r > \rho)$ are made up of a superposition of E and H waves that follow from Equations 9.147 through 9.151, with the dispersion relation given by Equation 9.149. We shall use a unified notation to represent the case of the plasma in free space and inside of a waveguide enclosure with perfectly conducting electric walls. The longitudinal fields, satisfying Equations 9.147 and 9.148, are

$$E_z^o = CF_{mm}(\Gamma r, \Gamma a)e^{jm\phi} \quad (10.159)$$

$$H_z^o = DF_{mm'}(\Gamma r, \Gamma a)e^{jm\phi} \quad (10.160)$$

where C and D are arbitrary, complex constants and the functions F_{mm} and $F_{mm'}$ are as given in Table 10.1. The function F is so chosen that in the case of the plasma in free space the

Table 10.1 Functional Relations Used in Equations 10.159 through 10.162

Plasma in	F_{mm}	Γ	ξ	ζ
Waveguide	H_{mm}	q	$-\frac{\gamma}{q^2}$	$\frac{k_0}{q}$
Free Space	K_m	T	$\frac{\gamma}{T}$	$-\frac{k_0}{T}$

$H_{mm}(x, y) = J_m(x)N_m(y) - N_m(x)J_m(y)$ (J_m ordinary Bessel function and N_m Neumann function)

$H_{mm'} = \frac{d}{dy} H_{mm}$

$K_m = K_m(Tr)$ (modified Bessel function of second kind)

$q^2 = \gamma^2 + k_0^2 = -T^2$

fields vanish for $r \to \infty$, and in the case of the plasma in the waveguide the tangential electric field at the wall vanishes. The transverse field components follow from Equations 9.150 and 9.151. We write down the ϕ components only:

$$E_\phi^o = \left[C\frac{jm}{r}\xi F_{mm}(\Gamma r, \Gamma a) + D\zeta Z_0 F'_{mm'}(\Gamma r, \Gamma a)\right] e^{jm\phi} \qquad (10.161)$$

$$H_\phi^o = \left[-C\zeta Y_0 F'_{mm}(\Gamma r, \Gamma a) + D\frac{jm}{r}\xi F_{mm'}(\Gamma r, \Gamma a)\right] e^{jm\phi} \qquad (10.162)$$

where ξ and ζ are also given in Table 10.1, and $Z_0 = 1/Y_0 = (\mu_0/\epsilon_0)^{\frac{1}{2}}$, is the free-space impedance.

The boundary conditions at the plasma surface $(r = \rho)$ shall be taken as the continuity of the tangential E and H fields. Matching the longitudinal fields (Equations 10.151, 10.152, 10.159, and 10.160) gives

$$CF_{mm} = AJ_m(p_1\rho) + BJ_m(p_2\rho) \qquad (10.163)$$

$$DF_{mm'} = Ah_1J_m(p_1\rho) + Bh_2J_m(p_2\rho) \qquad (10.164)$$

where the functions F are evaluated at $r = \rho$. Matching the ϕ components of E and H (Equations 10.153, 10.154, 10.161, and 10.162) and using Equations 10.163 and 10.164, we obtain

$$V_1A + V_2B = 0 \qquad (10.165)$$

$$W_1A + W_2B = 0 \qquad (10.166)$$

where

$$V_{1,2} = \left[\frac{jm}{\rho}(\ell_{2,1} - \xi) - \zeta Z_0 h_{1,2}\frac{F'_{mm'}}{F_{mm'}}\right] J_m(p_{1,2}\rho) + L_{2,1}p_{1,2}J_m'(p_{1,2}\rho) \tag{10.167}$$

$$W_{1,2} = \left[\frac{jm}{\rho}(y_{2,1} - \xi h_{1,2}) + \zeta Y_0 \frac{F'_{mm}}{F_{mm}}\right] J_m(p_{1,2}\rho) + Y_{2,1}p_{1,2}J_m'(p_{1,2}\rho) \tag{10.168}$$

The determinantal equation from Equations 10.165 and 10.166 is

$$V_1W_2 - V_2W_1 = 0 \tag{10.169}$$

Equation 10.169 and the dispersion relations, Equations 9.149 and 9.161, are in principle sufficient for determining γ, p_1, and p_2 as functions of frequency and plasma parameters. In practice, this is a formidable computational problem. We shall discuss some of the simple limiting solutions.

In addition to the resonances $(\gamma \to \infty)$ at the plasma and cyclotron frequencies, for which one of the p values remains finite, Equation 10.169 permits a surface-wave resonance at frequencies given by Equations 10.135 through 10.138 (see Figure 10.9). At these resonances both p values become large and approach the values given by the right-hand side of Equations 10.133 and 10.134, and the fields are concentrated near the plasma surface. For terms up to order $1/\gamma$, Equation 10.169 becomes $V_2W_1 = 0$. In this limit, Equations 10.135 and 10.136 follow.

The cutoff frequencies in the partially filled plasma waveguide have somewhat simpler determinantal equations. For $\gamma = 0$, the solutions inside the plasma separate into E and H waves, as given in Section 9.5.2. However, the p values are no longer frequency independent. The E-wave cutoff frequencies are determined by

$$p_e^2 = k_0^2 K_{\parallel} \tag{10.170}$$

$$\frac{p_e}{\omega\mu_0}\frac{J_m'(p_e\rho)}{J_m(p_e\rho)} = Y_0 \frac{H'_{mm}(k_0\rho,\ k_0a)}{H_{mm}(k_0\rho,\ k_0a)} \tag{10.171}$$

and the H-wave cutoff frequencies by

$$p_h^2 = k_0^2 \frac{K_r K_\ell}{K_\perp} \tag{10.172}$$

$$\frac{\omega\mu_0}{p_h^2}\left[\frac{K_\times}{K_\perp}\frac{jm}{\rho} + \frac{p_h J_m'(p_h\rho)}{J_m(p_h\rho)}\right] = Z_0 \frac{H'_{mm'}(k_0\rho,\ k_0a)}{H_{mm'}(k_0\rho,\ k_0a)} \tag{10.173}$$

We note that the H-wave cutoff frequencies for m positive and m negative are distinct.

In the low-frequency, MHD regime we have solutions that extend to zero frequency. Near zero frequency one of the p values remains finite and is as given by Equation 9.252. The other p value approaches zero as ω^2. The ϕ-independent solutions (m = 0) are E waves with dispersion determined by

$$\gamma^2 \approx -\frac{\omega^2}{u_\alpha^2}\left(1+\frac{u_\alpha^2}{c^2}\right)\left(1+\frac{p^2c^2}{\omega_p^2}\right) \tag{10.174}$$

$$\frac{J_0(p\rho)}{J_1(p\rho)} = -\frac{\omega_p^2\gamma^2}{\omega^2 p}\rho \ln\frac{a}{\rho} \tag{10.175}$$

We note that in this limit p is independent of frequency and a function of geometry only. The solutions with $m \neq 0$ are mixed E and H waves both inside and outside the plasma, with the H-wave part remaining necessary for matching the boundary conditions.

It is also interesting to note some of the general new features of the field solutions arising from the presence of the applied magnetic field. We note that the determinantal equation, Equation 10.169, is different for positive and negative m values, a fact indicating that the field solutions exhibit Faraday rotation. In the absence of the applied magnetic field, Equation 10.164 reduces to Equation 10.59 or Equation 10.46, both of which are functions of m^2 only. The isotropic plasma waveguide solutions (see Section 10.2.2) exhibit a surface-wave resonance at $\omega = \omega_p/\sqrt{2}$ and only a degenerate set of solutions at $\omega = \omega_p$. The presence of an applied magnetic field in the direction of propagation removes the degeneracy at $\omega = \omega_p$ and allows for resonances in propagation at the plasma and cyclotron frequencies. When the applied magnetic field is small or the density high, so that $\omega_{b-} \ll \omega_p$, we also find two surface-wave resonances; the high-frequency resonance, Equation 10.142, is approximately the same as in the isotropic plasma, and the low-frequency resonance, Equation 10.141, is at the electron-ion "hybrid" resonant frequency. As the applied magnetic field is raised or the density lowered, these surface-wave resonances may be inhibited; when $\omega_{b-} > \omega_p$, the high-frequency surface-wave disappears, and when $\omega_{b+} > \omega_p$, both surface-wave resonances cease to exist.

10.5 Quasi-Static Solutions for Slow Waves

One of the most interesting characteristics of plasma waveguides is that they can propagate slow waves ($\beta > k_0$) having phase velocities smaller than the velocity of light. In fact, for the simplest plasma model, described by the temperate and collisionless hydrodynamic equations (see Chapter 2), there are resonances ($\beta \to \infty$) in the propagation. Near these resonances the phase velocity of the wave can become extremely slow compared with the velocity of light ($\beta >> k_0$). An approximate field solution for these waves can be obtained by regarding the velocity of light to be essentially infinite and hence the fields essentially static. Under these conditions the magnetic field is small, the electric field is approximately curl-free, and its divergence is the first-order charge density of the plasma. Such a quasi-static analysis of wave propagation in LMG, single-particle, plasma waveguides was first carried out by Smullin and Chorney[32,48] and Trivelpiece and Gould.[33,49] †

Before we present the quasi-static analysis and its results, let us note the following. Near the resonances in the propagation the phase velocity of the wave can become arbitrarily close to zero. This clearly violates the assumptions on which the temperate-plasma model is based. We shall retain the temperate-plasma model and hence restrict our phase velocities to be large compared with the thermal velocities of the plasma particles. This restriction was of course implied throughout this and the previous chapter. Next, we note that the field solutions for slow waves are quasi-static only if both the propagation constant β and the transverse wave number p are large compared with k_0. In a dispersive medium slow waves ($\beta >> k_0$) do not necessarily imply large transverse wave numbers. Thus, near the resonances in propagation ($\beta >> k_0$), the transverse wave numbers p may remain finite (see Chapter 9) and dependent on plasma dimensions: For this case the condition $p >> k_0$ may usually be satisfied if the transverse dimensions of the plasma are small compared with a free-space wavelength at the frequency of the resonance. Even near resonances of surface waves where both β and p become large compared with k_0, the frequency variation of β near this resonance may depend upon the manner in which p becomes large. This will become clear when we compare the quasi-static dispersion relations with the dispersion relations of Chapters 9 and 10 for the resonance limit. Finally, we note that for the temperate plasma model the power flow associated with a wave is entirely electromagnetic (see Chapter 7). Hence, for completeness, the quasi-static wave analysis must also establish an approximate magnetic field.

10.5.1 Equations for the Electric Fields. We start by assuming that the first-order magnetic field is negligible. The equations that describe the electric field in the plasma are

$$\nabla \times \underset{\sim}{E} = 0 \qquad (10.176)$$

$$\nabla \cdot (\epsilon_0 \underset{\approx}{K} \cdot \underset{\sim}{E}) = 0 \qquad (10.177)$$

where the time dependence $\exp(j\omega t)$ is implied. Hence the electric field is derivable from a potential

† Quasi-static field solutions are also alluded to in a paper by Ya. B. Fainberg,[50] where reference is made to similar work in the U.S.S.R.

$$\underset{\sim}{E} = -\nabla\varphi \tag{10.178}$$

where φ must satisfy Equation 10.177. We assume that the plasma is homogeneous and may only partially fill the waveguide cross section. We take the waveguide axis in the z direction and assume solutions with $\exp(-\gamma z)$ dependence.

(a) Longitudinally Magnetized Plasma (Figure 9.1a)

In this case the applied magnetic field is along the z axis, and the normalized dielectric tensor is given by Equation 2.13. Using Equations 10.177, 10.178, and 2.13 we obtain

$$\nabla_T^2\varphi + p^2\varphi = 0 \tag{10.179}$$

$$p^2 = \gamma^2 \frac{K_\parallel}{K_\perp} \tag{10.180}$$

$$E_z = \gamma\varphi \tag{10.181}$$

$$\underset{\sim}{E}_T = -\nabla_T\varphi \tag{10.182}$$

where ∇_T is the two-dimensional gradient operator in the plane transverse to the z direction.

(b) Transversely Magnetized Plasma (Figure 9.1b)

In this case the applied magnetic field is along one of the coordinate axes in the plane transverse to the z axis, and the normalized dielectric tensor is given by Equation 9.279. We assume the space dependence along $\underset{\sim}{B}_0$ to be $\exp(-jp_B r_B)$, where r_B is the transverse coordinate along $\underset{\sim}{B}_0$, and we use the vector operator and field notation of Table 9.6. From Equations 10.177, 10.178, and 9.279 we find

$$\nabla_{T'}^2\varphi + p^2\varphi = 0 \tag{10.183}$$

$$p^2 = \gamma^2 - \frac{K_\parallel}{K_\perp} p_B^2 \tag{10.184}$$

$$E_z = -\gamma\varphi \tag{10.185}$$

$$\underset{\sim}{E}_T = -\nabla_T\varphi = -\nabla_{T'}\varphi + \underset{\sim}{i}_B jp_B\varphi \tag{10.186}$$

When the plasma does not fill the waveguide cross section, the space outside the plasma will be assumed to be free space. The quasi-static electric field in free space must satisfy

$$\nabla \times \underset{\sim}{E}^o = 0 \tag{10.187}$$

$$\nabla \cdot (\epsilon_0 \underset{\sim}{E}^o) = 0 \tag{10.188}$$

and hence

$$\underset{\sim}{E}^o = -\nabla\varphi^o \tag{10.189}$$

For solutions with z dependence $\exp(-\gamma z)$, Equations 10.188 and 10.189 give

$$\nabla_T^2 \varphi^o + \gamma^2 \varphi^o = 0 \qquad (10.190)$$

$$E_z^o = \gamma \varphi^o \qquad (10.191)$$

$$\underset{\sim}{E}_T^o = -\nabla_T \varphi^o \qquad (10.192)$$

The field solutions of Equations 10.179 through 10.192 will have to satisfy the following boundary conditions: (a) At the perfectly conducting electric wall, when the plasma fills the waveguide cross section,

$$\underset{\sim}{n} \times \underset{\sim}{E} = 0 \qquad (10.193)$$

where $\underset{\sim}{n}$ is a unit vector normal to the wall. (b) At the surface between the plasma and free space, when the waveguide cross section is only partially filled with plasma,

$$\underset{\sim}{n} \times (\underset{\sim}{E} - \underset{\sim}{E}^o) = 0 \qquad (10.194)$$

$$\underset{\sim}{n} \cdot (\epsilon_o \underset{\approx}{K} \cdot \underset{\sim}{E} - \epsilon_o \underset{\sim}{E}^o) = 0 \qquad (10.195)$$

where $\underset{\sim}{n}$ is a unit vector normal to the surface. The boundary conditions of Equations 10.194 and 10.195 imply a first-order surface charge density on the unperturbed boundary position, consistent with the force equation and continuity equation (see discussions in Sections 9.4.2 and 10.4). The fields outside the plasma at the waveguide walls must also satisfy

$$\underset{\sim}{n} \times \underset{\sim}{E}^o = 0 \qquad (10.196)$$

where $\underset{\sim}{n}$ is a unit vector normal to the waveguide wall.

10.5.2. Dispersion Characteristics. The dispersion equation for the quasi-static solutions of the LMG and TMG plasma waveguide are given, respectively, by Equations 10.180 and 10.184. When the waveguide is completely filled with plasma, the boundary condition of Equation 10.193 leads to positive and real p^2 values that are a function of geometry only (see footnote, page 139). For this case the dependence of γ upon frequency is entirely determined by Equations 10.180 and 10.184. In waveguides that are partially filled with a plasma the boundary conditions of Equations 10.194 through 10.196 lead to determinantal equations involving p, γ, and ω, which must be solved simultaneously with the dispersion relations. Some examples are given in the next section. Here we shall explore the dispersion characteristics for the plasma-filled waveguides and the surface-wave resonances for the partially filled plasma waveguides.

Plasma-Filled Waveguides. Consider first the simplest case when the plasma is isotropic, $K_\times = 0$, $K_\perp = K_\parallel$. For this case the dispersion relations become simply $\gamma^2 = p^2$, and we find no propagating wave solutions. This is as we should expect, since in the isotropic plasma-filled waveguide there are no slow-wave solutions (see Section 9.3), and the fast-wave solutions cannot be obtained from a quasi-static analysis. The cutoff wave (Equation 9.71) is properly described only at low frequencies, where k_0 becomes negligible compared to p, and even then only if $pc >> \omega_p$, that is, when the cutoff frequency of the free-space waveguide is large compared with the plasma frequency.

For the anisotropic plasma-filled waveguide we shall discuss the LMG and TMG cases separately.

(a) Longitudinally magnetized plasma waveguide

For the LMG plasma-filled waveguide, Equation 10.180 is

$$\gamma^2 = p^2 \frac{K_\perp}{K_\parallel} \tag{10.197}$$

where we regard p^2 as a positive, real constant determined by geometry and independent of frequency. Slow-wave solutions exist near the resonances of γ, and these occur in the vicinity of ω_p and $\omega_{b\pm}$. Near plasma resonance Equation 10.197 gives the correct dispersion relation for γ (see Equation 9.204). However, near the cyclotron resonances Equation 10.197 is correct only if

$$p^2 >> 2k_0^2 K_\parallel \tag{10.198}$$

(see Equation 9.219); that is, $p^2c^2 >> 2\left|\omega_{b\pm}^2 - \omega_p^2\right|$. The frequency dependence of γ near the resonances as predicted by the quasi-static dispersion relation, Equation 10.197, corresponds to the first column in Figure 9.8. The resonance characteristics at the cyclotron frequencies in the other two columns of Figure 9.8 are not predicted by the quasi-static dispersion relation.

Equation 10.197 for various values of ω_p and a particular value of p is shown in Figure 10.11. The fast-wave regions $(\beta \geq k_0)$ are shown dashed because Equation 10.197 is not valid there. We note that the finite frequency cutoffs predicted by Equation 10.197 are at $K_\perp = 0$. For $pc >> \omega_{b\pm}$, these cutoff frequencies are approximately equal to the H-wave cutoffs that have fields with $\nabla_T H_z$ normal to the perfectly conducting electric wall (see Figure 9.7). However, the fields at these cutoffs cannot be obtained even approximately from the quasi-static equations. The E-wave cutoffs (Equations 9.199) are not predicted by Equation 10.197. In the low-frequency regime, $\omega \to 0$, Equation 10.197 becomes

$$\gamma^2 \to \frac{\omega^2}{u_a^2} \frac{p^2c^2}{\omega_p^2} \left(1 + \frac{u_a^2}{c^2}\right) \tag{10.199}$$

This is approximately correct for the low-frequency E waves only if $pc >> \omega_p$ (see Equation 9.253). The MHD H waves (Equations 9.257 through 9.265) cannot be obtained at all from the quasi-static analysis.

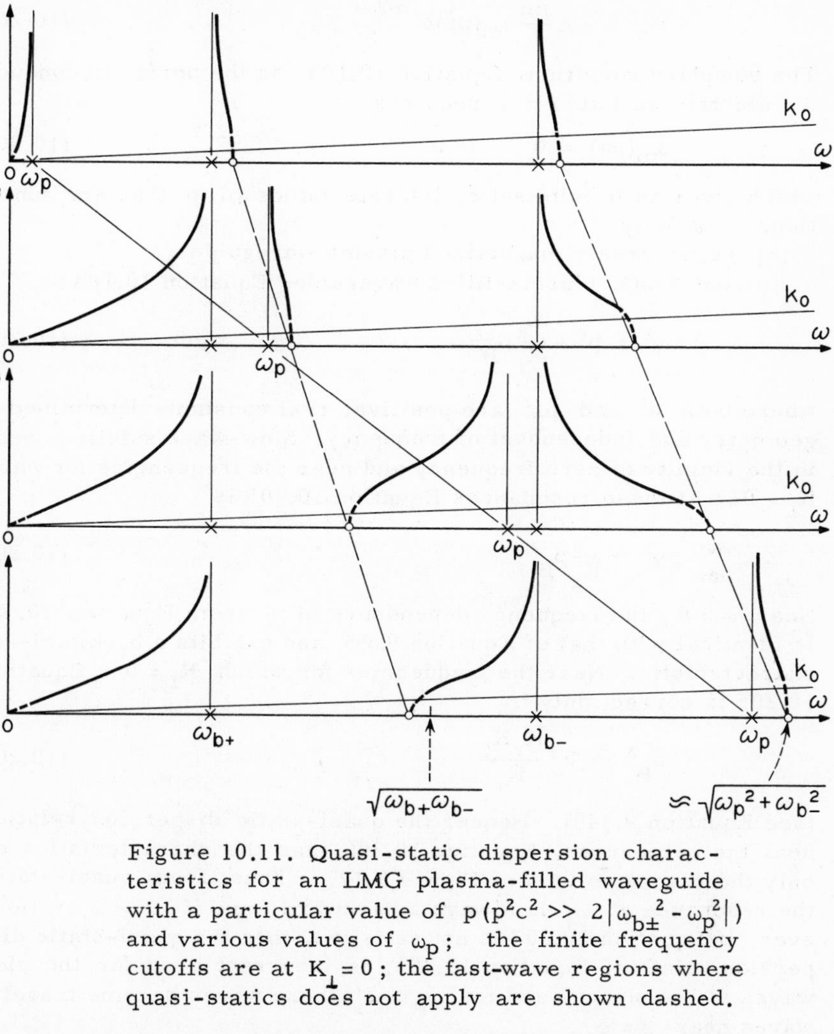

Figure 10.11. Quasi-static dispersion characteristics for an LMG plasma-filled waveguide with a particular value of $p\,(p^2c^2 >> 2|\omega_{b\pm}{}^2 - \omega_p{}^2|)$ and various values of ω_p; the finite frequency cutoffs are at $K_\perp = 0$; the fast-wave regions where quasi-statics does not apply are shown dashed.

As an example, consider the LMG, plasma-filled, circular waveguide. The solution to Equation 10.179, regular at $r = 0$, is

$$\varphi = AJ_m(pr)e^{jm\phi} \tag{10.200}$$

where J_m is the Bessel function of order m and A is an arbitrary complex constant. The electric field components follow from Equations 10.181 and 10.182:

$$E_z = A\gamma J_m(pr)e^{jm\phi} \tag{10.201}$$

$$E_r = -ApJ_m{}'(pr)e^{jm\phi} \tag{10.202}$$

$$E_\phi = -A\frac{jm}{r}J_m(pr)e^{jm\phi} \tag{10.203}$$

The boundary condition, Equation 10.193, on the perfectly conducting electric wall at $r = a$ requires

$$J_m(pa) = 0 \tag{10.204}$$

which gives an infinite set of discrete values of p that are functions of a only.

(b) Transversely magnetized plasma waveguide

For the TMG, plasma-filled waveguide, Equation 10.184 is

$$\gamma^2 = p^2 + \frac{K_\parallel}{K_\perp}p_B{}^2 \tag{10.205}$$

where both p^2 and $p_B{}^2$ are positive, real constants determined by geometry and independent of frequency. Slow-wave solutions occur in the vicinity of zero frequency and near the frequencies for which $K_\perp = 0$. At these resonances Equation 10.205 is

$$\gamma^2 \to \frac{K_\parallel}{K_\perp}p_B{}^2 \tag{10.206}$$

Near $\omega = 0$, the frequency dependence of γ from Equation 10.205 is identical with that of Equation 9.355, and exhibits a backward-wave characteristic. Near the frequencies for which $K_\perp = 0$, Equation 10.205 is correct only if

$$p_B{}^2 \gg k_0{}^2\frac{K_r K_\ell}{K_\parallel} \tag{10.207}$$

(see Equation 9.349). Hence, the quasi-static dispersion relation near the resonances, Equation 10.206, has the characteristics of only the second column of Figure 9.17. Thus, from quasi-statics the resonance at ω_1 is always a backward wave if $\omega_p > \omega_1$. However, if Equation 10.207 is not satisfied, then the quasi-static dispersion relation, Equation 10.206, is incorrect even for the slow waves at resonance, and for $\omega_p > \omega_1$ these waves become traveling waves near $\omega = \omega_1$.

The complete quasi-static dispersion relation, Equation 10.205, is sketched in Figure 10.12 for various values of ω_p. The fast-wave regions, where the quasi-static solution is not expected to be correct, are shown dashed. The cutoffs predicted by Equation 10.205 occur at

$$\left(\frac{p}{p_B}\right)^2 = -\frac{K_\parallel}{K_\perp} \tag{10.208}$$

Since p and p_B are real, the frequencies for the cutoffs are restricted to the ranges for which the right-hand side of Equation 10.20 is positive. A direct comparison of the cutoffs predicted by Equa-

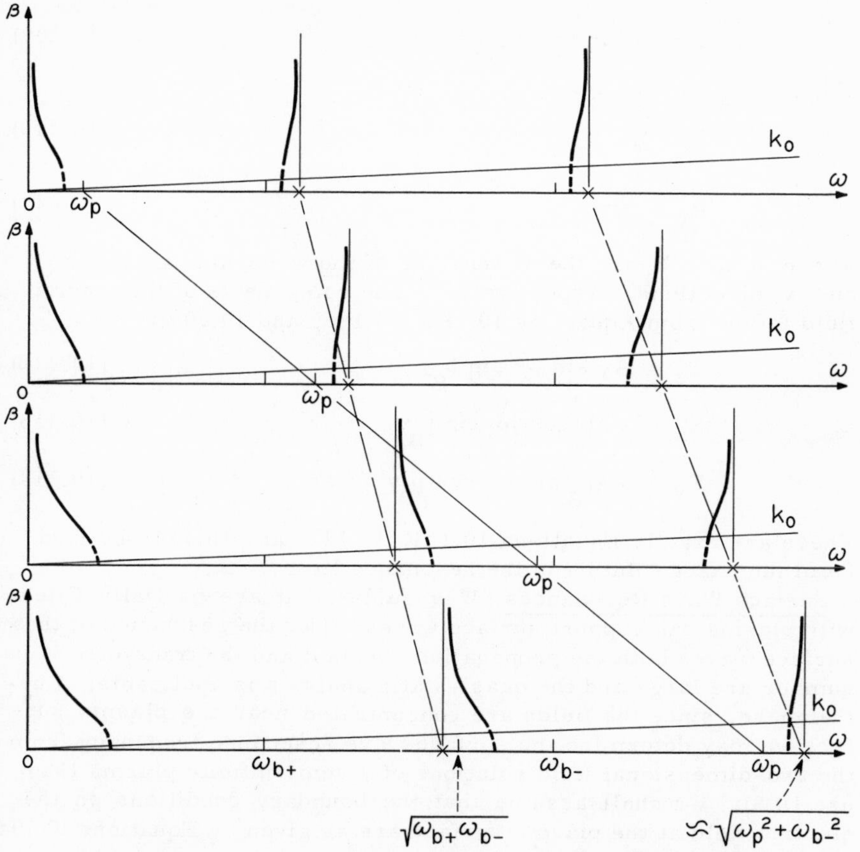

Figure 10.12. Quasi-static dispersion characteristics for a TMG plasma-filled waveguide with particular values of p and p_B (Equation 10.207), and various values of ω_p; the finite frequency cutoffs are in the regions where $K_{\|}/K_{\perp}$ is negative; the fast-wave regions where quasi-statics does not apply are shown dashed.

tion 10.208 with the true cutoffs in a closed, plasma-filled, TMG waveguide is not simple (see Section 9.7.3). However, for the case of either p or p_B equal to zero the cutoffs cannot be obtained from Equation 10.208, and the exact dispersion relation must be used (see Section 9.7.3).

To illustrate the quasi-static field solutions, consider a TMG plasma-filled waveguide of rectangular cross section (Figure 9.11a). The solution of Equation 10.183 that satisfies the boundary condition on the walls, Equation 10.193, is

$$\varphi = A \sin px \sin p_B y \qquad (10.209)$$

with

$$p = \frac{n\pi}{a} \qquad (10.210)$$

$$p_B = \frac{m\pi}{b} \qquad (10.211)$$

where a and b are the dimensions of the waveguide along the x and y directions, respectively. The components of the electric field follow from Equations 10.185, 10.186, and 10.209:

$$E_z = A\gamma \sin px \sin p_B y \qquad (10.212)$$

$$E_x = -Ap \cos px \sin p_B y \qquad (10.213)$$

$$E_y = -Ap_B \sin px \cos p_B y \qquad (10.214)$$

These are exactly Equations 10.111, 10.112, and 10.114 obtained from the exact solution in the resonance limit.

Surface Wave Resonances. Waveguides that are partially filled with plasma can support surface waves. Near the resonance of these surface waves both the propagation constant and the transverse wave number are large and the quasi-static analysis is applicable. Furthermore, since the fields are concentrated near the plasma surface, we may determine the surface-wave resonance frequency from the two-dimensional field solutions of a semi-infinite plasma (Figure 10.8). We shall assume that the boundary conditions on the electric field at the plasma surface are as given by Equations 10.194 and 10.195.

(a) Longitudinally Magnetized Plasma

Near the surface-wave resonance the fields in the plasma and in free space decay in the $\pm x$ direction, away from the surface (Figure 10.8a). From Equations 10.179 through 10.182 we obtain in the plasma

$$\varphi = Ae^{p_x x} \qquad (10.215)$$

$$E_z = Aj\beta e^{p_x x} \qquad (10.216)$$

$$E_x = -Ap_x e^{p_x x} \qquad (10.217)$$

where we have written $\gamma = j\beta$, and from Equation 10.180

$$-jp \to \beta\sqrt{\frac{K_\parallel}{K_\perp}} \equiv p_x \qquad (10.218)$$

with

$$\frac{K_{\parallel}}{K_{\perp}} > 0 \tag{10.219}$$

In the free space outside the plasma, Equations 10.190 through 10.192 give

$$\varphi^o = A^o e^{-\beta x} \tag{10.220}$$

$$E_z^o = A^o j\beta e^{-\beta x} \tag{10.221}$$

$$E_x^o = A^o \beta e^{-\beta x} \tag{10.222}$$

Applying the boundary conditions of Equations 10.194 and 10.195, we obtain the determinantal equation

$$-K_{\perp} p_x = \beta \tag{10.223}$$

Since β and p_x must be positive, the resonance can occur only for $K_{\perp} < 0$. Combining Equations 10.218, 10.219, and 10.223, we obtain

$$K_{\parallel} K_{\perp} = 1 \tag{10.224}$$

$$K_{\perp} < 0 \text{ and } K_{\parallel} < 0 \tag{10.225}$$

which are exactly Equations 10.135 and 10.136. Hence, the frequencies at which these resonances occur are given by Equations 10.137 and 10.138, and the conditions for their occurrence are illustrated in Figure 10.9.

(b) Transversely Magnetized Plasma

Consider first the case of Figure 10.8b, where $\underset{\sim}{B}_0$ is parallel to the plasma surface. Near resonance $(\gamma = j\beta \to \infty)$, the dispersion relation, Equation 10.184, is

$$p^2 \to -\beta^2 \tag{10.226}$$

and the solutions of Equations 10.183 through 10.192 are

$$\varphi = Ae^{\beta x} \tag{10.227}$$

$$E_z = Aj\beta e^{\beta x} \tag{10.228}$$

$$E_x = -A\beta e^{\beta x} \tag{10.229}$$

and

$$\varphi^o = A^o e^{-\beta x} \tag{10.230}$$

$$E_z^o = A^o j\beta e^{-\beta x} \tag{10.231}$$

$$E_x^o = A^o \beta e^{-\beta x} \tag{10.232}$$

At the plasma surface, $x = 0$, the tangential and normal electric fields must satisfy Equations 10.194 and 10.195, respectively:

$$A = A^o \tag{10.233}$$

$$(K_\times j\beta - K_\perp \beta)A = \beta A^o \tag{10.234}$$

Equations 10.233 and 10.234 are satisfied for

$$K_\ell = -1 \tag{10.235}$$

which is exactly Equation 10.148. The frequency at which this surface-wave resonance occurs is thus given by Equation 10.149.

Next we consider the surface-wave resonance for the case shown in Figure 10.8c, where $\underset{\sim}{B}_0$ is perpendicular to the plasma surface. The dispersion relation, Equation 10.184, near resonance ($\gamma = j\beta \to \infty$) is

$$-jp_B \to \beta\sqrt{\frac{K_\perp}{K_\parallel}} \equiv p_y \tag{10.236}$$

with

$$\frac{K_\perp}{K_\parallel} > 0 \tag{10.237}$$

We note that Equations 10.236 and 10.237 are reciprocal to Equations 10.218 and 10.219. The field solutions in the plasma are identical with Equations 10.215 through 10.217 with $p_x x$ replaced by $p_y y$. The field solutions in free space are identical with Equations 10.220 through 10.222 with x replaced by y. The boundary condition at the surface $y = 0$ requires

$$-K_\parallel p_y = \beta \tag{10.238}$$

which together with Equations 10.236 and 10.237 gives

$$K_\parallel K_\perp = 1 \tag{10.239}$$

$$K_\parallel < 0 \text{ and } K_\perp < 0 \tag{10.240}$$

Equations 10.239 and 10.240 are identical with Equations 10.224 and 10.225 and show that the surface-wave resonances occur at the same frequencies as in the LMG plasma (see Equations 10.137, 10.138, and Figure 10.9).

Until now, in all of our considerations of surface waves we have chosen the medium outside the plasma to be free space. The surface-wave resonances can be easily determined for the case when the outside medium is a homogeneous, isotropic dielectric and the boundary conditions are taken as

$$\underset{\sim}{n} \times (\underset{\sim}{E} - \underset{\sim}{E}^o) = 0 \tag{10.241}$$

$$\underset{\sim}{n} \cdot (\epsilon_0 \underset{\approx}{K} \cdot \underset{\sim}{E} - \epsilon \underset{\sim}{E}^o) = 0 \tag{10.242}$$

For the LMG plasma the determinantal equation, Equation 10.223, then has its right-hand side multiplied by ϵ/ϵ_0, and the resonances occur for

$$K_{\parallel} K_{\perp} = \frac{\epsilon}{\epsilon_0} \tag{10.243}$$

with $K_{\perp} < 0$ and $K_{\parallel} < 0$ as before. The low-frequency solution of Equation 10.243 is essentially the same as for $\epsilon = \epsilon_0$, and is given by Equation 10.137. The high-frequency solution is modified and becomes

$$\omega_{r2}^{\;2} \approx \frac{1}{\left(1 + \frac{\epsilon}{\epsilon_0}\right)} \left[\omega_p^{\;2} + \omega_{b-}^{\;2} \left(1 + \frac{\omega_{b-}}{\omega_{b+}}\right)\right] \tag{10.244}$$

instead of Equation 10.138. For the TMG plasma with $\underset{\sim}{B}_0$ perpendicular to the surface, the determinantal equation, Equation 10.238, has its right-hand side also multiplied by ϵ/ϵ_0, and the modifications in the resonant frequency are the same as in the LMG plasma, Equations 10.243 and 10.244. For the TMG plasma with $\underset{\sim}{B}_0$ parallel to the surface, the right-hand side of Equation 10.234 is multiplied by ϵ/ϵ_0, and Equation 10.235 becomes

$$K_{\ell} = -\frac{\epsilon}{\epsilon_0} \tag{10.245}$$

Hence, Equation 10.149 becomes modified, and the resonances occur at the solutions of

$$(\omega + \omega_{b-})(\omega - \omega_{b+}) = \frac{\omega_p^{\;2}}{\left(1 + \frac{\epsilon}{\epsilon_0}\right)} \tag{10.246}$$

For $\omega_p << \omega_{b+}\omega_{b-}$, the solution is essentially the same as for $\epsilon = \epsilon_0$, and is just above ω_{b+}. For $\omega_p >> \omega_{b-}$, the resonant frequency is lower than for $\epsilon = \epsilon_0$, and is slightly above $\omega_p/\sqrt{1 + \epsilon/\epsilon_0}$.

The surface-wave resonances considered so far were for the semi-infinite plasma. The quasi-static analysis is seen to predict accurately the frequencies at which these resonances occur. However, the frequency dependence of the propagation constant in the vicinity of these resonances may depend upon the exact manner in which the transverse wave numbers become large. This is most clearly evident in the case of the surface-wave resonance for a thin plasma slab above a conductor (Section 10.2.2). The exact determinantal equation for this case is given by Equation 10.74, which shows that near resonance $(\beta \to \infty)$ we must have $K_{\parallel} < -1$; therefore, the frequency dependence of β is that of a traveling wave, as shown in Figure 10.5f for large values of β. However, the quasi-static analysis starts with $T = \Gamma = \beta$, and hence, near resonance the determinantal equation (Equation 10.74 with $T = \Gamma = \beta$) then requires $K_{\parallel} > -1$, which falsely predicts a backward-wave characteristic for $\beta \to \infty$.

Partially Filled Longitudinally Magnetized Waveguide. As an example of a waveguide that is partially filled with a plasma we consider the LMG circularly concentric system of Figure 10.10a. The exact solution for this case was given in Section 10.4.2. In the quasi-static approximation the electric field inside the plasma is as given by Equations 10.200 through 10.203. The electric field outside the plasma cylinder follows from Equations 10.190 through 10.192. The solution of Equation 10.190 that satisfies the boundary condition on the perfectly conducting wall at $r = a$ is

$$\varphi^{o} = BG_{mm}(\beta r, \beta a)e^{jm\phi} \tag{10.247}$$

where

$$G_{mm}(\beta r, \beta a) = I_m(\beta r)K_m(\beta a) - K_m(\beta r)I_m(\beta a) \tag{10.248}$$

I_m and K_m are the modified Bessel functions of first and second kind, respectively, and B is an arbitrary, complex constant. Hence, the electric field outside the plasma is given by

$$E_z^{o} = Bj\beta G_{mm}e^{jm\phi} \tag{10.249}$$

$$E_r^{o} = -B\beta G'_{mm}e^{jm\phi} \tag{10.250}$$

$$E_\phi^{o} = -B\frac{jm}{r}G_{mm}e^{jm\phi} \tag{10.251}$$

where G'_{mm} is the derivative of G_{mm} with respect to βr. When Equations 10.201 through 10.203 and 10.249 through 10.251 are used to satisfy the boundary conditions at $r = \rho$, Equations 10.194 and 10.195, we obtain

$$\frac{B}{A} = \frac{J_m(p\rho)}{G_{mm}(r = \rho)} \tag{10.252}$$

and the determinantal equation

$$K_{\perp}p\rho\frac{J'_m(p\rho)}{J_m(p\rho)} - K_{\times}jm = \beta\rho\frac{G'_{mm}(r = \rho)}{G_{mm}(r = \rho)} \tag{10.253}$$

Equations 10.253 and 10.197 determine the wave numbers β and p as functions of frequency. It can be readily checked that resonances in propagation $(\beta \to \infty)$ can occur with either p finite or $p \to j\infty$; in the first case the resonances occur at ω_p, ω_{b+}, and ω_{b-}; in the second case they are the surface-wave resonances that occur at the frequencies given by Equations 10.137 and 10.138. Hence, in the partially filled, LMG waveguide, we have resonances at ω_p, ω_{b+}, ω_{b-}, and , depending upon the relative values of ω_p, ω_{b+}, and ω_{b-} (see Figure 10.9), either two, or one, or no surface-wave resonances.

10.5.3 The Magnetic Field and Power Flow. In Section 10.5.1 the quasi-static electric field was assumed to be coupled to the space charge of the plasma, and the magnetic field was assumed to be zero. The assumed wave solutions for the electric field gave the dispersion equations that were discussed in Section 10.5.2. However, the power flow associated with these waves is, for the temperate plasma (see Chapter 7), entirely electromagnetic. Therefore, a suitable, approximate magnetic field must also be established.

We consider the magnetic field to be coupled to the motion of the charges, that is, the currents, that arise from the interaction of the quasi-static electric field with the charges. Thus,

$$\nabla \times \underset{\sim}{H} = j\omega\epsilon_0 \underset{\approx}{K} \cdot \underset{\sim}{E} \tag{10.254}$$

$$\nabla \cdot \underset{\sim}{H} = 0 \tag{10.255}$$

where the right-hand side of Equation 10.254 will be assumed as known, and given by the solutions in Section 10.5.1. The solution for the magnetic field then consists of a particular and a homogeneous part. The particular solution can be readily found from equations for the longitudinal and transverse components of $\underset{\sim}{H}$ of Equations 10.254 and 10.255. For the LMG plasma we find

$$\underset{\sim}{H}_{pT} = Y_1 \nabla_T \varphi + Y_2 \underset{\sim}{i}_z \times \nabla_T \varphi \tag{10.256}$$

$$H_{pz} = Y_3 \gamma \varphi \tag{10.257}$$

where

$$Y_1 = \frac{j\omega\epsilon_0}{\gamma} \frac{K_\times K_\perp}{K_\perp - K_\parallel} \tag{10.258}$$

$$Y_2 = -\frac{j\omega\epsilon_0}{\gamma} K_\perp \tag{10.259}$$

$$Y_3 = -\frac{j\omega\epsilon_0}{\gamma} \frac{K_\times K_\parallel}{K_\perp - K_\parallel} \tag{10.260}$$

The homogeneous solution is determined from Equations 10.254 and 10.255 with the right-hand side equal to zero. This is found to be

$$\underset{\sim}{H}_{hT} = -\nabla_T \psi \tag{10.261}$$

$$H_{hz} = \gamma\psi \tag{10.262}$$

where ψ is a solution of

$$\nabla_T^2\psi + \gamma^2\psi = 0 \tag{10.263}$$

Hence, the total magnetic field is

$$\underset{\sim}{H}_T = Y_1\nabla_T\varphi + Y_2\underset{\sim}{i}_z \times \nabla_T\varphi - \nabla_T\psi \tag{10.264}$$

$$H_z = \gamma(Y_3\varphi - \psi) \tag{10.265}$$

The magnetic field outside the plasma is of the same form as Equations 10.264 and 10.265 but with $K_\times = 0$, $K_\perp = K_\parallel$, and $\varphi = \varphi^0$ of Equation 10.190. The boundary conditions, Equations 10.193 through 10.195, completely determine ψ. We note that because of the anisotropy of the medium two transverse eigenvalues (p^2 and γ^2) and two transverse wave functions (φ and ψ) are required for meeting all the boundary conditions. Analogous equations can be derived for the magnetic field in the case of a TMG plasma waveguide.

The power flow associated with the quasi-static electric and magnetic fields can now be determined. Using Equations 10.182 and 10.264, together with the boundary conditions on the perfectly conducting walls, we find

$$P = \int_A \operatorname{Re} \tfrac{1}{2}\underset{\sim}{E}_T \times \underset{\sim}{H}_T^* \cdot \underset{\sim}{i}_z \, da = \tfrac{1}{2} \operatorname{Re} \omega\epsilon_0\gamma \int_A K_\parallel |\varphi|^2 \, da \tag{10.266}$$

where $K_\parallel$ may be a function of the transverse coordinates.

Appendix A

EVALUATION OF INTEGRALS FOR THE CONDUCTIVITY

We shall indicate here how the first two diagonal terms of σ' (Equation 6.14) are evaluated for a Maxwellian distribution:

$$f_0 = \left(\frac{m}{2\pi eT}\right)^{\frac{3}{2}} \exp\left(-\frac{w^2+u^2}{\frac{2eT}{m}}\right)$$

$$= \left(\frac{1}{2\pi v_T^2}\right)^{\frac{3}{2}} \exp\left(-\frac{w^2+u^2}{2v_T^2}\right) \tag{A.1}$$

$$\left.\begin{aligned} \frac{\partial f_0}{\partial w} &= -\frac{w}{(2\pi)^{\frac{3}{2}} v_T^5} \exp\left(-\frac{w^2+u^2}{2v_T^2}\right) \\ \frac{\partial f_0}{\partial u} &= -\frac{u}{(2\pi)^{\frac{3}{2}} v_T^5} \exp\left(-\frac{w^2+u^2}{2v_T^2}\right) \end{aligned}\right\} \tag{A.2}$$

where

$$v_T^2 = \frac{eT}{m} \tag{A.3}$$

From Equations 6.14 and A.1, we have

$$\sigma'_{\substack{r\\ \ell}} = \frac{Ne^2}{2m\omega_b} \frac{1}{(2\pi)^{\frac{3}{2}} v_T^5} \int_{-\infty}^{\infty} \exp\left(-\frac{u^2}{2v_T^2}\right) du \int_0^{\infty} w^3 \exp\left(-\frac{w^2}{2\pi v_T^2}\right) dw$$

$$\times \int_0^{2\pi} d\phi \int_{-\infty}^{\phi} \exp\{-j[(a \mp 1)(\phi - \phi') - b(\sin\phi - \sin\phi')]\}\, d\phi' \tag{A.4}$$

In the integral over ϕ and ϕ',

$$I = \int_0^{2\pi} d\phi \int_{-\infty}^{\phi} \exp\{-j[(a \mp 1)(\phi - \phi') - b(\sin\phi - \sin\phi')]\}\, d\phi'$$

$$= \int_0^{2\pi} d\phi \int_{-\infty}^{\phi} \exp\left\{-j\left[(a \mp 1)(\phi - \phi') - 2b \sin\left(\frac{\phi - \phi'}{2}\right) \cos\left(\frac{\phi - \phi'}{2}\right)\right]\right\} d\phi' \tag{A.5}$$

we change variables by setting

$$\left.\begin{aligned} \phi - \phi' &= 2x \qquad & \phi &= x + y \\ \phi + \phi' &= 2y \qquad & \phi' &= y - x \end{aligned}\right\} \tag{A.6}$$

$$d\phi\, d\phi' = \begin{vmatrix} \dfrac{\partial \phi}{\partial x} & \dfrac{\partial \phi}{\partial y} \\[2ex] \dfrac{\partial \phi'}{\partial x} & \dfrac{\partial \phi'}{\partial y} \end{vmatrix} dx\, dy = 2dx\, dy \tag{A.7}$$

The regions of integration in the (ϕ, ϕ') and (x, y) planes are shown in Figure A.1. Hence

$$I = 2\int_0^{\infty} dx \int_{-x}^{-x+2\pi} \exp\{-j[(a \mp 1)2x - 2b \sin x \cos y]\}\, dy$$

$$= 2\int_0^{\infty} dx \int_0^{2\pi} \exp\{-j[(a \mp 1)2x - 2b \sin x \cos y]\}\, dy \tag{A.8}$$

because the integrand is periodic. Using Equation A.8 in Equation A.4, we have

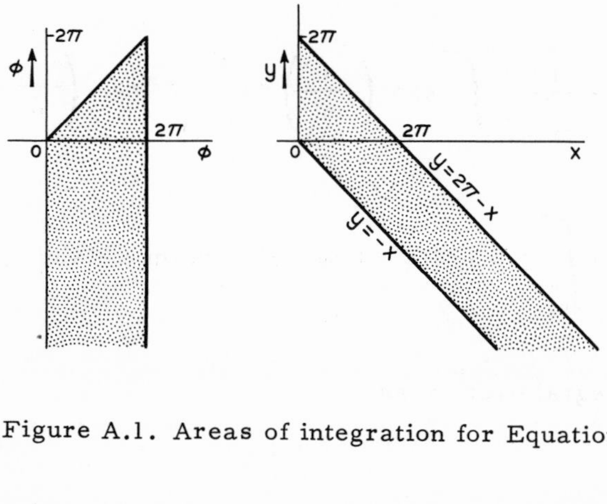

Figure A.1. Areas of integration for Equation A.7.

$$= \frac{Ne^2}{m\omega_b}(2\pi)^{-\frac{3}{2}} v_T^{-5} \int_0^\infty dx \int_{-\infty}^\infty \exp\left(-\frac{u^2}{2v_T^2} + j\frac{2uk_\parallel}{\omega_b}x\right) du \int_0^\infty w^3 \exp\left(-\frac{w^2}{2v_T^2}\right) dw$$

$$\times \int_0^{2\pi} \exp\left\{-j\left[\left(\frac{\omega - j\nu}{\omega_b} \mp 1\right) 2x - 2b \sin x \cos y\right]\right\} dy \qquad (A.9)$$

Now

$$\int_0^{2\pi} \exp(2jb \sin x \cos y)\, dy = 2\pi J_0(2b \sin x)$$

$$= 2\pi J_0\left(2\frac{wk_\perp}{\omega_b} \sin x\right) \qquad (A.10)$$

and

$$\int_{-\infty}^\infty \exp\left(-\frac{u^2}{2v_T^2} + 2j\frac{k_\parallel x}{\omega_b}\right) du = \sqrt{2\pi}v_T \exp\left(-\frac{2k_\parallel^2 v_T^2}{\omega_b^2}x^2\right) \qquad (A.11)$$

so that

$$\sigma'_{\substack{r\\ \ell}} = \frac{Ne^2}{m\omega_b}\frac{1}{v_T^4}\int_0^\infty dx \exp\left\{-j\left[\left(\frac{\omega - j\nu}{\omega_b} \mp 1\right) 2x\right] - \frac{2k_\parallel^2 v_T^2}{\omega_b^2}x^2\right\}$$

$$\times \int_0^\infty w^3 J_0\left(2\frac{k_\perp w}{\omega_b} \sin x\right) \exp\left(-\frac{w^2}{2v_T^2}\right) dw \qquad (A.12)$$

Using the formula

$$\left.\begin{aligned} &\int_0^\infty z^{v+1} J_v(az) \exp(-p^2 z^2)\, dz = \frac{a^v}{(2p^2)^{v+1}} \exp\left(-\frac{a^2}{4p^2}\right) \\ &\qquad \mathrm{Re}\,(2v + 2) > 0; \qquad |\arg p| < \frac{\pi}{4} \end{aligned}\right\} \qquad (A.13)$$

we find for the integral over w

$$\int_0^\infty w^3 J_0\left(\frac{2k_\perp w}{\omega_b}\sin x\right)\exp\left(-\frac{w^2}{2v_T^2}\right)dw$$

$$= v_T^3\frac{\partial}{\partial v_T}\int_0^\infty w\exp\left(-\frac{w^2}{2v_T^2}\right)J_0\left(2\frac{k_\perp w}{\omega_b}\sin x\right)dw$$

$$= v_T^3\frac{\partial}{\partial v_T}\left\{v_T^2\exp\left[-2\left(\frac{k_\perp v_T}{\omega_b}\right)^2\sin^2 x\right]\right\}$$

$$= 2v_T^4\left[1-2\left(\frac{k_\perp^2 v_T^2}{\omega_b^2}\right)\sin^2 x\right]\exp\left[-2\left(\frac{k_\perp v_T}{\omega_b}\right)^2\sin^2 x\right] \tag{A.14}$$

Hence

$$\sigma_{\substack{r\\ \ell}}' = \frac{Ne^2}{m\omega_b}\,2\int_0^\infty \exp\left\{-j\left[\left(\frac{\omega-j\nu}{\omega_b}\mp 1\right)2x-\frac{k_\parallel^2 v_T^2}{\omega_b^2}x^2-2\frac{k_\perp^2 v_T^2}{\omega_b^2}\sin^2 x\right]\right\}$$

$$\times\left(1-2\frac{k_\perp^2 v_T^2}{\omega_b^2}\sin^2 x\right)dx \tag{A.15}$$

Define

$$\left.\begin{aligned}\mu_\parallel &= \frac{k_\parallel^2 eT}{m\omega_b^2} = \frac{\delta n^2}{\beta^2}\cos^2\theta\\ \mu_\perp &= \frac{k_\perp^2 eT}{m\omega_b^2} = \frac{\delta n^2}{\beta^2}\sin^2\theta\end{aligned}\right\} \tag{A.16}$$

where† $\delta = eT/mc^2$, and set $2x = y$. This yields

$$\sigma_{\substack{r\\ \ell}}' = \frac{Ne^2}{m\omega_b}\left(1+\mu_\perp\frac{\partial}{\partial\mu_\perp}\right)$$

$$\times\int_0^\infty \exp\left\{-j\left[\left(\frac{\omega-j\nu}{\omega_b}\mp 1\right)y-\tfrac{1}{2}\mu_\parallel y^2-\mu_T(1-\cos y)\right]\right\}dy \tag{A.17}$$

† This definition of δ differs by γ from that in Equation 5.12.

Also since

$$\exp(\mu_\perp \cos y) = \sum_{i=-\infty}^{\infty} I_i(\mu_\perp) \exp(ji\,y) \tag{A.18}$$

where $I_i(\mu_\perp)$ is the modified Bessel function, we finally obtain

$$\sigma'_{\substack{r\\ \ell}} = \frac{Ne^2}{m\omega_b}\left(1 + \mu_\perp \frac{\partial}{\partial \mu_\perp}\right) \exp(-\mu_\perp) \sum_{i=-\infty}^{\infty} I_i(\mu_\perp)$$

$$\times \int_0^\infty \exp\left[-j\left(\frac{\omega - j\nu}{\omega_b} \mp 1 - i\right)x - \tfrac{1}{2}\mu_{\|}x^2\right] dx \tag{A.19}$$

In a similar manner we find

$$\sigma'_{r\ell} = \sigma'_{\ell r} = \frac{Ne^2}{m\omega_b} \mu_\perp \frac{\partial}{\partial \mu_\perp} \exp(-\mu_\perp) \sum_{i=-\infty}^{\infty} I_i(\mu_\perp)$$

$$\times \int_0^\infty \exp\left[-j\left(\frac{\omega - j\nu}{\omega_b} - i\right)x - \tfrac{1}{2}\mu_{\|}x^2\right] dx \tag{A.20}$$

$$\sigma'_{\substack{rp\\ \ell p}} = -\sqrt{2}\,\frac{Ne^2}{m\omega_b}\sqrt{\mu_{\|}\mu_\perp}$$

$$\times \int_0^\infty \exp\left[-j\left(\frac{\omega - j\nu}{\omega_b} \mp \frac{1}{2}\right)x - \tfrac{1}{2}\mu_{\|}x^2 - \mu_\perp(1 - \cos x)\right] x \sin\frac{x}{2}\, dx \tag{A.21}$$

$$\sigma'_{rp} = \sigma'_{pr}$$

$$\sigma'_{\ell p} = \sigma'_{p\ell}$$

$$\sigma'_p = \frac{Ne^2}{m\omega_b}\left(1 + 2\mu_{\|} \frac{\partial}{\partial \mu_{\|}}\right) \exp(-\mu_\perp) \sum_{i=-\infty}^{\infty} I_i(\mu_\perp)$$

$$\times \int_0^\infty \exp\left[-j\left(\frac{\omega - j\nu}{\omega_b} - i\right)x - \tfrac{1}{2}\mu_{\|}x^2\right] dx \tag{A.22}$$

so that

$$
\sigma' = \begin{bmatrix} \sigma_r & \sigma_{r\ell} & \sigma_{rp} \\ \sigma_{r\ell} & \sigma_{\ell} & \sigma_{\ell p} \\ \sigma_{rp} & \sigma_{\ell p} & \sigma_p \end{bmatrix} \tag{A.23}
$$

If we transform back to Cartesian coordinates,

$$
\begin{aligned}
\underset{\approx}{\sigma} &= \underset{\approx}{U}^{-1} \cdot \underset{\approx}{\sigma}' \cdot \underset{\approx}{U} \\
&= \frac{1}{2} \begin{bmatrix} \sigma_r + \sigma_{\ell} + 2\sigma_{r\ell} & -j(\sigma_r - \sigma_{\ell}) & \sqrt{2}\,(\sigma_{rp} + \sigma_{\ell p}) \\ j(\sigma_r - \sigma_{\ell}) & \sigma_r + \sigma_{\ell} - 2\sigma_{r\ell} & -j\sqrt{2}\,(\sigma_{rp} - \sigma_{\ell p}) \\ \sqrt{2}\,(\sigma_{rp} + \sigma_{\ell p} & j\sqrt{2}\,(\sigma_{rp} - \sigma_{\ell p}) & 2\sigma_p \end{bmatrix} \\
&= \begin{bmatrix} \sigma_{xx} & -\sigma_{xy} & \sigma_{xz} \\ \sigma_{xy} & \sigma_{yy} & \sigma_{yz} \\ \sigma_{xz} & -\sigma_{yz} & \sigma_{zz} \end{bmatrix}
\end{aligned} \tag{A.24}
$$

Appendix B

EXPANSION OF THE DIELECTRIC TENSOR FOR LOW TEMPERATURES

When the temperature T is sufficiently low that $\mu = \mu_{\parallel} + \mu_{\perp}$ is much less than unity (wavelength much larger than the electron Larmor radius), it is possible to expand the various integrals in σ in a series in μ. As an example, we consider the expansion of σ_p given by Equation A.22

$$\sigma_p = \frac{Ne^2}{m\omega_b}\left(1 + 2\mu_{\parallel}\frac{\partial}{\partial\mu_{\parallel}}\right) \exp(-\mu_{\perp}) \sum_{i=-\infty}^{\infty} I_i(\mu_{\perp})$$

$$\times \int_0^{\infty} \exp\left[-j\left(\frac{\omega - \nu}{\omega_b} - i\right)x - \tfrac{1}{2}\mu_{\parallel}x^2\right] dx \qquad \text{(B.1)}$$

The integral is evaluated as follows

$$\int_0^{\infty} \exp\left[-j\left(\frac{\omega - \nu}{\omega_b} - i\right)x - \tfrac{1}{2}\mu_{\parallel}x^2\right] dx = \int_0^{\infty} \exp(-j\zeta x - \tfrac{1}{2}\mu_{\parallel}x^2)\, dx$$

$$\approx \int_0^{\infty} (1 - \mu_{\parallel}x^2) \exp(-j\zeta x)\, dx$$

$$= \left(1 + \mu_{\parallel}\frac{\partial}{\partial\zeta^2}\right) \int_0^{\infty} \exp(-j\zeta x)\, dx$$

$$= -j\frac{1}{\zeta}\left(1 + \frac{\mu_{\parallel}}{\zeta^2}\right) \qquad \text{(B.2)}$$

where

$$\zeta = \frac{\omega - j\nu}{\omega_b} - i \qquad \text{(B.3)}$$

This manner of evaluating the integral is approximately correct provided that $\mu_{\parallel}^2 << \zeta^2$ or $n_{\parallel}^2\delta < \beta^2|1 - \beta - j\nu/\omega|$. This yields

$$\sigma_p = \frac{Ne^2}{jm\omega_b}\left(1 + 2\mu_{\parallel}\frac{\partial}{\partial\mu_{\parallel}}\right)\exp(-\mu_\perp)\sum_{i=-\infty}^{\infty} I_i(\mu_\perp)\frac{1}{\zeta}\left(1 + \frac{\partial\mu_{\parallel}}{\zeta^2}\right)$$

$$= \frac{Ne^2}{jm\omega_b}\exp(-\mu_\perp)\sum_{i=-\infty}^{\infty} I_i(\mu_\perp)\frac{1}{\zeta}\left(1 + \frac{3\mu_{\parallel}}{\zeta^2}\right) \qquad \text{(B.4)}$$

Assuming, in turn, that $\mu_\perp << 1$, and expanding further we find that only the $i = 0$ and $i = \pm 1$ terms in the infinite sum contribute to terms first-order in $\mu_\perp$. The $i = 0$ term yields

$$\sigma_{p_{i=0}} = \frac{Ne^2}{jm\omega_b}(1 - \mu_\perp)\left(\frac{\omega_b}{\omega - j\nu}\right)\left[1 + \frac{3\mu_{\parallel}\omega_b^2}{(\omega - j\nu)^2}\right]$$

$$= \frac{Ne^2}{jm\omega_b}\left[1 - \mu_\perp + \frac{3\mu_{\parallel}\omega_b^2}{(\omega - j\nu)^2}\right] \qquad \text{(B.5)}$$

The $i = \pm 1$ term yields

$$\sigma_{p_{i=\pm 1}} = \frac{Ne^2}{jm\omega_b}\,\frac{\mu_\perp}{2}\,\frac{\omega_b}{(\omega - j\nu \mp 1)} \qquad \text{(B.6)}$$

Adding the terms in Equations B.5 and B.6, we find

$$K_p = 1 + \frac{\sigma_p}{j\omega\epsilon_0}$$

$$= 1 - \alpha^2 - \alpha^2\beta^2\left(3\mu_{\parallel} + \frac{\mu_\perp}{1 - \beta^2}\right)$$

$$= 1 - \alpha^2 - \alpha^2\delta\left(3n^2\cos^2\theta + \frac{n^2\sin^2\theta}{1 - \beta^2}\right) \qquad \text{(B.7)}$$

where

$$\alpha^2 = \frac{\omega_p^2}{\omega^2}, \qquad \beta = \frac{\omega_b}{\omega} \qquad \text{(B.8)}$$

and where for simplicity we set $\nu = 0$.

Proceeding in this manner with the other elements of the conductivity tensor, we obtain the expressions given in Equations 6.16 and 6.17.

REFERENCES

Part I

1. I. E. Tamm, Am. Scientist, 47, 169 (1959).
2. W. P. Allis, "Motions of Ions and Electrons," Handbuch der Physik, Vol. 21, S. Flügge, Ed., Springer-Verlag, Berlin (1956), pp. 383-444.
3. W. P. Allis, op. cit., pp. 420-429.
4. G. A. Deschamps and W. L. Weeks, Report AFCRC-TN-60-468, Air Force Cambridge Research Center, Bedford, Mass. (April 26, 1960).
5. E. Åström, Arkiv Fysik, 2, 443 (1951).
6. A. Sommerfeld, Optics, Lectures on Theoretical Physics, Vol. 4 (translated from the German by O. Laporte and P. A. Moldauer), The Academic Press, Inc., New York (1954).
7. H. Bremmer, Handbuch der Physik, Vol. 16, S. Flügge, Ed., Springer-Verlag, Berlin (1958), p. 570.
8. K. G. Budden, Radio Waves in the Ionosphere, Cambridge University Press, Cambridge (1961).
9. J. A. Ratcliffe, The Magneto-ionic Theory and Its Applications to the Ionosphere; a Monograph, Cambridge University Press, Cambridge (1959).
10. H. G. Booker, Proc. Roy. Soc. (London), A, 147, 352 (1934).
11. P. L. Auer, H. Hurwitz, Jr., and R. D. Miller, Phys. Fluids, 1, 501 (1958).
12. A. A. Vlasov, J. Exptl. Theoret. Phys. (U.S.S.R.), 8, 291 (1938).
13. D. Bohm and E. P. Gross, Phys. Rev., 75, 1851, 1864 (1949); 79, 992 (1950).
14. B. D. Fried and R. W. Gould, Phys. Fluids, 4, 139 (1961).
15. N. Herlofson, Nature, 165, 1020 (1950);
    H. C. van der Hulst, Problems of Cosmical Aerodynamics, Central Air Documents Office, Dayton, Ohio (1951), Chapter 6.
16. N. G. Van Kampen, Physica, 21, 949 (1955); 23, 641 (1957).
17. E. G. Harris, J. Nuclear Energy, C 2, 138 (1961).
18. I. B. Bernstein and S. K. Trehan, Nuclear Fusion, 1, 3 (1960).
19. L. Landau, J. Exptl. Theoret. Phys. (U.S.S.R.), 16, 574 (1946).
20. A. Peskoff, "Waves in a Relativistic Plasma," Ph.D. Thesis, Department of Physics, M.I.T., Cambridge (September, 1960).

Part II

21. P. A. Sturrock, "A variational principle and an energy theorem for small-amplitude disturbances of electron beams and of electron-ion plasmas," Ann. Phys., 4, 306-324 (July, 1958);
    D. L. Bobroff, H. A. Haus, and J. W. Klüver, "On the Small Signal Power Theorem of Electron Beams," J. Appl. Phys., 33, 2932-2942 (October, 1962).
    A. Bers and P. Penfield, Jr., "Conservation Principles for Plasmas and Relativistic Electron Beams," IRE Trans. on Electron Devices, ED-9, 12-26 (January, 1962).

22. C. G. Montgomery, R. H. Dicke, and E. M. Purcell, Eds., Principles of Microwave Circuits, Radiation Laboratory Series, Vol. 8, McGraw-Hill Book Company, Inc., New York (1948), Sections 3.7 and 8.3.

23. L. D. Landau and E. M. Lifshitz, Electrodynamics of Continuous Media, Pergamon Press, Oxford and New York, and Addison-Wesley Publishing Company, Reading, Mass. (1960), Section 77.

L. Brillouin, Wave Propagation and Group Velocity, Academic Press, Inc., New York (1960).

24. R. B. Adler, L. J. Chu, and R. M. Fano, Electromagnetic Energy Transmission and Radiation, John Wiley & Sons, Inc., New York (1960), Section 7.2.2.

25. P. Chorney, "Power and Energy Relations in Bidirectional Waveguides," Proceedings of the Symposium on Electromagnetics and Fluid Dynamics of Gaseous Plasma, New York, N.Y., April 4, 5, 6, 1961, Vol. 11, Polytechnic Press, Polytechnic Institute of Brooklyn, Brooklyn, N.Y. (1962), pp. 195-210.

26. J. F. Denisse and J. L. Delcroix, Théorie des Ondes dans les Plasmas, Dunod, Paris (1961), Chapter 5.

27. A. Bers, "Properties of Waves in Time- and Space-Dispersive Media," Quarterly Progress Report No. 65, Research Laboratory of Electronics, M.I.T., Cambridge, Mass. (April 15, 1962), pp. 89-93; "Energy and Power in Media with Temporal and Spacial Dispersion," Quarterly Progress Report No. 66, Research Laboratory of Electronics, M.I.T., Cambridge, Mass. (July 15, 1962), pp. 111-116.

28. C. G. Montgomery, R. H. Dicke, and E. M. Purcell, Eds., op. cit., Sections 2.11 to 2.15.

F. E. Borgnis and C. H. Papas, "Electromagnetic Waveguides and Resonators," Handbuch der Physik, Vol. 16, S. Flügge, Ed., Springer-Verlag, Berlin (1958), pp. 285-422.

29. W. W. Balwanz, Interaction Between Electromagnetic Waves and Flames; Part 6 - Theoretical Plots of Absorption, Phase Shift, and Reflection, NRL Report 5388, U.S. Naval Research Laboratory, Washington, D.C. (September 23, 1959).

30. R. B. Adler, Properties of Guided Waves on Inhomogeneous Cylindrical Structures, Technical Report No. 102, Research Laboratory of Electronics, M.I.T., Cambridge, Mass. (May 27, 1949); also Proc. IRE., 40, 339-348 (March, 1952).

31. W. C. Hahn, "Small Signal Theory of Velocity-Modulated Electron Beams," Gen. Elec. Rev. 42, 258-270 (June, 1939).

32. L. D. Smullin and P. Chorney, "Properties of Ion Filled Waveguides," Proc. IRE, 46, 360-361 (January, 1958).

33. A. W. Trivelpiece and R. W. Gould, "Space Charge Waves in Cylindrical Plasma Columns," J. Appl. Phys., 30, 1784-1793 (November, 1959).

34. R. W. Newcomb, "The Hydromagnetic Wave Guide," Magnetohydrodynamics: a Symposium, R. K. M. Landshoff, Ed., Stanford University Press, Palo Alto, Calif. (1957), p. 109.

35. R. Gajewski, "Magnetohydrodynamic Waves in Wave Guides," Phys. Fluids, 2, 633-641 (November, 1959).

36. H. Suhl and L. R. Walker, "Topics in Guided-Wave Propagation Through Gyromagnetic Media," Bell System Tech. J., 33 (1954); Part I - "The Completely Filled Cylindrical Guide," 579-659 (May, 1954); Part II - "Transverse Magnetization and Non-Reciprocal Helix," 939-986 (July, 1954); Part III - "Perturbation Theory and Miscellaneous Results," 1133-1194 (September, 1954).

37. A. A. Th. M. Van Trier, "Guided Electromagnetic Waves in Anisotropic Media," Appl. Sc. Research (Netherlands), B3, 305-371 (1953).

38. S. C. Brown, Basic Data of Plasma Physics, The Technology Press of M.I.T. and John Wiley & Sons, Inc., New York (1959), Chapter 3.

39. E. Jahnke and F. Emde, Tables of Functions with Formulae and Curves, 4th ed., Dover Publications, New York (1945), Chapter 11.
E. Jahnke, F. Emde, and F. Lösch, Tables of Higher Functions, 6th ed., B. G. Teubner Verlagsgesellschaft, Stuttgart (1960).

40. W. Magnus, F. Oberhettinger, and F. G. Tricomi, Higher Transcendental Functions, Vol. I, A. Erdélyi, Ed., McGraw-Hill Book Company, Inc., New York (1953), Chapter 6.

41. P. S. Epstein, "On the Possibility of Electromagnetic Surface Waves," Proc. Natl. Acad. Sci. U.S., 40, 1158-1165 (1954).
W. O. Schumann, "Wellen längs homogener Plasmaschichten," Sitzber. math.-naturw. Kl. bayer. Akad. Wiss. München, 255-279 (1948).

42. B. Agdur and B. Enander, "Resonances of a Microwave Cavity Partially Filled with a Plasma," J. Appl. Phys., 33, 575-581 (February, 1962).

43. A. W. Trivelpiece, Slow Wave Propagation in Plasma Waveguides, Technical Report No. 7, Electron Tube and Microwave Laboratory, California Institute of Technology, Pasadena, Calif. (May, 1958).

44. A. A. Oliner and T. Tamir, "Backward Waves on Isotropic Plasma Slabs," J. Appl. Phys., 33, 231-233 (January, 1962).
S. F. Paik, "A Backward Wave in Plasma Waveguide," Proc. IRE, 50, 462-463 (April, 1962).
W. O. Schumann, "Über die Entstehung einer 'Backward Wave' in einem nichtmagnetisierten, von Luft begrenzten Plasmazylinder," Z. angew. Phys., 12, 4, 145-148 (April, 1960).

45. P. S. Epstein, "Theory of Propagation in a Gyromagnetic Medium," Revs. Modern Phys., 28, 3-17 (January, 1956).

46. W. O. Schumann, "Über Wellenausbreitung im Plasma zwischen zwei unendlich gut leitenden Ebenen in Richtung eines aufgeprägten äusseren Magnetfeldes," Z. angew. Phys., 8, 482-485 (1956).

47. A. I. Akhiezer, Y. B. Fainberg, A. G. Sitenko, K. Stepanov, V. Kurilko, M. Gorbatenko, and U. Kirochkin, "High-frequency Plasma Oscillations," Proc. 2nd U.N. Conference on Peaceful Uses of Atomic Energy, Vol. 31, United Nations, Geneva (1958), pp. 99-111;
Ya. B. Fainberg and M. F. Gorbatenko, "Electromagnetic Waves in a Plasma Situated in a Magnetic Field," Soviet Phys.-Tech. Phys., 4, 487-500 (November, 1959).
W. O. Schumann, "Über die Ausbreitung elektrischer Wellen längs einer dielektrisch begrenzten Plasmaschicht mit einem longitudinalen Magnetfeld," Z. angew. Phys., 10, 26-31 (1958).

48. L. D. Smullin and P. Chorney, "Propagation in Ion Loaded Waveguides," Proceedings of the Symposium on Electronic Waveguides, New York, N.Y., April 8, 9, 10, 1958, Vol. 8, Polytechnic Press, Polytechnic Institute of Brooklyn, Brooklyn, N.Y. (1958), pp. 229-247.

49. R. W. Gould and A. W. Trivelpiece, "A New Mode of Wave Propagation on Electron Beams," Proceedings of the Symposium on Electronic Waveguides, New York, N.Y., April 8, 9, 10, 1958, Vol. 8, Polytechnic Press, Polytechnic Institute of Brooklyn, Brooklyn, N.Y., (1958), pp. 215-228.

50. Ya. B. Fainberg, "The Use of Plasma Waveguides as Accelerating Structures in Linear Accelerators," CERN Symposium on High Energy Accelerators and Pion Physics, Proceedings, Vol. 1, Geneva, (1956), pp. 84-90.

SELECTED BIBLIOGRAPHY

Books

Alfvén, H., Cosmical Electrodynamics, Clarendon Press, Oxford (1950), Chapter 4.

Budden, K. G., Radio Waves in the Ionosphere, Cambridge University Press, Cambridge (1961).

Denisse, J. F., and J. L. Delcroix, Théorie des Ondes dans les Plasmas, Dunod, Paris (1961).

Ginzburg, V. L., Propagation of Electromagnetic Waves in Plasma, translated from the Russian by Royer and Roger, Gordon and Breach, New York (1961).

Ratcliffe, J. A., The Magneto-Ionic Theory and Its Application to the Ionosphere, Cambridge University Press, Cambridge (1959).

Spitzer, L., Physics of Fully Ionized Gases, Chapter 3, 2nd Edition, Interscience Publishers, Inc., New York (1962).

Stix, T. H., Theory of Plasma Waves, Monographs in Advanced Physics, McGraw-Hill Book Company, Inc., New York (in press).

Papers

Agdur, B., "Notes on the Propagation of Guided Microwaves Through an Electron Gas in the Presence of a Static Magnetic Field," Proceedings of the Symposium on Electronic Waveguides, New York, N.Y., April 8, 9, 10, 1958, Vol. 8, Polytechnic Press, Polytechnic Institute of Brooklyn, Brooklyn, N.Y. (1958), pp. 177-197.

Akhiezer, A. I., Y. B. Fainberg, A. G. Sitenko, K. N. Stepanov, V. Kurilko, M. Gorbatenko, and U. Kirochkin, "High-frequency Plasma Oscillations," Proc. 2nd U.N. Conference on Peaceful Uses of Atomic Energy, Vol. 31, United Nations, Geneva (1958), p. 99.

Akhiezer, A. I., G. Ya. Lyubarskii, and R. V. Polovin, "Simple Waves in Magnetohydrodynamics," J. Tech. Phys. (U.S.S.R.), 29, No. 8 (1959); Soviet Phys.-Tech. Phys., 4, 849 (1960).

Akhiezer, A. I., and R. V. Polovin, "On the Oscillations of a Plasma in Crossed Electric and Magnetic Fields," J. Tech. Phys. (U.S.S.R.), 22, 1794 (1952).

Anderson, N. S., "Longitudinal Magneto-Hydrodynamic Waves," J. Acoust. Soc. Am., 25, 529 (1953).

Appleton, E. V., "Geophysical Influences on the Transmission of Wireless Waves," Proc. Phys. Soc. (London), 37, 16D, (1924).

Appleton, E. V., and G. Builder, "The Ionosphere as a Doubly Refracting Medium," Proc. Phys. Soc. (London), 45, 208 (1933).

Åström, E., "On Waves in an Ionized Gas," Arkiv Fysik, 2, 443 (1951).

Åström, E., "Magneto-Hydrodynamic Waves in a Plasma," Nature, 165, 1019 (1950).

Auer, P. L., H. Hurwitz, Jr., and R. D. Miller, "Collective Oscillations in a Cold Plasma," Phys. Fluids, 1, 501 (1958).

Bailey, V. A., "Plane Waves in an Ionized Gas with Static Electric and Magnetic Fields Present," Australian J. Sci. Research, A1, 351 (1948).

Baños, A., "Magneto-Hydrodynamic Waves in Incompressible and Compressible Fluids," Proc. Roy. Soc. (London), A, 233, 350 (1955).

Beard, D. B., "Cyclotron Radiation from Magnetically Confined Plasmas," Phys. Fluids, 2, 366 (1960).

Bekefi, G., J. L. Hirshfield, and S. C. Brown, "Kirchhoff's Radiation Law for Plasmas with Non-Maxwellian Distributions," Phys. Fluids, 4, 173 (1961).

Bernstein, I. B., "Waves in a Plasma in a Magnetic Field," Phys. Rev., 109, 10 (1958).

Bernstein, I. B., "Plasma Oscillations Perpendicular to a Constant Magnetic Field," Phys. Fluids, 3, 489 (1960).

Bernstein, I. B., and R. M. Kurslud, "Ion Wave Instabilities," Phys. Fluids, 3, 937 (1960).

Bernstein, I. B., and S. K. Trehan, "Plasma Oscillations," Nuclear Fusion, 1, 3 (1960).

Berz, F., "On the Theory of Plasma Waves," Proc. Phys. Soc. (London), 69B, 939 (1956).

Bohm, D., and E. P. Gross, "Theory of Plasma Oscillations: A. Origin of Medium-like Behavior," Phys. Rev., 75, 1851 (1949).

Bohm, D., and E. P. Gross, "Theory of Plasma Oscillations: B. Excitation and Damping of Oscillations," Phys. Rev., 75, 1864 (1949).

Booker, H. G., "Some General Properties of the Formulae of the Magneto Ionic Theory," Proc. Roy. Soc. (London), A, 147, 352 (1934).

Booker, H. G., "The Application of the Magneto-Ionic Theory to the Ionosphere," Proc. Roy. Soc. (London), A, 150, 267 (1935).

Braginskiy, S. I., "The Modes of Plasma Oscillation in a Magnetic Field," Dokl. Akad. Nauk S.S.S.R., 115, 475 (1957); Soviet Physics-Doklady 2, 345 (1957).

Buchsbaum, S. J., "Resonance in a Plasma with Two Ion Species," Phys. Fluids, 3, 418 (1960).

Buchsbaum, S. J., L. L. Mower, and S. C. Brown, "Interaction Between Cold Plasma and Guided Electromagnetic Waves," Phys. Fluids, 3, 806 (1960).

Budden, K. G., "A Reciprocity Theorem on the Propagation of Radio Waves via the Ionosphere," Proc. Cambridge Phil. Soc., 50, 604 (1954).

Case, K. M., "Plasma Oscillations," Ann. Phys., 7, 349 (1959).

Darwin, C. G., "The Refractive Index of an Ionized Medium," Proc. Roy. Soc. (London), A, 146, 17 (1934); A, 182, 152 (1943).

Dawson, J., "On Landau Damping," Phys. Fluids, 4, 869 (1961).

Dawson, J., and C. Oberman, "Oscillations of a Finite Cold Plasma in a Strong Magnetic Field," Phys. Fluids, 2, 103 (1958).

Dawson, J., and C. Oberman, "High-Frequency Conductivity and the Emission and Absorption Coefficients of a Fully Ionized Plasma," Phys. Fluids, 5, 517 (1962).

Denisov, N. G., "Propagation of Electromagnetic Signals in an Ionized Gas," J. Exptl. Theoret. Phys. (U.S.S.R.), 21, 1354 (1951).

Denisov, N. G., "On the Absorption of Radio Waves in Resonance Regions of Heterogeneous Plasma," Radiotekh. i Elektron., 4, 388 (1959).

Dnestrovskii, Yu. N., and D. P. Kostomarov, "Electromagnetic Waves in a Half-Space Filled with a Plasma," J. Exptl. Theoret. Phys. (U.S.S.R.), 39, 845 (1960); Soviet Phys.-JETP, 12, 587 (1961).

Dnestrovskii, Yu. N., and D. P. Kostomarov, "Dispersion Equation for an Ordinary Wave Moving in a Plasma Across an External Static Magnetic Field," J. Exptl. Theoret. Phys. (U.S.S.R.), 40, 1404 (1961); Soviet Phys.-JETP, 13, 986 (1961).

Dnestrovskii, Yu. N., and D. P. Kostomarov, "The Dispersion Equation for an Extraordinary Wave Moving in a Plasma Across an External Magnetic Field," J. Exptl. Theoret. Phys. (U.S.S.R.), 41, 1527 (1961); Soviet Phys.-JETP, 14, 1089 (1962).

Doyle, P. H., and J. Neufeld, "On the Behavior of Plasma at Ionic Resonance," Phys. Fluids, 2, 390 (1959).

Drummond, J. E., "Basic Microwave Properties of Hot Magnetoplasmas," Phys. Rev., 110, 293 (1958).

Drummond, J. E., "Microwave Propagation in Hot Magnetoplasmas," Phys. Rev., 112, 1460 (1958).

Drummond, W. E., and M. N. Rosenbluth, "Cyclotron Radiation from a Hot Plasma," Phys. Fluids, 3, 45 (1960); 3, 491 (1960).

Dungey, J. W., "Derivation of the Dispersion Equation for Alfvén's Magneto-Hydrodynamic Waves from Bailey's Electromagneto-Ionic Theory," Nature, 167, 1029 (1951).

Engelhardt, A. G., and A. A. Dougal, "Dispersion of Ion Cyclotron Waves in Magnetoplasmas," Phys. Fluids, 5, 29 (1962).

Epstein, P. S., "Theory of Wave Propagation in a Gyromagnetic Medium," Revs. Modern Phys., 28, 3 (1956).

Ferraro, V. C. A., "Hydromagnetic Waves in a Rare Ionized Gas and Galactic Magnetic Fields," Proc. Roy. Soc. (London), A, 233, 310 (1955).

Frank-Kamenetskii, D. A., "Natural Oscillations of a Bound Plasma," J. Exptl. Theoret. Phys. (U.S.S.R.), 39, 669-679 (1960); Soviet Phys.-JETP, 12, 469-475 (1961).

Fried, B. D., "Mechanism for Instability of Transverse Plasma Waves," Phys. Fluids, 2, 337 (1959).

Fried, B. D., M. Gell-mann, J. D. Jackson and H. W. Wyld, "Longitudinal Plasma Oscillations in an Electric Field," J. Nuclear Energy, Part C, 1, 190 (1960).

Fried, B. D., and R. W. Gould, "Longitudinal Ion Oscillations in a Hot Plasma," Phys. Fluids, 4, 139 (1961).

Furutsu, K., "On the Group Velocity, Wave Path and Their Relations to the Poynting Vector of the Electromagnetic Field in an Absorbing Medium," J. Phys. Soc. Japan, 7, 458 (1952).

Gajewski, R., "Magnetohydrodynamic Waves in Wave Guides," Phys. Fluids, 2, 633 (1959).

Gajewski, R., and O. K. Mawardi, "Hydromagnetic Resonators," Phys. Fluids, 3, 820 (1960).

Gallet, R., "Propagation and Production of Electromagnetic Waves in a Plasma," Nuovo cimento Supp., 13, 234 (1959).

Gershman, B. N., "On the Spreading of Electromagnetic Signals in an Ionized Gas," J. Tech. Phys. (U.S.S.R.), 22, 101 (1952).

Gershman, B. N., "Kinetic Theory of Magnetohydrodynamic Waves," J. Exptl. Theoret. Phys. (U.S.S.R.), 24, 453 (1953).

Gershman, B. N., "On the Propagation of Electro-Magnetic Waves in a Plasma Located in a Magnetic Field, Taking into Account Thermal Motion of Electrons," J. Exptl. Theoret. Phys. (U.S.S.R.), 24, 659 (1953).

Gershman, B. N., "Note on Waves in a Homogeneous Magnetoactive Plasma," J. Exptl. Theoret. Phys. (U.S.S.R.), 31, 707 (1956); Soviet Phys.-JETP, 4, 582 (1957).

Gershman, B. N., "Nonresonant Absorption of Electromagnetic Waves in a Magnetoactive Plasma," J. Exptl. Theoret. Phys. (U.S.S.R.), 37, 695 (1959); Soviet Phys.-JETP, 10, 497 (1960).

Gershman, B. N., "Gyromagnetic Resonance Absorption of Electromagnetic Waves in Plasma," J. Exptl. Theoret. Phys. (U.S.S.R.), 38, 912 (1960); Soviet Phys.-JETP, 11, 657 (1960).

Gershman, B. N., "Group Velocity of Plasma Waves in the Presence of a Magnetic Field," Izv. Vysshikh. Uchebh. Zavedenii, Radiofiz., 3, 146 (1960).

Gershman, B. N., "The Problem of Propagation of Electromagnetic Waves in a Weakly Relativistic Magnetoactive Plasma," Izv. Vysshikh. Uchebh. Zavedenii, Radiofiz., 3, 534 (1960).

Gershman, B. N., V. L. Ginzburg, and N. G. Denisov, "Propagation of Electromagnetic Waves in Plasma." Uspekhi Fiz. Nauk, 61, 561 (1957).

Gershman, B. N., and M. S. Kovner, "On the Characteristics of Quasi-Transverse Propagation of Magnetohydrodynamic Waves in Plasma," Izv. Vysshikh. Uchebh. Zavedenii, Radiofiz., 1, 19 (1958).

Gershman, B. N., and M. S. Kovner, "On Certain Characteristics of Wave Propagation Associated with Collisions in a Magnetoactive Plasma," Izv. Vysshikh. Uchebh. Zavedenii, Radiofiz., 2, 28 (1959).

Gerson, N. C., and S. L. Seaton, "Generalized Magneto-Ionic Theory," J. Franklin Inst., 246, 483 (1948).

Ginzburg, V. L., "On the Theory of the Propagation of Electromagnetic Waves in Magnetoactive Media," J. Exptl. Theoret. Phys. (U.S.S.R.), 18, 487 (1948).

Ginzburg, V. L., "On Magnetohydrodynamic Waves in Gases," J. Exptl. Theoret. Phys. (U.S.S.R.), 21, 788 (1951).

Ginzburg, V. L., and V. V. Zhelenznyakov, "On Absorption and Emission of Electromagnetic Waves in a Magnetoactive Plasma," Izv. Vysshikh. Uchebh. Zavedenii, Radiofiz., 1, 59 (1958).

Golant, V. E., and A. P. Zhilinskii, "Propagation of Electromagnetic Waves Through Waveguides Filled with Plasma," J. Tech. Phys. (U.S.S.R.), 30, 15 (1960); Soviet Phys.-Tech. Phys., 5, 12 (1960).

Golant, V. E., A. P. Zhilinskii, M. V. Krisvosheev, and L. I. Chernova, "Propagation of Microwaves Through Waveguides Filled with Plasma in the Positive Column of a Discharge. II," J. Tech. Phys. (U.S.S.R.), 31, 63 (1961); Soviet Phys.-Tech. Phys., 6, 44 (1961).

Golant, V. E., A. P. Zhilinskii, M. V. Krivosheev, and G. P. Nekrutkina, "Propagation of Microwaves Through Waveguides Filled with Plasma in the Positive Column of a Discharge," J. Tech. Phys. (U.S.S.R.), 31, 23 (1961); Soviet Phys.-Tech. Phys., 6, 38 (1961).

Gordeyev, G. V., "Plasma Oscillations in a Magnetic Field," J. Exptl. Theoret. Phys. (U.S.S.R.), 23, 660 (1952); 27, 24 (1954).

Graf, K. A., and M. P. Bachynski, "Electromagnetic Waves in a Bounded, Anisotropic Plasma," Can. J. Phys., 10, 887 (1962).

Harris, E. G., "Unstable Plasma Oscillations in a Magnetic Field," Phys. Rev. Letters, 2, 34 (1959).

Harris, E. G., "Plasma Instabilities Association with Anisotropic Energy Distributions," J. Nuclear Energy, C2: Plasma Physics, 138 (1961).

Hartree, D. R., "Propagation of Electromagnetic Waves in a Refracting Medium in a Magnetic Field," Proc. Cambridge Phil. Soc., 27, 143 (1931).

Hauser, W., "On the Theory of Anisotropic Obstacles in Cavities," Quart. J. Mech. Appl. Math., 11, 112 (1958).

Hauser, W., "Variational Principles for Guided Electromagnetic Waves in Anisotropic Materials," Quart. Appl. Math., 16, 259 (1958).

Hauser, W., "On the Theory of Anisotropic Obstacles in Waveguides," Quart. J. Mech. Appl. Math., 11, 427 (1958).

Herlofson, N., "Magneto-Hydrodynamic Waves in a Compressible Fluid Conductor," Nature, 165, 1020 (1950).

Hines, C. O., "Wave Packets, the Poynting Vector and Energy Flow. Part I: Non-dissipative (anisotropic) Homogeneous Media; Part II: Group Propagation Through Dissipative Isotropic Media; Part III: Packet Propagation Through Dissipative Anisotropic Media; Part IV: Poynting and Macdonald Velocities in Dissipative Anisotropic Media (conclusion)," J. Geophys. Research, 56, 63, 107, 207, 535 (1951).

Huxley, L. G. H., "The Propagation of Electromagnetic Waves in an Ionized Atmosphere," Phil. Mag., 25, 148 (1938).

Huxley, L. G. H., "The Propagation of Electromagnetic Waves in an Atmosphere Containing Free Electrons," Phil. Mag., 29, 313 (1940).

Hwa, R. C., "Effects of Electron-Electron Interactions on Cyclotron Resonances in Gaseous Plasmas," Phys. Rev., 110, 307 (1958).

Imre, K., "Oscillations in a Relativistic Plasma," Phys. Fluids, 5, 459 (1962).

Jackson, J. D., "Longitudinal Plasma Oscillations," J. Nuclear Energy, Part C, 1, 171 (1960).

Karplus, R., "Radiation of Hydromagnetic Waves," Phys. Fluids, 3, 800 (1960).

Kelso, J. M., "On the Coupled Wave Equations of Magneto-Ionic Theory," J. Geophys. Research, 58, 431 (1953).

Kitsenko, A. B., and K. N. Stepanov, "Instability of a Plasma with an Anisotropic Distribution of Ion and Electron Velocities," J. Exptl. Theoret. Phys. (U.S.S.R.), 38, 1840 (1960); Soviet Phys.-JETP, 11, 1323 (1960).

Kitsenko, A. B., and K. N. Stepanov, "Cyclotron Instability in a Plasma," J. Tech. Phys. (U.S.S.R.), 31, 176 (1961); Soviet Phys.-Tech. Phys., 6, 127 (1961).

Kovrizhnykh, L. M., "Oscillations of an Electron-Ion Plasma," J. Exptl. Theoret. Phys. (U.S.S.R.), 37, 1692 (1959); Soviet Phys.-JETP, 10, 1198 (1960).

Kunkel, W. B., "Remarks on Magnetoacoustic Waves," Phys. Fluids, 5, 867 (1962).

Landau, L. D., "Electron Plasma Oscillations," J. Phys. (U.S.S.R.), 10, 25 (1946).

Leontovich, M., "Generalization of the Kramers-Kronig Formulas to Media with Spatial Dispersion," J. Exptl. Theoret. Phys. (U.S.S.R.), 40, 907 (1961); Soviet Phys.-JETP, 13, 634 (1961).

Levitskii, S. M., and N. S. Baranchuk, "On the Propagation of Electromagnetic Waves Along a Plasma Rod," Izv. Vysshikh. Uchebh. Zavedenii, Radiofiz., 3, 725 (1960).

Lieboff, R. L., "Long-Wavelength Phenomena in a Plasma," Phys. Fluids, 5, 963 (1962).

Lighthill, M. J., "Studies on Magneto-Hydrodynamic Waves and Other Anisotropic Wave Motions," Phil. Trans. Roy. Soc. (London), A, 252, 397 (1960).

Mower, L., "Conductivity of Warm Plasma," Phys. Rev., 116, 16 (1959).

Oehrl, W., "Die Ausbreitung langsamer elektromagnetischer Wellen längs inhomogener Plasmaschichten," Z. angew. Phys., 9, 164 (1957).

Oster, L., "Linearized Theory of Plasma Oscillations," Revs. Modern Phys., 32, 141 (1960).

Pai, S. I., "Wave Motions of Small Amplitude in a Fully Ionized Plasma Without External Magnetic Field," Revs. Modern Phys., 32, 882 (1960).

Pai, S. I., "Wave Motions of Small Amplitude in a Fully Ionized Plasma under Applied Magnetic Field," Phys. Fluids, 5, 234 (1962).

Perel', V. I., and S. M. Eliashberg, "Absorption of Electromagnetic Waves in a Plasma," J. Exptl. Theoret. Phys. (U.S.S.R.), 41, 886 (1961); Soviet Phys.-JETP, 14, 633 (1962).

Piddington, J. H., "The Four Possible Waves in a Magnetoionic Medium," Phil. Mag., 46, 1037 (1955).

Piddington, J. H., "Hydromagnetic Waves in Ionized Gases," Monthly Notices Roy. Astron. Soc., 115, 671 (1955).

Platzman, P. M., and S. J. Buchsbaum, "Effect of Collisions of the Landau Damping of Plasma Oscillations," Phys. Fluids, 4, 1288 (1961).

Polovin, R. V., "Contribution to the Theory of Simple Magnetohydrodynamic Waves," J. Exptl. Theoret. Phys. (U.S.S.R.), 39, 463 (1960); Soviet Phys.-JETP, 12, 326 (1961).

Postnov, G. A., "Wave Propagation in Isotropic Plasma Waveguide," Radiotekh. i Elektron., 5, 1598 (1960) [Translated in Radio Engineering and Electronics, 5, 1598 (1960)].

Pradhan, T., "Plasma Oscillations in Steady Magnetic Field: Circularly Polarized Electromagnetic Modes," Phys. Rev., 107, 1222 (1957).

Rosenbluth, M. N., and N. Rostoker, "Scattering of Electromagnetic Waves by a Nonequilibrium Plasma," Phys. Fluids, 5, 776 (1962).

Rukhadze, A. A., and V. P. Silin, "Electrodynamics of Media with Spatial Dispersion," Soviet Phys.-Uspekhi, 4, 459 (1961). Uspekhi Fiz. Nauk, 74, 223 (1961).

Rytov, S. M., "Some Theorems Concerning the Group Velocity of Electromagnetic Waves," J. Exptl. Theoret. Phys. (U.S.S.R.), 17, 930, (1947).

Sagdeyev, R. Z., and V. D. Shafranov, "Absorption of High-frequency Electromagnetic Waves in High-temperature Plasma," Proc. 2nd U.N. Conference on the Peaceful Uses of Atomic Energy, Vol. 31, United Nations, Geneva (1958), p. 118.

Sagdeyev, R. Z., and V. D. Shafranov, "On the Instability of a Plasma with an Anisotropic Distribution of Velocities in a Magnetic Field," J. Exptl. Theoret. Phys. (U.S.S.R.), 39, 181 (1960); Soviet Phys.-JETP, 12, 130 (1961).

Scarf, F. L., "Landau Damping and the Attenuation of Whistlers," Phys. Fluids, 5, 6 (1962).

Sen, H. K., and A. A. Wyler, "On the Generalization of the Appleton-Hartree Magnetoionic Formulas," J. Geophys. Research, 65, 3931 (1960).

Shafranov, V. D., "Propagation of an Electromagnetic Field in a Medium with Spatial Dispersion," J. Exptl. Theoret. Phys. (U.S.S.R.), 34, 1475 (1958); Soviet Phys.-JETP, 7, 1019 (1958).

Silin, V. P., "Of the Electromagnetic Properties of Relativistic Plasma," J. Exptl. Theoret. Phys. (U.S.S.R.), 38, 1577 (1960); Soviet Phys.-JETP, 11, 1136 (1960).

Silin, V. P., "Kinetic Equation for Rapidly Varying Processes," J. Exptl. Theoret. Phys. (U.S.S.R.), 38, 1771 (1960); Soviet Phys.-JETP, 11, 1277 (1960).

Silin, V. P., "Electromagnetic Properties of a Relativistic Plasma. II.," J. Exptl. Theoret. Phys. (U.S.S.R.), 40, 616 (1960); Soviet Phys.-JETP, 13, 430 (1962).

Silin, V. P., "High-Frequency Dielectric Constant of a Plasma," J. Exptl. Theoret. Phys. (U.S.S.R.), 41, 861 (1961); Soviet Phys.-JETP, 14, 617 (1962).

Sitenko, A. G., and Yu. A. Kirochkin, "Excitation of Waves in a Plasma," J. Tech. Phys. (U.S.S.R.), 29, 801 (1959); Soviet Phys.-Tech. Phys., 4, 723 (1960).

Sitenko, A. G., and K. N. Stepanov, "On the Oscillations of an Electron Plasma in a Magnetic Field," J. Exptl. Theoret. Phys. (U.S.S.R.), 31, 642 (1956); Soviet Phys.-JETP, 4, 512 (1957).

Stepanov, K. N., "Kinetic Theory of Magnetohydrodynamic Waves," J. Exptl. Theoret. Phys. (U.S.S.R.), 34, 1292 (1958); Soviet Phys.-JETP, 7, 892 (1958).

Stepanov, K. N., "Low-Frequency Oscillations of a Plasma in a Magnetic Field," J. Exptl. Theoret. Phys. (U.S.S.R.), 35, 1155 (1958); Soviet Phys.-JETP, 8, 808 (1959).

Stepanov, K. N., "Cyclotron Absorption of Electromagnetic Waves in a Plasma," J. Exptl. Theoret. Phys. (U.S.S.R.), 38, 265 (1960); Soviet Phys.-JETP, 11, 192 (1960).

Stix, T. H., "Oscillations of a Cylindrical Plasma," Phys. Rev., 106, 1146 (1957).

Stix, T. H., "Generation and Thermalization of Plasma Waves," Phys. Fluids, 1, 308 (1958).

Stix, T. H., "Absorption of Plasma Waves," Phys. Fluids, 3, 19 (1960).

Tanenbaum, B. S., "Dispersion Relations in a Stationary Plasma," Phys. Fluids, 4, 1262 (1961).

Taylor, M., "The Appleton-Hartree Formula and Dispersion Curves for the Propagation of Electromagnetic Waves Through an Ionized Medium in the Presence of an External Magnetic Field. Part I. Curves for Zero Absorption," Proc. Phys. Soc. (London), 45, 245 (1933).

Taylor, M., "The Appleton-Hartree Formula and Dispersion Curves for the Propagation of Electromagnetic Waves Through an Ionized Medium in the Presence of an External Magnetic Field. Part II. Curves with Collisional Friction," Proc. Phys. Soc. (London), 46, 408 (1934).

Tonks, L., and I. Langmuir, "Oscillations in Ionized Gases," Phys. Rev., 33, 195 (1929).

Trubnikov, B. A., "Plasma Radiation in a Magnetic Field," Dokl. Acad. Nauk S.S.S.R., 118, 913 (1958).

Trubnikov, B. A., and V. S. Kudryatsov, "Plasma Radiation in a Magnetic Field," Proc. 2nd U.N. Conference on the Peaceful Uses of Atomic Energy, Vol. 31, United Nations, Geneva (1958), p. 93

Van Kampen, N. G., "On the Theory of Stationary Waves in Plasmas," Physica, 21, 949 (1955).

Van Kampen, N. G., "The Dispersion Equation for Plasma Waves," Physica, 23, 641 (1957).

Vlasov, A. A., "Vibrational Properties of an Electron Gas," J. Exptl. Theoret. Phys. (U.S.S.R.), 8, 291 (1938).

Volkov, T. F., "Ion Oscillations in Plasma," J. Exptl. Theoret. Phys. (U.S.S.R.), 37, 422 (1959); Soviet Phys.-JETP, 10, 302 (1960).

Weibel, E. S., "Spontaneously Growing Transverse Waves in a Plasma Due to an Anisotropic Velocity Distribution," Phys. Rev. Letters, 2, 83 (1959).

Westfold, K. C., "The Interpretation of the Magneto-Ionic Theory," J. Atmospheric and Terrest. Phys., 1, 152 (1951).

Whale, H. A., and J. P. Stanley, "Group and Phase Velocities from the Magnetoionic Theory," J. Atmospheric and Terrest. Phys., 1, 82 (1950).

Yakimenko, V. L., "Oscillations in a Cold Plasma Containing Two Ion Species," J. Tech. Phys. (U.S.S.R.), 32, 168 (1962); Soviet Phys.-Tech. Phys., 7, 117 (1962).

Zhelensnyakov, V. V., "Interaction of Electromagnetic Waves in Plasma, I and II," Izv. Vysshikh. Uchebh. Zavedenii, Radiofiz., 1, 32 (1959); 2, 858 (1959).

INDEX